DIATOMIC INTERACTION POTENTIAL THEORY

Volume 1

Fundamentals

DIATOMIC INTERACTION POTENTIAL THEORY

Jerry Goodisman

DEPARTMENT OF CHEMISTRY
SYRACUSE UNIVERSITY
SYRACUSE, NEW YORK

Volume 1

Fundamentals

ACADEMIC PRESS   New York and London   1973

A Subsidiary of Harcourt Brace Jovanovich, Publishers

ACADEMIC PRESS, INC.
111 Fifth Avenue, New York, New York 10003

United Kingdom Edition published by
ACADEMIC PRESS, INC. (LONDON) LTD.
24/28 Oval Road, London NW1

Library of Congress Cataloging in Publication Data

Goodisman, Jerry.
    Diatomic interaction potential theory.

    (Physical chemistry, a series of monographs)
    Includes bibliographies.
    CONTENTS: v. 1.   Fundamentals.–v.   2.   Applications.
    1.   Quantum chemistry.   I.   Title.   II.   Series.
QD462.G65   541'.28   72-9985
ISBN 0–12–290201–7 (v. 1)

PRINTED IN THE UNITED STATES OF AMERICA

Contents

Chapter I Introduction to Potential Curves

Chapter II Qualitative Discussion of Potential Energy Curves

Chapter III Methods of Calculation

Preface

The calculation of the energy of a diatomic system as a function of internuclear separation is a problem which has a long history and has generated an enormous amount of literature. Due in part to advances in computational hardware and software in the past few years, quantum chemists can now produce reliable interaction potentials for diatomic systems in their ground states. The situation for excited states and for polyatomic systems is less satisfactory, but there is hope that it will shortly improve.

These two volumes cover the theoretical material involved in calculations for diatomic systems in their ground states, with attention given to the variety of the approaches one may use. The first volume contains mostly basic and general material; the second includes more in the way of specific descriptions of modern calculations.

The problem is defined in Chap. I, Vol. 1. A discussion is given of the nature of an interatomic interaction potential or potential energy curve, including its relation to reality (experiment). Chapter II presents a general discussion of its shape. Chapter III treats the main approaches to schemes of calculation: variation theory, perturbation theory, the virial and Hellmann–Feynman theorems, local energy principles, and quantum statistical theories. In Chapter I of Volume 2, the calculation of the interaction potential for large and small values of the internuclear distance R (separated and united atom limits) is considered. Chapter II treats the methods used for intermediate values of R, which in principle means *any* values of R. The Hartree–Fock and configuration interaction schemes described here have been the most used of all the methods. Semiempirical theories and methods constitute the subject of the last chapter of Volume 2.

The level of treatment throughout, it is hoped, is sufficiently elementary for the material to be understood after an introductory quantum mechanics course. By means of this book, the reader should be able to go from that degree of preparation to the current literature.

Work on this book started about five years ago; its proximate cause was my participation in a special topics graduate course, with Prof. D. Secrest and Prof. J. P. Toennies, at the University of Illinois. The course largely dealt with scattering experiments and their relation to potential energy curves. At that time I was struck by the fact that there was much material which was common knowledge among those involved in quantum chemical calculations but unfamiliar to students, even those with good course backgrounds. Thus, one goal of the present book is to make that material conveniently available to students and others interested in the subject, and to introduce them to the current literature.

For those interested in the theory of quantum chemical calculations, I want to provide in one place as much information as I can on the varied methods which are available. For those interested in potential curves or in quantum chemistry, but not particularly interested in calculating potential curves, I hope this book will be a guide to what has been going on, as well as an aid in reading the literature.

The subject has been limited to diatomic interactions, and, still further, to diatomic ground-state interactions. Of course, the limitation on the calculations discussed does not mean that the methods of calculation have no other applicability. I hope that the general discussions, particularly in the first volume, will be of interest to those who care about other systems. Some of the methods may even find their greatest applicability to those other systems. However, only by severely limiting my subject could I hope to attain some measure of completeness of coverage. Even so, I have had to give very limited space to certain topics. I have not discussed calculations specifically applicable to one- and two-electron systems (another book should be written on these, as was done for the corresponding atoms); I have slighted relativistic and magnetic effects; I have given unjustly brief coverage to many-body theory. There are undoubtedly other sins of omission.

Nevertheless, I believe that this book gives a balanced picture of the enormous amount of work that has been done on ground state diatomic potential curves. While limitations of certain methods have been discussed, I hope the main point is still clear: after many years of effort, reliable potential curves can now be generated for most systems of interest.

Notes on Notation and Coordinate Systems

As an aid in keeping formulas more legible, a Dirac-like bracket notation is employed frequently, without necessarily implying notions of states, representations, and so on. The triangular bracket $\langle \Phi_i/\Phi_k \rangle$ means the product of Φ_i^* and Φ_k, integrated over the entire configuration space, which must be the same for the two functions. This means integration over all spatial coordinates and sums over spin coordinates. The arguments of Φ_i and Φ_k need not be stated, although sometimes they are. The "factors" in this "scalar product" may be considered separately. Thus $|\Phi\rangle$ and Φ are equivalent, both being wave functions.

The adjoint wavefunction is written $\langle \Phi |$, but this is also used to denote an operator in the following sense: $\langle \Phi |$ multiplying $|\Psi\rangle$ gives the bracket $\langle \Phi | \Psi \rangle$, a number which is computed by multiplying Φ^* by Ψ and integrating over the configuration space. Thus, $\langle \Phi |$ may be interpreted as the operation of multiplication by Φ^*, followed by integration. If the functions ψ_i form a complete set,

$$\left[\sum_i | \psi_i \rangle \langle \psi_i | \right] \phi \rangle = | \phi \rangle$$

so that the operator in the square bracket is the identity operator.

Italics are used for operators, e.g., F or h. We use $\langle \Psi | F | X \rangle$ equivalently to $\langle \Psi | FX \rangle$: thus F operates on the function X and the scalar product of the result with Ψ is then taken. Operators are assumed to operate to the right in all bracket expressions. Thus,

$$\langle \Phi | P | \Psi \rangle = \langle P^\dagger \Phi | \Psi \rangle = \langle \Psi | P^\dagger \Phi \rangle^*$$

where $P^\dagger$ is the adjoint of P.

In general, we use bold face for matrices, e.g., $\mathbf{M}$ is the matrix with elements M_{ij}. The determinant of the matrix $\mathbf{M}$ is written $| \mathbf{M} |$ or $\det[\mathbf{M}]$

and the trace is written Tr[**M**]. The dagger (†) is used to indicate the adjoint of matrices, so $(\mathbf{M}^\dagger)_{ij}$, the (i, j) element of the matrix $\mathbf{M}^\dagger$, is M_{ji}^*. The transpose is similarly denoted by T: $(\mathbf{M}^T)_{ij} = M_{ji}$. The transpose or adjoint of a column vector is a row vector and vice versa.

The expression "matrix element of the operator P between states (or functions) Ψ_i and X" means the integral

$$\int \Psi_i^* PX \, d\tau = \langle \Psi_i \mid P \mid X \rangle.$$

The "scalar product" of X_i and X_j, $\langle X_i \mid X_j \rangle$, is sometimes referred to as an element of the overlap matrix. In these integrals or brackets an integration, or summation where appropriate, is implied over all coordinates, unless there is a specific remark to the contrary.

"Real part of" is denoted by Re, e.g., Re(ϕ). Curly brackets are used to refer to all the members of a set of functions, as in: "we orthonormalize the $\{\phi_i\}$ among themselves."

Several different coordinate systems are used in our discussions. Consider the two nuclei separated by a distance R, and a Cartesian coordinate system located at the midpoint, with the Z axis along the internuclear axis.

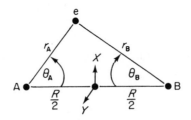

The position of an electron can be specified by its distances from the nuclei, plus the angle φ between the plane containing the ABe triangle and a reference plane containing the Z axis. Alternatively, one can use the angles θ_A and θ_B together with φ, or one of the sets (r_A, θ_A, φ) and (r_B, θ_B, φ). The last two are of course just spherical polar coordinates centered on one nucleus or the other.

It is convenient for many purposes to use coordinates involving r_A and r_B, together with φ. Most often, one uses rather than r_A and r_B themselves, the confocal ellipsoidal coordinates

$$\xi = (r_A + r_B)/R$$

$$\eta = (r_A - r_B)/R.$$

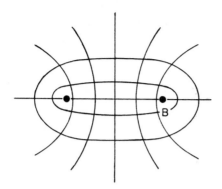

ξ and η are dimensionless. The surfaces of constant ξ are ellipsoids of revolution with foci at the nuclei. The value of ξ can run from 1 (corresponding to the line between the nuclei) to ∞. The surfaces of constant η are hyperboloids of revolution with foci at the nuclei. The value of η is between -1 (corresponding to the extension of the internuclear axis to the left of A) and $+1$ (corresponding to the extension of the axis to the right of B). $\eta = 0$ is the plane bisecting the internuclear axis.

The volume element for this coordinate system is

$$d\tau = \left(\frac{R}{2}\right)^3 (\xi^2 - \eta^2) \, d\xi \, d\eta \, d\varphi$$

and the Laplacian is

$$\Delta = \left(\frac{2}{R}\right)^2 (\xi^2 - \eta^2)^{-1} \left[\frac{\partial}{\partial\xi} (\xi^2 - 1) \frac{\partial}{\partial\xi} + \frac{\partial}{\partial\eta} (1 - \eta^2) \frac{\partial}{\partial\eta} \right.$$
$$\left. + \frac{\xi^2 - \eta^2}{(\xi^2 - 1)(1 - \eta^2)} \frac{\partial^2}{\partial\varphi^2} \right].$$

All distances are proportional to R for given values of the coordinates ξ and η. Thus

$$r_A = \frac{R}{2} (\xi + \eta), \qquad x = \frac{R}{2} [(\xi^2 - 1)(1 - \eta^2)]^{1/2} \cos\varphi,$$

$$r_B = \frac{R}{2} (\xi - \eta); \qquad y = \frac{R}{2} [(\xi^2 - 1)(1 - \eta^2)]^{1/2} \sin\varphi,$$

$$z = \frac{R}{2} \xi\eta.$$

and the distance between two points (ξ_1, η_1) and (ξ_2, η_2) is

$$r_{12}^2 = \left(\frac{R}{2}\right)^2 \{([(\xi_1{}^2 - 1)(1 - \eta_1{}^2)]^{1/2} - [(\xi_2{}^2 - 1)(1 - \eta_2{}^2)]^{1/2})^2$$
$$+ (\xi_1\eta_1 - \xi_2\eta_2)^2\}.$$

Contents of Volume 2

DIATOMIC INTERACTION POTENTIAL THEORY

Volume 1

Fundamentals

Chapter *I*   Introduction to Potential Curves

A. Introduction

This book discusses the theoretical approaches to the following problem: Calculate the wave function and energy for the lowest state of a system of N electrons moving in the field of two fixed point charges (the nuclei of a diatomic system) separated by a distance R. The energy, which we denote here as E_{el}, depends parametrically on R. From a series of calculations for different values of R, we produce a function $E_{el}(R)$, to which we generally add the internuclear repulsion $Z_A Z_B / R$ (Z_A and Z_B denote the nuclear charges) to produce what we shall refer to as a potential energy curve or interaction potential. For large R, it approaches the energy of the separated atoms or ions; for small R it becomes infinite because of the internuclear repulsion.

A first interpretation of the potential energy curve is that it represents the potential energy that the nuclei experience as R changes. Then the description of the nuclear motion reduces to the problem of two mass points interacting with this potential energy. This picture is approximately correct, and a useful first approximation to which corrections may be calculated. It would be exact if the nuclear masses were infinite, rather than simply very large (ratio of 10^4 or higher) compared to electronic masses. In reality, the nuclei are never fixed, and their motion must be discussed simultaneously with that of the electrons. It can be said that

1

they move slowly compared with the electrons, or crudely, that the electrons can adjust to changes of nuclear position, so that we may consider the electronic problem for various values of R. To the extent that this is true, the notion of a potential energy curve is reasonable and useful. This model will be expressed more correctly later in terms of the adiabatic theorem.

In order to show in more detail the relation between the clamped-nucleus potential curve, with whose calculation we are concerned, and the true nuclear–electronic wave function, we shall present the treatment of Born and Oppenheimer and a more general treatment, sometimes referred to as the Born separation. Then some discussion of the experimental measurement of interaction potentials will be given to help clarify the relation of the subject matter of this book to reality.

In Chapter II, we consider some simple ideas related to the form of the potential curve, which we denote by $U(R)$. Most of what we have to say here was known in the earliest years of the application of quantum mechanics to chemistry, and is semiquantitative, involving little or no calculation. This is in contrast to the main subject matter of this book—the recent progress made in calculating accurate potential curves. However, the qualitative discussions in Sections A, B, and C of Chapter II give the framework around which the more detailed work was done. Section D of Chapter II presents most of the functional forms that have been used for $U(R)$, based in part on the ideas discussed in the preceding sections. The title of Chapter II, Section A, "Very Large Internuclear Distances," refers to values of R large enough for neglect of overlap between the electronic wave functions of the interacting atoms (we do not discuss relativistic effects in this book). The calculation of $U(R)$ for such values of R is discussed in detail in Section A of Chapter I (Volume 2). Similarly, details of the calculations corresponding to the qualitative discussions of Section B of Chapter II are found in Chapter I, Section B of Vol. 2. In the latter two sections, we consider values of R too small for the neglect of interatomic overlap, so that whe are concerned with the formation of a diatomic system from the separated atoms. The ideas associated with the valence bond theory are most useful here. The concepts of the molecular orbital approach are discussed in Chapter II, Section C. As the valence bond picture is related to the separated atom (large R) limit, the molecular orbital approach has a kinship to the united atom (small R) limit. Calculations based on the united atom concepts are discussed in Chapter I, Section C (Vol. 2). However, calculations in the molecular orbital framework require in addition most of Chapter II (Vol. 2) for discussion, since they make up the bulk of the work done on the calculation of diatomic interactions.

The remaining chapters, beginning with Chapter III, bring us to the detailed presentation of methods for the calculation of $U(R)$, the main subject matter of the book. In Chapter III we introduce the different quantum mechanical methods that have been used, and discuss the basic theory involved. Material related to the variational principle is given in most detail (Chapter III, Section A) because of the predominant importance of variational calculations. We consider the use of the variational principle and some of the commonly used variational functions for problems in the electronic structure of diatomic systems. Perturbation theory, also extensively used, is taken up in Chapter III, Section B. The emphasis is on perturbation formalisms that have been applied to the calculation of $U(R)$. Most of the important applications of variational and perturbation methods are discussed in Chapters I and II (Vol. 2). In addition to these main methods, we feel it is important to present some of the other approaches which have been used. For these less employed methods, most applications, as well as basic formulas, are included in Chapter III, rather than being left to later chapters. Section C of that chapter discusses the virial theorem and calculations based on it. Section D does the same for the Hellmann–Feynman theorem. In Section E we take up methods which can be related to local energy principles, in which the wave function ψ is chosen to make the local energy as constant as possible rather than to minimize the expectation value of the Hamiltonian H. The local energy $H\psi/\psi$ is independent of the arguments of ψ if ψ is an exact eigenfunction. The local energy methods relate naturally to methods in which discrete sums over points are substituted for integrals. Chapter III, Section F discusses calculations based on quantum statistical theories, which involve treating the electrons at each point in the diatomic system as a Fermi gas.

In Chapter I (Vol. 2), we reenter the mainstream of modern quantum chemistry. The calculations discussed in the first two chapters of Vol. 2 represent the bulk of the work on the calculation of $U(R)$. The subject matter of Chapter I (Vol. 2) involves methods whose main utility would seem to be for particular regions of R. In Chapter I, Section A (Vol. 2), we consider interactions between nonoverlapping atomic systems. The interactions are often divided into electrostatic, inductive, and dispersive parts, according to the form of the perturbation theory terms that describe them. Most of Chapter I, Sections A.4–A.8 (Vol. 2) is concerned with the evaluation of the dispersive interactions, since great progress has been made on this old problem in recent years and much work has been done. In Chapter I, Section B (Vol. 2), we consider interaction of atoms when R is too small to neglect overlap ("exchange forces"). We include calculations according to the valence bond theory (Subsections 4 and 5) and also the recently popular and unsolved problem of developing perturbation theories for exchange (Subsections 2 and 3). Chapter I (Vol. 2), Section C is concerned with calculations, presumably useful for small R, which are based on the united atom.

The methods for intermediate R, which actually may be applied for all values of R of interest to us, constitute the subject matter of Chapter II (Vol. 2). Here the self-consistent field and configuration interaction calculations are discussed, with their variants and improvements. Most modern calculations of $U(R)$ involve these methods, and the amount of published work is extremely large. In Chapter II, Section A (Vol. 2), the single determinant function for a diatomic system is considered. The properties of a single determinant and of the spin orbitals from which it is constructed are discussed, as are the Hartree–Fock methods and their implementation. The derivation of wave functions built from several determinants, both symmetry projected single determinants and configuration interactions, are discussed in Vol. 2, Chapter II, Section B. In Section C, the derivation of the potential $U(R)$ from calculated energies for a series of values of R, such as are produced by the methods discussed in Sections A and B, is considered, as well as the correction of the results of Hartree–Fock calculations to give accurate potential curves. Finally, many of the results for $U(R)$ for various diatomic systems, produced by the methods discussed, are summarized (Subsections 3–4), hopefully giving an idea of what has been done and of the "state of the art" at this writing.

Chapter III in Vol. 2 is devoted to calculations that make use of both experimentally measured and calculated quantities. In Section A semiempirical molecular orbital and valence bond theories which have been used for calculation of $U(R)$ are considered, and methods employing pseudopotential and atoms-in-molecules theories are discussed. These have in common the fact that they attempt to simplify, by the introduction of available experimental data, calculations such as those discussed in Chapter II (Vol. 2). In contrast, Section B deals with simple models based on a physical picture that do not attempt to mimic *ab initio* calculations. Many of these methods lead to relations between potential constants. A discussion of such relations, from simple models as well as other sources, is also found in Section B.

The General Bibliography, at the end of both Volumes 1 and 2, gives some references to books closely related to the material of this volume, and to bibliographies of calculations. To avoid the detailed presentation of much elementary material, we have used capital letters in the text to refer to a number of basic books on quantum chemistry. These books are listed in the General Bibliography.

B. Separation of Nuclear from Electronic Motions

1. Center of Mass and Relative Motion

Before separation of nuclear from electronic motions, it is convenient to separate the translational motion of the molecule as a whole, which may be done exactly. A transformation of coordinates which accomplishes this (there are others) has been given by Fröman [1, 2]. We start with a system of coordinate axes fixed in space. The Hamiltonian for the N-electron diatomic system is

$$H' = -\frac{\hbar^2}{2m} \sum_{i=1}^{N} \boldsymbol{\nabla}_i' \cdot \boldsymbol{\nabla}_i' - \frac{\hbar^2}{2m_A} \boldsymbol{\nabla}_A' \cdot \boldsymbol{\nabla}_A' - \frac{\hbar^2}{2m_B} \boldsymbol{\nabla}_B' \cdot \boldsymbol{\nabla}_B' + V \quad (1)$$

where the terms are electronic kinetic energy, kinetic energy of nuclei A and B, and potential energy. Here, V is a sum of Coulombic interactions, depending on interparticle coordinates only, m is the electronic mass, and m_A and m_B denote the nuclear masses. The total mass is denoted by M:

$$M = m_A + m_B + Nm. \quad (2)$$

The primes refer to coordinates with respect to space-fixed axes. The $3N + 6$ coordinates used in (1) are first replaced by a new set as follows:

$$\mathbf{C} = M^{-1}\left[m_A \mathbf{r}_A' + m_B \mathbf{r}_B' + m \sum_{i=1}^{N} \mathbf{r}_i' \right] \quad (3a)$$

$$\mathbf{r}_i = \mathbf{r}_i' - (m_A + m_B)^{-1}[m_A \mathbf{r}_A' + m_B \mathbf{r}_B'] \quad (3b)$$

$$\mathbf{r}_A = \mathbf{r}_A' - (m_A + m_B)^{-1}[m_A \mathbf{r}_A' + m_B \mathbf{r}_B'] \quad (3c)$$

$$\mathbf{r}_B = \mathbf{r}_B' - (m_A + m_B)^{-1}[m_A \mathbf{r}_A' + m_B \mathbf{r}_B'] \quad (3d)$$

Here, $\mathbf{C}$ gives the coordinates of the center of mass of the molecule, and the new (unprimed) coordinates for electrons and nuclei are relative to the center of mass of the nuclei. Fröman proceeds slightly differently, since his procedure is designed to handle polyatomics as well. Note that the electronic coordinate axes, while referred to an origin translating with the molecule, still have their directions parallel to the initial, space-fixed axes. The set (3) is redundant, and $\mathbf{r}_A$ and $\mathbf{r}_B$ may be replaced by

$$R = \mathbf{r}_A - \mathbf{r}_B = \mathbf{r}_A' - \mathbf{r}_B' \quad (4)$$

In these new coordinates, the Hamiltonian is

$$H = -\frac{\hbar^2}{2M}\boldsymbol{\nabla}_c \cdot \boldsymbol{\nabla}_c - \frac{\hbar^2}{2m}\sum_i \boldsymbol{\nabla}_i^2$$
$$+ V - \frac{\hbar^2}{2(m_A + m_B)}\left(\sum_i \boldsymbol{\nabla}_i\right)\cdot\left(\sum_i \boldsymbol{\nabla}_i\right) - \frac{\hbar^2}{2\mu}\boldsymbol{\nabla}_R^2 \qquad (5)$$

Here, μ is the reduced mass of the nuclei:

$$\mu = m_A m_B/(m_A + m_B) \qquad (6)$$

Clearly, the eigenfunction of H may be expressed as a product of a translational eigenfunction and a function describing the rotational and internal motion, which is an eigenfunction of the Hamiltonian

$$H_0 - \frac{\hbar^2}{2(m_A + m_B)}\left(\sum_i \boldsymbol{\nabla}_i\right)\left(\sum_i \boldsymbol{\nabla}_i\right) - \frac{\hbar^2}{2\mu}\boldsymbol{\nabla}_R^2 \qquad (7)$$

Here,

$$H_0 = -\frac{\hbar^2}{2m}\sum_i \boldsymbol{\nabla}_i^2 + V \qquad (8)$$

is the fixed-nucleus Hamiltonian, which would obtain for infinite nuclear mass.

The last term in (7) is the nuclear kinetic energy. The term preceding it is the "mass polarization," arising from the constraints introduced by the separation of the center of mass motion. Sometimes, one writes

$$-\frac{\hbar^2}{2(m_A + m_B)}\left(\sum_i \boldsymbol{\nabla}_i\right)\cdot\left(\sum_i \boldsymbol{\nabla}_i\right)$$
$$= -\frac{\hbar^2}{2(m_A + m_B)}\left(\sum_i \boldsymbol{\nabla}_i^2 + 2\sum_{i<j} \boldsymbol{\nabla}_i \cdot \boldsymbol{\nabla}_j\right) \qquad (9)$$

and combines the first part with the kinetic energy operator in (8):

$$-\frac{\hbar^2}{2m}\sum_i \boldsymbol{\nabla}_i^2 - \frac{\hbar^2}{2(m_A + m_B)}\sum_i \boldsymbol{\nabla}_i^2 = -\frac{\hbar^2}{2m^*}\sum_i \boldsymbol{\nabla}_i^2$$

where m^* is a reduced mass for the electrons:

$$m^* = \frac{m(m_A + m_B)}{m + m_A + m_B}.$$

This corresponds to a change in units in the electronic Schrödinger equation.

The usual system of atomic units takes $e = \hbar = m = 1$, giving the Bohr radius $a_0 = \hbar^2/me^2$ as unit of length and e^2/a_0 as unit of energy. Here we replace m by m^*, scaling the energy unit correspondingly. The cross terms in (9) may be referred to as "true mass polarization." We shall return to them later. For polyatomic systems, similar terms arise for the nuclei [1].

The separation of the center of mass used here is not the only one. Others are discussed in the literature, and may be convenient for certain purposes [3–6]. Pack and Hirschfelder [6] have recently discussed three in detail, in which electronic coordinates are taken relative to (a) the center of mass of the nuclei, (b) the geometric center of the nuclei, and (c) the individual nuclei (different for different electrons).

2. Born–Oppenheimer Separation

After separation of the motion of the center of mass, the wave function in the relative coordinates describes the electronic motion as well as the motion of the nuclei relative to their center of mass. If the system corresponds to a stable molecule, so that interparticle distances are limited in size, we may classify the motions described by the wave function as rotational, vibrational, and electronic. We expect that electronic energies are large compared with those associated with the nuclei and that energies of nuclear rotation are small compared with the energies of nuclear vibration. The Born–Oppenheimer treatment makes this quantitative, and provides a formal treatment of the separation of nuclear from electronic motions for the case of a molecule.

We start with several order of magnitude estimates. The energy of an electron is roughly the energy needed to remove it from the molecule, i.e., an ionization potential. A typical ionization potential for an atom is 5–10 eV, and this quantity should also hold for a molecule. Alternatively, we may use the uncertainty principle: The size of a molecule, which we denote by d, is of the order of angstroms, and this is the distance over which the electron is localized, i.e., the uncertainty in its position. Then the uncertainty in its momentum is of size $\hbar/d$. The average momentum is zero, so for the average square momentum

$$\langle p^2 \rangle \sim (\hbar/d)^2$$

($\sim$ means "of the size of"). Then the kinetic energy of the electron, which is the same size as its total energy, obeys

$$(2m)^{-1}\langle p^2 \rangle \sim (2m)^{-1}(\hbar/d)^2 \sim 5 \text{ eV}.$$

The energy of nuclear rotation is the square of the angular momentum, say $\hbar$, divided by twice the moment of inertia, which is something like Md^2, where M is a typical nuclear mass. To estimate the energy of nuclear vibration, we use the harmonic oscillator formula and write it as $\hbar\omega$, where ω, the vibration frequency, is $(k/M)^{1/2}$. The force constant k,

$$k = (d^2E/dR^2)_0,$$

is now estimated by noting that changing R by a distance comparable to d essentially dissociates a molecule, and requires an energy comparable to the ionization potential, say 5 eV. Thus

$$\tfrac{1}{2}kd^2 \sim (2m)^{-1}(\hbar/d)^2$$

which leads to

$$\hbar\omega \sim \frac{\hbar}{M^{1/2}}\, \frac{\hbar/d^2}{m^{1/2}}$$

To summarize, we have roughly

$$E^{\text{el}} : E^{\text{vib}} : E^{\text{rot}} : : (\hbar/d)^2(2m)^{-1} : (\hbar/d)^2(mM)^{-1/2} : (\hbar/d)^2(2M)^{-1} \quad (10)$$

Since nuclear masses are, say, 10^4 electronic masses, we have

$$E^{\text{el}} : E^{\text{vib}} : E^{\text{rot}} : : 1 : 10^{-2} : 10^{-4}.$$

The preceding arguments may be extended to estimate the size of a typical nuclear vibrational displacement, say x, compared to d. We have

$$E^{\text{vib}} \sim \tfrac{1}{2}kx^2$$

which gives $x^2 \sim (m/M)^{1/2}d^2$, i.e., x is smaller than a typical electronic displacement by $(m/M)^{1/4}$. (In a rotational motion, the displacements of the nuclei are of size d.) Note that letting m/M become zero corresponds to fixing the nuclei. If we want to treat the effect of nuclear motions as a perturbation, a reasonable parameter of smallness would be

$$\eta \equiv (m/M)^{1/4} \sim 10^{-1}.$$

We now review the treatment of Born and Oppenheimer [7, 8], which develops the Hamiltonian in powers of η, and calculates the energy as a series in η by perturbation theory. We expect vibrational energy to enter in second order and rotational energy in fourth order. The Hamiltonian is now written explicitly, except for the mass-polarization terms, for a case more general than the diatomic

$$H = -\frac{\hbar^2}{2m} \sum_{i=1}^{N} \nabla_i^2 - \sum_K \frac{\hbar^2}{2\mu_K} \nabla_K^2 + V(\mathbf{r}, \mathbf{R}) \tag{11}$$

Here, $\mathbf{r}$ represents electronic coordinates and $\mathbf{R}$ nuclear coordinates. K runs over some set of relative nuclear coordinates; for the diatomic, the second term in (11) is just $-(\hbar^2/2\mu) \nabla_\mathbf{R}^2$. The $3v$ nuclear coordinates for v nuclei are replaced by three center of mass coordinates (removed), three rotational coordinates (two for a linear molecule), and $3v - 6$ internuclear coordinates ($3v - 5$ for a linear molecule) associated with vibration. Denote the rotational coordinates by θ and the vibrational coordinates by ξ.

The reduced mass μ_K may be written as $\eta^{-4}m_K$, where m_K is of size m. Then the Hamiltonian becomes

$$H = H_0 + \eta^4 H_1 \tag{12}$$

where H_0 refers to fixed nuclei. Both H_0 and H_1 contain dependence on nuclear coordinates and hence on η, as we see later. Since we will obtain the energy as a power series in η, all occurrences of η must be made explicit. The size of the coordinates ξ depends on η, so, for the ith one, we write

$$\Delta\xi_i = \xi_i - \xi_{i0} = \eta\zeta_i \tag{13}$$

with the new coordinates ζ_i to be of the size of electronic or rotational coordinates. The ξ_{i0} are convenient origins, to be specified later. It is convenient to write

$$H_1 = (-\hbar^2/2m)(T_{\xi\xi} + T_{\xi\theta} + T_{\theta\theta}) \tag{14}$$

where $T_{\xi\xi}$ involves two differentiations with respect to ξ coordinates, $T_{\xi\theta}$ one with respect to ξ and one with respect to θ, and $T_{\theta\theta}$ two with respect to θ. We must use the ζ coordinates, whose size is independent of η, so, for example, in $T_{\xi\xi}$ we must write

$$\frac{\partial^2}{\partial\xi_i \partial\xi_j} = \eta^{-2} \frac{\partial^2}{\partial\zeta_i \partial\zeta_j} \tag{15}$$

In considering H_0 we note that V depends on the nuclear vibrational coordinates:

$$V(\mathbf{r}, \mathbf{R}) = V(\mathbf{r}, \xi_0) + \sum_i \left(\frac{\partial V}{\partial\xi_i}\right)_\theta (\xi_i - \xi_{i0}) + \cdots \tag{16}$$

Then (13) shows that we must write

$$H_0 = H_{00} + \eta H_{01} + \eta^2 H_{02} + \cdots \qquad (17)$$

where

$$H_{00} = -\frac{\hbar^2}{2m} \sum_{i=1}^{N} \nabla_i^2 + V(r, \xi_0) \qquad (18a)$$

$$\eta H_{01} = \sum_{i=1}^{N} \left(\frac{\partial V}{\partial \xi_i}\right)_0 (\xi_i - \xi_{i0}) \qquad (18b)$$

and so on.

The program is first to consider the fixed-nucleus problem, i.e., to solve the eigenvalue equation

$$H_0(r, \xi)\varphi_n(r, \xi) = U_n(\xi)\varphi_n(r, \xi), \qquad (19)$$

and then to use the results in treating the full problem. The parametric dependence of the Hamiltonian on ξ means that the eigenfunctions and eigenvalues will have a similar parametric dependence. $U_n(\xi)$ will be referred to as the electronic energy surface or (for a diatomic) the potential energy curve. The n labels the different eigenstates (electronic quantum number). Here, H_0 is expressed in a power series in η, so we may write

$$\varphi_n = \varphi_n^{(0)} + \eta\varphi_n^{(1)} + \eta^2\varphi_n^{(2)} + \cdots \qquad (20a)$$

and

$$U_n = U_n^{(0)} + \eta U_n^{(1)} + \eta^2\varphi_n^{(2)} + \cdots \qquad (20b)$$

and use perturbation theory (EYRING, Sect. 7a, PILAR, Sect. 10-4, LEVINE, Sect. 9.2). The equations are

$$H_{00}\varphi_n^{(0)} = U_n^{(0)}\varphi_n^{(0)} \qquad (21)$$

$$(H_{00} - U_n^{(0)})\varphi_n^{(1)} + (H_{01} - U_n^{(1)})\varphi_n^{(0)} = 0 \qquad (22)$$

$$(H_{00} - U_n^{(0)})\varphi_n^{(2)} + (H_{01} - U_n^{(1)})\varphi_n^{(1)} + (H_{02} - U_n^{(2)})\varphi_n^{(0)} = 0 \qquad (23)$$

$$\vdots$$

We assume these to be solved. Note that $\varphi_n^{(0)}$ is the electronic eigenfunction for the nuclei fixed at the positions ξ_{i0}, and $U_n^{(0)}$ is the corresponding energy. From (22), we have

$$U_n^{(1)} = \int \varphi_n^{(0)*} H_{01} \varphi_n^{(0)} \, d\tau_e \qquad (24)$$

where the integral is over electronic coordinates and we have taken $\varphi_n^{(0)}$ normalized. From (23), multiplication by $\varphi_n^{(0)*}$ and integration gives

$$\int \varphi_n^{(0)*} H_{01} \varphi_n^{(1)} \, d\tau_e + \int \varphi_n^{(0)*} H_{02} \varphi_n^{(0)} \, d\tau_e = U_n^{(1)} \int \varphi_n^{(0)*} \varphi_n^{(1)} \, d\tau_e + U_n^{(2)}. \quad (25)$$

Now we return to the full problem, finding eigenfunctions and eigenvalues for $H_0 + \eta^4 H_1$. A typical term in $T_{\xi\xi}$ is

$$g_{ij}(\xi) \frac{\partial^2}{\partial \xi_i \, \partial \xi_j}.$$

It can be shown [7] that g_{ij} does not depend on θ. It may be expanded

$$g_{ij}(\xi) = g_{ij}(\xi_0) + \sum_k \left(\frac{\partial g_{ij}}{\partial \xi_k} \right)_\theta (\xi_k - \xi_{k0}) + \cdots$$

Going over to scaled coordinates (the ζ), we have

$$T_{\xi\xi} = \eta^{-2}(T_{\zeta\zeta}^0 + \eta T_{\zeta\zeta}^1 + \eta^2 T_{\zeta\zeta}^2 + \cdots) \quad (26a)$$

Here, $T_{\zeta\zeta}^0$ consists of terms like $g_{ij}(\xi_0) \, \partial^2/\partial \zeta_i \, \partial \zeta_j$, $T_{\zeta\zeta}^1$ terms like $(\partial g_{ij}/\partial \xi_k)_0 \zeta_k$ $\partial^2/\partial \zeta_i \, \partial \zeta_j$, and so on. Similarly, we will have

$$T_{\xi\theta} = \eta^{-1}(T_{\zeta\theta}^0 + \eta T_{\zeta\theta}^1 + \eta^2 T_{\zeta\theta}^2 + \cdots) \quad (26b)$$

and

$$T_{\theta\theta} = T_{\theta\theta}^0 + \eta T_{\theta\theta}^1 + \eta^2 T_{\theta\theta}^2 + \cdots \quad (26c)$$

All the η-dependence in the Hamiltonian is now explicit. Regrouping terms,

$$H = H_{00} + \eta H_{01} + \eta^2(H_{02} + T_{\zeta\zeta}^0) + \eta^3(H_{03} + T_{\zeta\zeta}^1 + T_{\zeta\theta}^0)$$
$$+ \eta^4(H_{04} + T_{\zeta\zeta}^2 + T_{\zeta\theta}^1 + T_{\theta\theta}^0) + \cdots \quad (27)$$

and we want to solve

$$H\Psi_n = E_n \Psi_n. \quad (28)$$

If we write

$$\Psi_n = \Psi_n^{(0)} + \eta \Psi_n^{(1)} + \eta^2 \Psi_n^{(2)} + \cdots \quad (29a)$$

$$E_n = E_n^{(0)} + \eta E_n^{(1)} + \eta^2 E_n^{(2)} + \cdots \quad (29b)$$

and substitute into (28), we may again invoke perturbation theory.

The terms in η^0 give

$$H_{00}\Psi_n^{(0)} = E_n^{(0)}\Psi_n^{(0)} \tag{30}$$

which means [see Eq. (21)]

$$\Psi_n^{(0)} = \varphi_n^{(0)}(x;\xi_{i0})\chi_n^{(0)}(\zeta,\theta) \tag{31}$$

where $\chi_n^{(0)}$ is so far an arbitrary function of nuclear coordinates, and

$$E_n^{(0)} = U_n^{(0)} \tag{32}$$

This is the fixed-nucleus electronic energy, with the nuclear positions given by ξ_{i0}. The first-order equation is

$$(H_{00} - E_n^{(0)})\Psi_n^{(1)} + (H_{01} - E_n^{(1)})\Psi_n^{(0)} \tag{33}$$

Multiply by $\varphi_n^{(0)*}$ and integrate over electronic coordinates. Using (21), (32), and the fact that $\chi_n^{(0)}$ does not vanish, (33) implies

$$\int \varphi_n^{(0)*}\left[\sum_i \left(\frac{\partial V}{\partial \xi_i}\right)_0 \zeta_i - E_n^{(1)}\right]\varphi_n^{(0)}\,d\tau_e = 0 \tag{34}$$

The ζ_i being independent coordinates, this can be true only if $E_n^{(1)} = 0$. As expected, there are no energy terms going as η—the vibrational energy goes as η^2. Equation (34) also implies that

$$\int \varphi_n^{(0)*} H_{01}\varphi_n^{(0)}\,d\tau_e = 0$$

or $U_n^{(1)} = 0$. This means that the positions "0" must represent an extremum in the fixed-nucleus electronic energy surface, limiting the present treatment to stable molecules. The first-order equation (33), since there is no differentiation with respect to nuclear coordinates, has the general solution

$$\Psi_n^{(1)} = \chi_n^{(0)}\varphi_n^{(1)} + \chi_n^{(1)}\varphi_n^{(0)} \tag{35}$$

with $\chi_n^{(1)}$ another arbitrary function of ζ and θ, and $\varphi_n^{(1)}$ the solution to (22) with $U_n^{(1)} = 0$. The second term in (35) is a general solution to the homogeneous equation

$$(H_{00} - E_n^{(0)})\Psi = 0$$

The second-order equation is

$$(H_{00} - U_n^{(0)})\Psi_n^{(2)} + H_{01}(\chi_n^{(0)}\varphi_n + \chi_n^{(1)}\varphi_n^{(0)}) + (H_{02} + T_{\zeta\zeta}^0 - E_n^{(2)})(\chi_n^{(0)}\varphi_n^{(0)}) = 0 \tag{36}$$

On multiplication by $\varphi_n^{(0)*}$ and integration over electronic coordinates, with use of (25) and $U_n^{(1)} = 0$, we obtain an equation determining $\chi_n^{(0)}$:

$$(T_{\zeta\zeta}^0 + U_n^{(2)} - E_n^{(2)})\chi_n^{(0)} = 0 \tag{37}$$

Here

$$U_n^{(2)} = \sum_{i,j} \left(\frac{\partial^2 U_n}{\partial \xi_i \, \partial \xi_j}\right)_0 \zeta_i \zeta_j$$

$$T_{\zeta\zeta}^0 = -\sum_{i,j} (g_{ij})_0 \, \frac{\partial^2}{\partial \zeta_i \, \partial \zeta_j}$$

It is seen that, at least to this order, U_n plays the role of a potential energy for nuclear motion. Since the ζ_i are independent, a transformation to new coordinates is possible, say

$$q_k = \sum_i c_i{}^k \zeta_i.$$

The coefficients may be chosen to throw $U_n^{(2)}$ into the form $\sum_i a_i q_i{}^2$ and make $T_{\zeta\zeta}^0 = \sum_i b_i \, \partial^2/\partial q_i{}^2$ (EYRING, Sect. 14f, PILAR, Sect. 15-5, LEVINE, Chap. 5). We may write

$$\chi_n^{(0)} = \sigma_{ns}(\zeta)\varrho_{ns}(\theta) \tag{38}$$

with ϱ_{ns} arbitrary for the moment and $\sigma_{ns}(\zeta)$ a product of harmonic oscillator functions in the $\{q_i\}$. The second-order energy takes the form

$$E_n^{(2)} = \sum_i (j_i + \tfrac{1}{2})\hbar\omega_i \tag{39}$$

The ω_i are the normal frequencies associated with the q_i, and the j_i are integers. The set of these vibrational quantum numbers is supposed to be specified by s. It should be noted that the transformation to the q_i which diagonalize $U_n^{(2)}$ and $T_{\zeta\zeta}^0$ requires U_n positive definite. This is guaranteed for a stable molecule provided that the origin of nuclear coordinates is taken to be the position of the minimum fixed-nucleus energy. The vibrational wave function gives terms in η^2 in the energy, as expected. A particular solution to (36) is $\chi_{ns}^{(1)}\varphi_{ns}^{(1)} + \chi_{ns}^{(0)}\varphi_{ns}^{(2)}$, where

$$(H_{00} - U_n^{(0)})\varphi_n^{(2)} + H_{01}\varphi_n^{(1)} + (H_{02} - U_n^{(2)})\varphi_n^{(0)} = 0$$

as may be seen by inspection. To this we add an arbitrary solution to the homogeneous equation

$$(H_{00} - U_n^{(0)})\psi = 0.$$

Thus we have

$$\Psi_n^{(2)} = \chi_{ns}^{(0)}\varphi_n^{(2)} + \chi_{ns}^{(1)}\varphi_n^{(1)} + \chi_{ns}^{(2)}\varphi_n^{(0)}. \tag{40}$$

$\chi_{ns}^{(1)}$ and $\chi_{ns}^{(2)}$ are determined from the third- and fourth-order equations. It may also be shown [7] that $E_{ns}^{(3)}$ vanishes, and that $E_{ns}^{(4)}$ includes the rotational energy, which thus enters as η^4. Couplings between different kinds of motions may also be elucidated [7].

Through second order (sometimes called the "harmonic approximation") our wave function is

$$\begin{aligned}
\Psi_{ns} &= \Psi_{ns}^{(0)} + \eta\Psi_{ns}^{(1)} + \eta^2\Psi_{ns}^{(2)} \\
&= \chi_{ns}^{(0)}[\varphi_n^{(0)} + \eta\varphi_n^{(1)} + \eta^2\varphi_n^{(2)}] + \eta\chi_{ns}^{(1)}[\varphi_n^{(0)} + \eta\varphi_n^{(1)}] + \eta^2\chi_{ns}^{(2)}\varphi_n^{(0)}.
\end{aligned}$$

The energy can be calculated from this correctly through η^5 terms. (This is general: The wave function to order n of perturbation theory suffices to give the energy through order $2n + 1$.) With the neglect of terms of higher order than second in η, the wave function may be written as follows:

$$\Psi_{ns} = [\varphi_n^{(0)} + \eta\varphi_n^{(1)} + \eta^2\varphi_n^{(2)}][\chi_{ns}^{(0)} + \eta\chi_{ns}^{(1)} + \eta^2\chi_{ns}^{(2)}]. \tag{41}$$

This amounts to a factoring of electronic and nuclear motions. The first bracket in (41) is, to terms in η^3, the fixed-nucleus electronic wave function as a function of nuclear coordinates, with which we will be concerned.

Recently, Moshinsky and Kittel [9] have tested the Born–Oppenheimer approximation for a simple system: two heavy particles and one light one interacting with harmonic forces. This problem is soluble exactly as well as in the Born–Oppenheimer approximation, permitting comparison of the wave functions and energies.

3. Adiabatic Treatment

An alternative to the perturbation treatment is possible. It is particularly advantageous when calculations are to be carried out. Suppose the eigenfunctions of H_0, and their eigenvalues, are computed for all $\mathbf{R}$, where $\mathbf{R}$ represents a set of nuclear coordinates.

$$H_0(r; \mathbf{R})\varphi_n(r; \mathbf{R}) = U_n(\mathbf{R})\varphi_n(r; \mathbf{R}) \tag{42}$$

Here, H_0 includes the electronic kinetic energy operator, all interparticle potential energy terms, and perhaps such corrections as mass polarization, which do not destroy the arguments to follow. We have a set of energies

U_n, $n = 0, 1, 2, \ldots$, for each $\mathbf{R}$. Here, we are particularly interested in $n = 0$. For a diatomic U_n can depend only on the internuclear distance R. As $R \to \infty$, each U_n must approach a sum of energies of states of the separated atoms. For small R, each U_n must approach the energy of some state of the united atom (formed by imagining the nuclei to coalesce) plus the internuclear repulsion energy (which goes infinite as $1/R$). By connecting corresponding U_n for adjacent values of R, we generate a set of potential curves, providing we can follow each over all R from $R = 0$ to $R = \infty$. Since the cylindrical symmetry (and the inversion symmetry for a homonuclear molecule) is maintained for all R, the eigenstates can be labeled according to their behavior under the symmetry operations of the group, and this behavior must be unchanged on changing R. There is thus no danger of confusing states of different symmetry, nor states of the same symmetry, provided that they are separated in energy. Furthermore, states of the same symmetry, whose energies differ for some R, are prevented from crossing or having the same energy at a nearby R by a theorem to be discussed in the next section. Then the correlation can be carried out uniquely.

Having generated the $U_n(\mathbf{R})$, we now look for an approximate solution of the full Schrödinger equation of the form

$$\Psi_A = \varphi_n(r; \mathbf{R})\chi_n(\mathbf{R}). \tag{43}$$

Since $H = H_0 + T_{\mathrm{nuc}}$, the latter (nuclear kinetic energy operator) involving double differentiations with respect to nuclear coordinates, we have

$$H\Psi_A = U_n(\mathbf{R})\Psi_A + [T_{\mathrm{nuc}}\chi_{ns}]\varphi_n + [T_{\mathrm{nuc}}\varphi_n]\chi_{ns} + \text{cross terms}. \tag{44}$$

The cross terms have one differentiation on χ and one on φ. If both they and the third term on the right side of (44) can be neglected, Ψ_A will be an eigenfunction of H when χ_{ns} is chosen to obey

$$[T_{\mathrm{nuc}} + U_n(\mathbf{R})]\chi_{ns} = E_A\chi_{ns}. \tag{45}$$

Here, the fixed-nucleus energy becomes the potential for nuclear motion. Note that we need not assume that U_n leads to an equilibrium configuration.

We call E_A the adiabatic energy because of the relation to the adiabatic theorem. As $\mathbf{R}$ changes, one can follow any electronic eigenstate and speak of its energy as a function of $\mathbf{R}$. Thus the motion of the nuclei, considered classically, leads to a time-dependent electronic Hamiltonian. The adiabatic theorem [10] states that if this Hamiltonian changes slowly enough compared to electronic transition frequencies no transitions will be induced. Let the electronic system be in some eigenstate at time t_1 when $\mathbf{R} = \mathbf{R}_1$. Then at time t_2, when $\mathbf{R} = \mathbf{R}_2$, it will almost certainly be in the eigenstate which correlates with the original one as H_0 changes from $H_0(\mathbf{R}_1)$ to $H_0(\mathbf{R}_2)$. Furthermore, if the nuclear configuration returns (slowly) to $\mathbf{R}_1$, the electronic state will be what it was initially. This is what is expressed by writing the wave function in the form (43). The terms we wanted to drop in (44) are related to nuclear velocities. The adiabatic behavior will not hold for small electronic transition frequencies (degeneracy or near degeneracy) and/or high nuclear velocities.

Since Ψ_A of (43), with χ_n obeying (45), is an approximate wave function, the expectation value of the full Hamiltonian over Ψ_A will be an upper bound to the true energy of the ground state. This expectation value is not E_A because of the terms in (44) dropped in going to (45). It can, in fact, be shown [11] that E_A for the lowest state is a *lower* bound to the energy. Letting Ψ be the exact wave function and E the true energy, we have

$$E \int \Psi^* \Psi \, d\tau_n \, d\tau_e = \int \Psi^*(H_0 + T_{\text{nuc}})\Psi \, d\tau_n \, d\tau_e, \qquad (46)$$

the integration being over both nuclear and electronic coordinates. Because $\varphi_0(r; \mathbf{R})$ is the lowest eigenstate of $H_0(R)$ for each R,

$$U_0(\mathbf{R}) = \frac{\int \varphi_0^* H_0 \varphi_0 \, d\tau_e}{\int \varphi_0^* \varphi_0 \, d\tau_e} \leq \frac{\int \Psi^* H_0 \Psi \, d\tau_e}{\int \Psi^* \Psi \, d\tau_e} \qquad (47)$$

In the last member, Ψ is evaluated for some fixed value of $\mathbf{R}$. Similarly, because E_A is the lowest eigenvalue of the Hamiltonian of Eq. (45),

$$E_A \int \Psi^* \Psi \, d\tau_n \leq \int \Psi^* T_{\text{nuc}} \Psi \, d\tau_n + \int \Psi^* U_0(\mathbf{R})\Psi \, d\tau_n \qquad (48)$$

Here, the electronic coordinates are fixed, and the integration is over nuclear coordinates. We now integrate both sides of (48) over electronic coordinates. In the last term, we use (47) in the form

$$\int \left[\int \Psi^* \Psi \, d\tau_e U_0(\mathbf{R}) \right] d\tau_n \leq \int \left[\int \Psi^* H_0 \Psi \, d\tau_e \right] d\tau_n$$

Thus we have

$$E_A \int \Psi^* \Psi \, d\tau_n \, d\tau_e \leq \int \Psi^* [T_{\text{nuc}} + H_0]\Psi \, d\tau_n \, d\tau_e \qquad (49)$$

or $E_A \leqq E$: we have a *lower* bound to the true energy. If we add to E_A what is obtained by multiplication of the last two sets of terms in (44) by $\Psi_A{}^*$ and integration over all coordinates, we obtain an upper bound. We would then be calculating the expectation value of the full Hamiltonian over Ψ_A (see page 18).

Equation (42), solved for each $\mathbf{R}$, gives a set of electronic eigenfunctions and energies. The functions are assumed to form a complete orthonormal set. Then an exact wave function (for fixed $\mathbf{R}$) may be expanded in them:

$$\Psi_s(r; \mathbf{R}) = \sum_i \chi_{is}(\mathbf{R})\varphi_i(r; \mathbf{R}). \tag{50}$$

This is a generalization of Eq. (43), which takes only one term in the expansion. It is sometimes referred to as the Born separation [12], and gives, in principle, a program for the calculation of corrections to the wave function Ψ_A. Substituting (50) into the eigenvalue equation for H, multiplying by some $\varphi_j(r; \mathbf{R})^*$, and integrating over electronic coordinates gives

$$\sum_i \int \varphi_j{}^*[U_i(\mathbf{R}) + T_{\mathrm{nuc}}]\chi_{is}(\mathbf{R})\varphi_i(r; \mathbf{R})\, d\tau_e = E_s\chi_{js}(\mathbf{R}).$$

Explicitly,

$$T_{\mathrm{nuc}} = -\sum_K \frac{\hbar^2}{2\mu_K} \frac{\partial^2}{\partial R_K{}^2}$$

so that we have

$$\left[-\sum_K \frac{\hbar^2}{2\mu_K} \frac{\partial^2}{\partial R_K{}^2} + U_j(\mathbf{R}) - E_s\right]\chi_{js}(\mathbf{R})$$

$$= \sum_K \frac{\hbar^2}{2\mu_K}\left[\sum_i \int \varphi_j{}^* \frac{\partial^2\varphi_i}{\partial R_K{}^2}\, d\tau_e\chi_{is} + 2\sum_i \int \varphi_j{}^* \frac{\partial\varphi_i}{\partial R_K}\, d\tau_e \frac{\partial\chi_{is}}{\partial R_K}\right]. \tag{51}$$

For φ_j real,

$$\int \varphi_j{}^* \frac{\partial\varphi_j}{\partial R_K}\, d\tau_e = \frac{1}{2} \frac{\partial}{\partial R_K} \int \varphi_j{}^*\varphi_j\, d\tau_e = 0$$

because φ_j is normalized to unity independently of $\mathbf{R}$. Also, the contribution for $i = j$ ("diagonal correction for nuclear motion") in the first term on the right of (51), may be brought to the left side. The χ_j satisfy the coupled equations

$$\left[H_{Aj} - \sum_K \frac{\hbar^2}{2\mu_K} \int \varphi_j{}^* \frac{\partial^2\varphi_j}{\partial R_K{}^2}\, d\tau_e - E_s\right]\chi_{js}(\mathbf{R})$$

$$= \sum_i^{(i \neq j)} \sum_K \frac{\hbar^2}{2\mu_K}\left[\int \varphi_j{}^* \frac{\partial^2\varphi_i}{\partial R_K{}^2}\, d\tau_e\chi_{is} + 2\int \varphi_j{}^* \frac{\partial\varphi_i}{\partial R_K}\, d\tau_e \frac{\partial\chi_{is}}{\partial R_K}\right]. \tag{52}$$

Here,

$$H_{Aj} = -\sum_K \frac{\hbar^2}{2\mu_K} \frac{\partial^2}{\partial R_K^2} + U_j(\mathbf{R})$$

the adiabatic Hamiltonian used in Eq. (52).

If the "coupling terms" or "off-diagonal corrections for nuclear motion" on the right of (52) are neglected, χ_{js} satisfies the eigenvalue equation

$$\left[H_{Aj} - \sum_K \frac{\hbar^2}{2\mu_K} \int \varphi_j{}^* \frac{\partial^2 \varphi_j}{\partial R_K^2} \, d\tau_e \right] \chi_{js}(\mathbf{R}) = E_s \chi_{js}(\mathbf{R}) \qquad (53)$$

and a single term in the expansion (50) suffices. The Hamiltonian in (53) differs from that in (45) by the diagonal correction for nuclear motion, so E_s differs from E_A and χ_{js} from χ_{js} of equation (45). While the correction appears as an additional potential energy term, it is one that depends on the nuclei, through the reduced masses μ_K. Since E_s is the expectation value of the true Hamiltonian over the wave function

$$\Psi' = \chi_{js}(\mathbf{R})\varphi_j(r; \mathbf{R}), \qquad (54)$$

it is an upper bound to the true energy of the lowest state. A similar theorem holds for excited vibrational energies within the ground electronic state [13]. It seems reasonable that (54) should be used in preference to (43), since it includes at least partially the breakdown of the simple adiabatic picture just sketched. The effect of the additional term may be calculated from the χ_{js} of (45) by perturbation theory, since it is small. By use of a closure or average energy approximation (LEVINE, Sect. 9-11; Sect. III.B.1 below), even the effect of the off-diagonal terms can be incorporated into a local potential for nuclear motion, preserving the simple picture. Usually, however, we distinguish between adiabatic and nonadiabatic corrections. The latter, if important, make it impossible to describe the nuclear motion in terms of particles moving under the influence of an interaction potential.

Solution of (53) produces a set of nuclear wave functions χ_{js} which may be considered complete. Then we may form the wave functions

$$\Psi_k = \varphi_k(r; \mathbf{R})\chi_{js}(\mathbf{R}), \qquad j = 1, \ldots, \infty, \quad s = 1, \ldots, \infty$$

which almost diagonalize the Hamiltonian and treat the off-diagonal terms by perturbation theory [3]. Note that the electronic state from which the nuclear wave functions are derived is always the same and may be arbitrarily chosen. It is also interesting to note that if one takes a variational trial function of the form (54), where φ_j obeys the eigenvalue equation, Eq. (42), and chooses $\chi_{js}(\mathbf{R})$ to make the expectation value of the Hamiltonian stationary, one finds [13a] that $\chi_{js}(\mathbf{R})$ must obey (53).

For scattering problems, the use of the conventional adiabatic states becomes inconvenient when one considers certain types of reactive collisions. O'Malley [14] has proposed the use of a different set of states, variously called quasi stationary, resonant, or diabatic. These states do not diagonalize the electronic Hamiltonian, and hence do not obey the noncrossing rule of the next section.

4. Noncrossing Rule

As already mentioned, we are interested in the sets of electronic energies corresponding to a series of nuclear configurations. In order to speak of potential curves, we must be able to identify each energy for one value of $\mathbf{R}$ with one for a nearby value of $\mathbf{R}$. If two states of the same symmetry could have the same energy at some nuclear configuration (a "crossing" of potential energy curves) their identities would be lost. The theorem discussed in this section prevents such an occurrence.

We first give the proof of von Neumann and Wigner [15]. Suppose we compute matrix elements, in some basis, of a Hamiltonian which depends parametrically on parameters $\{\mathbf{R}\}$. If the Hamiltonian is invariant under some symmetry group, we need consider only the basis functions of some one symmetry. The matrix of the Hamiltonian may be written in the form

$$\mathbf{H} = \mathbf{U}\,\mathbf{E}\,\mathbf{U}^\dagger \tag{55}$$

where $\mathbf{E}$ is the diagonal matrix of the eigenvalues ($\mathbf{H}$ being Hermitian, $\mathbf{E}$ is real) and $\mathbf{U}$ the unitary matrix which diagonalizes $\mathbf{H}$. If the basis is n-dimensional, $\mathbf{E}$ has n elements and $\mathbf{U}$ has n^2. A unitary matrix of dimension n in general requires n^2 real parameters for its complete specification, but $\mathbf{U}$ is arbitrary to the extent of postmultiplication by any unitary matrix $\mathbf{V}$ which commutes with $\mathbf{E}$. Here $\mathbf{V}$ is block-diagonal; the blocks correspond to the degeneracies in $\mathbf{E}$, and each block is unitary. Thus, if we demand that there be f different eigenvalues and that the eigenvalue E_i be g_i-fold degenerate ($\sum_{i=1}^{f} g_i = n$), $\mathbf{V}$ consists of f blocks and contains $\sum_{i=1}^{f} g_i^2$ real parameters. The specification of $\mathbf{U}$ then demands only $n^2 - \sum_{i=1}^{f} g_i^2$ real parameters, and the specification of $\mathbf{H}$ demands

$$N = n^2 - \sum_{i=1}^{f} g_i^2 + f \tag{56}$$

parameters. For no degeneracy (all $g_i = 1$) $N = n^2$; for one double degeneracy $N = n^2 - 3$. Therefore, in order to have a degeneracy for one value of **R** and not for another, at least three parameters must be available. In the case of the electronic Hamiltonian for a diatomic molecule, there is only one parameter available—the internuclear distance R. Thus there can be no degeneracies at some one R which disappear for another R, i.e., no crossings of potential curves.

This means that the potential energy curves seem to exert a mutual repulsion as they approach each other. This is seen more clearly by the following demonstration.

Suppose the eigenvalue problem has been solved for R_1, where the fixed-nucleus Hamiltonian is H_1, and we try to solve for a nearby value R_2 where the Hamiltonian is H_2. As a basis for the latter problem, let us use the eigenfunctions of H_1. The difference between H_1 and H_2 may be considered as a small perturbation. Suppose further that at R_1 the two states φ_a and φ_b have nearly equal energies, i.e., the difference between their energies is much smaller than between either and energies of other states. Then it is a good approximation to consider only these two when studying the energy of either at R_2. The Schrödinger equation leads to the secular determinant

$$\begin{vmatrix} E_a + V_{aa} - E & V_{ab} \\ V_{ba} & E_b + V_{bb} - E \end{vmatrix} = 0. \qquad (57)$$

Here, $E_a = \langle \varphi_a | H_1 | \varphi_a \rangle$, $V_{ab} = \langle \varphi_a | H_2 - H_1 | \varphi_b \rangle$, $V_{ba} = \langle \varphi_b | H_2 - H_1 | \varphi_a \rangle$, $E_b = \langle \varphi_b | H_1 | \varphi_b \rangle$, and φ_a and φ_b are orthonormal. The two energies for $R = R_2$ are the solutions to a quadratic equation, and will be equal if

$$(E_a + V_{aa} - E_b - V_{bb})^2 + 4 | V_{ab} |^2 = 0.$$

This can occur only if $E_a + V_{aa} = E_b + V_{bb}$ and $V_{ab} = 0$ simultaneously. By choosing a value for the single parameter R_2 it will not be possible to satisfy both conditions. Thus, if the energies for two states at $R = R_1$ are sufficiently close together (so that the effect of consideration of other states would be small) there is no nearby value of R for which their energies will become equal. Therefore the curves of the functions $E_a(R)$ and $E_b(R)$ cannot cross.

The argument breaks down when $V_{ab} = 0$. Then the roots are $E_a + V_{aa}$ and $E_b + V_{bb}$ (no interaction between the states a and b), and these may be equal. In general, $V_{ab} = 0$ only when φ_a and φ_b are of different symmetry. (For diatomics, $H_2 - H_1$ has the full cylindrical symmetry.) An exception is the case where the functions φ_a and φ_b are each products of functions for different coordinates, say ξ and η. If the perturbation is a sum of terms in these coordinates,

$$\langle \varphi_a \mid H_2 - H_1 \mid \varphi_b \rangle = \langle \varphi_a'(\xi)\varphi_a''(\eta) \mid V_\xi(\xi) + V_\eta(\eta) \mid \varphi_b'(\xi)\varphi_b''(\eta) \rangle$$

$$= \langle \varphi_a' \mid V_\xi \mid \varphi_b' \rangle \langle \varphi_a'' \mid \varphi_b'' \rangle + \langle \varphi_a' \mid \varphi_b' \rangle \langle \varphi_a'' \mid V_\eta \mid \varphi_b'' \rangle$$

which will vanish when φ_a' is orthogonal to φ_b' and φ_a'' orthogonal to φ_b''. This will occur when the Hamiltonian is separable in the form

$$H = H_\xi(\xi) + H_\eta(\eta)$$

but something similar arises when it is of the form

$$H = [H_\xi(\xi) + H_\eta(\eta)]/[f_\xi(\xi) + f_\eta(\eta)]$$

The latter is applicable [16] to the one-electron diatomic molecular ion. Crossings of potential curves corresponding to states of the same symmetry do occur here. But it is still possible to correlate and follow potential curves by investigation of the nodal properties.

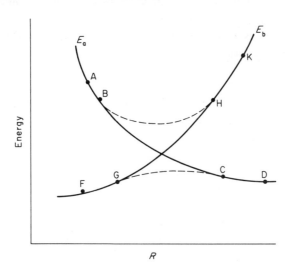

Figure 1.

The noncrossing rule refers to the exact potential curves. Approximate functions may give potential curves that cross. Such a case is shown in Fig. 1, where E_a and E_b (solid lines) are energies (as a function of R) calculated with approximate functions $\tilde{\varphi}_a$ and $\tilde{\varphi}_b$ which depend parametrically on R. For a value of R for which $\tilde{\varphi}_a$ and $\tilde{\varphi}_b$ have approximately the same energy, better wave functions and energies are obtained by taking a linear combination of them and using the variational principle:

$$\psi = c_a\tilde{\varphi}_a + c_b\tilde{\varphi}_b. \tag{58}$$

Let S_{aa}, S_{bb}, etc. be the overlap integrals between the functions, and $\tilde{H}_{aa}$, $\tilde{H}_{ab}$, etc. matrix elements of H, the Hamiltonian for some particular value of R: $E_a = \tilde{H}_{aa}/S_{aa}$, $E_b = \tilde{H}_{bb}/S_{bb}$. Then (58) leads to the secular equation

$$f(E) = \begin{vmatrix} \tilde{H}_{aa} - ES_{aa} & \tilde{H}_{ab} - ES_{ab} \\ \tilde{H}_{ba} - ES_{ba} & \tilde{H}_{bb} - ES_{bb} \end{vmatrix} = 0 \tag{59}$$

This may be written as a quadratic equation in E. Since $f(E)$ is positive for $E \to \pm\infty$, it is a parabola in E concave upward. The value of $f(E)$ is easily seen to be negative for $E = \tilde{H}_{aa}/S_{aa}$ and $E = \tilde{H}_{bb}/S_{bb}$. The roots, where $f(E) = 0$, therefore lie above the higher and below the lower of these two quantities, which are, of course, the original approximate energies. There is thus a "repulsion" which prevents the crossing, as shown in the diagram, where the full lines indicate the original energies and the dotted lines the final energies, using (58). Far from the crossing the effect of mixing the approximate wave functions becomes small and the roots are essentially $E = E_a$ and $E = E_b$.

Where E_a is much lower than E_b, to the right in Fig. 1, the lower eigenvalue has a wave function essentially $\tilde{\varphi}_a$. On the other side of the crossing region, the wave function of the lower curve is essentially $\tilde{\varphi}_b$. In going through the crossing region, the nature of the function corresponding to the lower root changes gradually from pure $\tilde{\varphi}_a$ to pure $\tilde{\varphi}_b$. This can be seen by examing the explicit solutions for energy and wave function, derived from writing the secular equation as a quadratic. It can be said that the crossing which exists for the approximate potential curves is removed in the first approximation (58), but that the zeroth approximation $\psi = \tilde{\varphi}_a$ or $\psi = \tilde{\varphi}_b$ is good except near the region of the crossing.

5. *Separation of Rotation*

For a diatomic molecule, the nuclear wave function χ_{ns} of Eq. (45) (where certain coupling terms have been neglected) may be written as a product of vibrational and rotational wave functions by going over to spherical coordinates for the nuclei. We may write

$$T_{\text{nuc}} = -\frac{\hbar^2}{2\mu R^2}\left[\frac{\partial}{\partial R}\left(R^2\,\frac{\partial}{\partial R}\right) + \frac{1}{\sin\theta}\,\frac{\partial}{\partial\theta}\left(\sin\theta\,\frac{\partial}{\partial\theta}\right) + \frac{1}{\sin^2\theta}\,\frac{\partial^2}{\partial\varphi^2}\right] \tag{60}$$

Here, R is the length of the internuclear axis and θ and φ its orientation in space. Evidently the eigenfunction can be written as a product of a radial (vibrational) wave function and a function of θ and φ. The latter is an eigenfunction of the angular part of the Laplacian with eigenvalue $J(J+1)$, J an integer. The radial wave function χ_R obeys

$$\left\{\frac{-\hbar^2}{2\mu R^2}\left[\frac{\partial}{\partial R}\left(R^2\,\frac{\partial}{\partial R}\right)\right] + U_n(R) + \frac{J(J+1)\hbar^2}{2\mu R^2}\right\} = E_R\chi_R \tag{61}$$

Thus the effective potential for nuclear motion is the fixed-nucleus potential plus a repulsive centrifugal term for $J > 0$. However, the terms representing interaction between nuclear and electronic motions, dropped in obtaining (45), make this simple picture invalid. The separation of rotational motion cannot be done exactly [6].

In an exact treatment, J would not be a good quantum number. What is conserved is not the nuclear angular momentum but the total angular momentum, which includes an electronic contribution. The operator for the square of the total momentum is [17]

$$M^2 = -\hbar^2\left\{\frac{1}{\sin\theta}\,\frac{\partial}{\partial\theta}\left(\sin\theta\,\frac{\partial}{\partial\theta}\right)\right.$$
$$\left. + \frac{1}{\sin^2\theta}\left(\frac{\partial}{\partial\varphi} - i\,\frac{L_z}{\hbar}\,\cos\theta\right)^2 - \frac{L_z^2}{\hbar^2}\right\} \tag{62}$$

L_z is the electronic angular momentum operator along the internuclear axis. There are actually coupling terms between nuclear rotation and electronic motion which were neglected in (45). They may be displayed by considering the transformation, for the electronic coordinates, from a system of axes with directions fixed in space to one rotating with the nuclear frame. One transformation that accomplishes this is given by [19]

$$\begin{pmatrix} x_i \\ y_i \\ z_i \end{pmatrix} = \begin{pmatrix} \cos\theta\cos\varphi & \cos\theta\sin\varphi & -\sin\theta \\ -\sin\theta & \cos\varphi & 0 \\ \sin\theta\cos\varphi & \sin\theta\sin\varphi & \cos\theta \end{pmatrix} \begin{pmatrix} x_i' \\ y_i' \\ z_i' \end{pmatrix} \tag{63}$$

Here, the primed coordinates refer to the space-oriented system and the unprimed coordinates refer to the system rotating with the internuclear axis. The z axis is along R, the y axis perpendicular to the z and z' axes, and the x axis is chosen to form a right-handed system. We expect the description of the electronic wave function to be simpler in the new coordinates, if certain small terms, representing the breakdown of the Born–Oppenheimer picture or the inability of the electrons to follow the nuclear tumbling, are neglected.

We must introduce the new coordinates into the Hamiltonian of Eq. (7). There will be no change in H_0 [Eq. (8)], as expected, since the Laplacian is invariant to rotation. Nor will the mass polarization terms be altered. However, the differentiations with respect to nuclear coordinates now lead to additional terms because of the implicit dependence of the electronic coordinates on θ and φ. Differentiation with respect to θ is replaced by $\partial/\partial\theta - i(L_y/\hbar)$, where the partial derivative indicates differentiation with respect to the explicit occurrence of θ only, and L_y is the operator for the y component of electronic angular momentum in the molecule-fixed coordinate system. Similarly, the differentiation with respect to φ becomes

$$\partial/\partial\varphi + iL_x/\hbar \sin\theta - iL_z/\hbar \cos\theta.$$

Then the additional terms from the Laplacian, responsible for coupling between electronic and rotational motion, are $C_1 + C_2 + C_3$, where

$$C_1 = \frac{1}{R^2}\left\{\frac{L_x{}^2 + L_y{}^2}{\hbar^2} - \cot^2\theta\,\frac{L_z{}^2}{\hbar^2} - 2i\,\frac{\cot\theta}{\sin\theta}\,\frac{L_z}{\hbar}\right\} \qquad (64a)$$

$$C_2 = \frac{L_+}{R^2\hbar}\left\{-\frac{\partial}{\partial\theta} + \cot\theta\,\frac{L_z}{\hbar} + \frac{i}{\sin\theta}\,\frac{\partial}{\partial\varphi}\right\} \qquad (46b)$$

$$C_3 = \frac{L_-}{R^2\hbar}\left\{\frac{\partial}{\partial\theta} + \cot\theta\,\frac{L_z}{\hbar}\,\frac{i}{\sin\theta}\,\frac{\partial}{\partial\varphi}\right\} \qquad (64c)$$

Fisk and Kirtman [3] and Meath and Hirschfelder [18] have given similar expressions for different coordinate systems.

The Hamiltonian is now

$$H = -\frac{\hbar^2}{2m}\sum_i \boldsymbol{\nabla}_i{}^2 + V$$

$$-\frac{\hbar^2}{2\mu R^2}\left[\frac{\partial}{\partial R}\left(R^2\frac{\partial}{\partial R}\right) + \frac{1}{\sin\theta}\frac{\partial}{\partial\theta}\left(\sin\theta\frac{\partial}{\partial\theta}\right) + \frac{1}{\sin^2\theta}\frac{\partial^2}{\partial\varphi^2}\right]$$

$$-\frac{\hbar^2}{2\mu}[C_1 + C_2 + C_3] \qquad (65)$$

It may be shown straightforwardly that M^2 [Eq. (62)] commutes with H, so that the eigenfunctions of H are also eigenfunctions of M^2 with eigenvalues $J(J+1)\hbar^2$. Furthermore, if C_2 and C_3 may be neglected, L_z commutes with H, so that we can specify the component of electronic angular momentum along the internuclear axis. This quantity is, of course, conserved in the fixed-nuclei case. Furthermore, the eigenfunction (after neglect of C_2 and C_3) may be factored [17] into rotational and vibronic wave functions, the first depending on θ, φ, and the angle of rotation about the figure axis. The vibronic wave function is an eigenfunction of

$$-\frac{\hbar^2}{2m}\sum_i \boldsymbol{\nabla}_i^2 + V - \frac{\hbar^2}{2\mu R^2}\frac{\partial}{\partial R}\left(R^2\frac{\partial}{\partial R}\right) + \frac{\hbar^2}{2\mu R^2}J(J+1)$$

$$-\frac{\hbar^2}{2\mu R^2}(L_x^2 + L_y^2 + L_z^2)$$

Were it not for the last term, one could speak of vibrational motion of the nuclei under the influence of the fixed-nucleus electronic energy plus a centrifugal term, as in (61), with a change in the meaning of J. It may be a good approximation to calculate the effect of the last term by first-order perturbation theory.

The coupling of electronic and rotational motions is generally treated as a perturbation after classification as one of Hund's cases [20, 21] and we shall not discuss it here. Note that, for some of these, the spin angular momentum must be included in the electronic angular momentum above [17]. For further discussion of the coupling terms neglected in separating out rotational motion, see Pack and Hirschfelder [6].

6. *Correcting for Nuclear–Electronic Interaction*

We have indicated the origins of terms which couple the electronic motion to the translation, rotation, and vibration of the molecule. All of these make inexact the picture of nuclei moving in a potential given by the solution to the electronic Schrödinger equation for nuclei fixed. However, some of the terms may be treated in such a way as to maintain the idea of nuclei moving according to some interaction potential. For example, it is easy to include the mass-polarization terms in the electronic Hamiltonian, as emphasized by Fröman [1], or to calculate their effect by perturbation theory. The result is what Fröman calls a "reduced electronic energy." Thus, without modifying the idea of an interaction potential, we can come closer to the actual effective interaction. Certain corrections arising from the separation of the rotational motion, from relativity, etc., can be similarly treated. Hirschfelder and Meath [18, p. 18] suggest that the effect of nonadiabatic corrections due to rotation can be minimized by choosing an effective angular momentum quantum number in the centrifugal potential, which is supposed to include some of the nonnuclear contributions to the angular momentum.

It seems that, in many cases, nonadiabatic effects may be ignored [2, 22]. Among the exceptions are muonic molecules, where the difference in masses between the different particles involved is not as great as for ordinary molecules. A treatment for a three-particle system, related to the Born separation discussed earlier, has been given by Hunter and co-workers [23]. The relation to the usual treatment when one particle is much lighter than the other two is indicated. Since the treatment is variational, an upper bound on the energy must be obtained. The effect of nonadiabaticity is less than 10^{-6} hartrees, i.e., several tenths of a wavenumber, for ordinary molecules. Naturally, both adiabatic and nonadiabatic effects of coupling between nuclear and electronic motions are most important for H_2^+, H_2, and related molecules, because of the small nuclear masses. At the same time, extremely accurate experimental and theoretical results are available to test the various approximations. A detailed discussion of the effects of coupling is beyond the scope of this book. The reader is referred to the recent and growing literature [24, 25].

Even where an adiabatic treatment is not applicable, the adiabatic or fixed-nuclei curves may be useful. If a set of product-type wave functions is used to treat the exact problem [as in Eq. (50)], perturbation theory arguments show that trouble arises when several such functions have almost the same energy. Then we can no longer speak of nuclei interacting according to some potential. However, we can usually describe the situation in terms of a small number of adiabatic states [4, 26, 27].

Suppose we have two approximate potential curves ABCD and FGHK which cross, as in Fig. 1. Except near the crossing point, they are supposed to be good approximations to the exact fixed-nuclei curves. Near the crossing, both curves must be taken into account simultaneously, producing a "repulsion" and new nonintersecting potential curves, ABHK and FGCD, for which vibrational levels may be calculated. If a vibrational level of ABHK and one from FGCD have approximately the same energy, their interaction must be considered; the adiabatic picture is meaningless. It may be more convenient to consider the interaction of vibrational levels corresponding to the nonadiabatic curves ABCD and FGHK [20, pp. 296–297]. Under certain circumstances, the nuclei behave as if ABCD and

FGHK were the relevant curves, i.e., the system goes from one adiabatic curve to another. London's treatment [27] for chemical reactions shows that the adiabatic viewpoint is valid for small velocities and/or large energy separations between the adiabatic curves, while the nonadiabatic curves are relevant for large velocities. However, the definition of "nonadiabatic curves" leads to some problems [14].

It must be noted that, in principle, we can dispense entirely with the notion of potential curves. Without attempting to separate the electronic and nuclear motions, we can try to find the complete wave function, a function of both nuclear and electronic coordinates. In particular, we can use a variational method with a trial function like (50), but with a finite number of terms [19, 28]. The completely nonadiabatic approach has been discussed by Kołos and Wołniewicz [19] and applied to H_2, D_2, and T_2 [29]. They contend that this approach is in some ways more efficient than the separation of nuclear and electronic motions and formation of product functions.

While this may be true for the highly accurate calculations for the energy levels of two-electron diatomics, with which Kołos and Wołniewicz were concerned, it is apparently not true in general. Calculations of molecular energies almost universally proceed by way of the potential curve, obtained by solving the electronic problem for fixed nuclei. There are a number of reasons. Partly, it is because many well-understood and well-tested techniques are available for the calculation of electronic wave functions, developed over the years. In addition, the notion of the potential curve is, in almost all cases, an excellent approximation, which also seems to make good physical sense. As we shall see in the next section, a wide variety of experimental results are understood in terms of this concept. Unless we seek only numbers for comparison with experiment, we probably would separate electronic and nuclear motion at the end of a calculation, even if not at the beginning. We feel strongly that the nonadiabatic approach will see only limited application in the future to the systems which concern us here.

REFERENCES

1. A. Fröman, *J. Chem. Phys.* **36**, 1490 (1962).
2. A. Dalgarno and R. McCarroll, *Proc. Roy. Soc. Ser. A* **237**, 383 (1956).
3. G. A. Fisk and B. Kirtman, *J. Chem. Phys.* **41**, 3516 (1964).
4. D. W. Jepsen and J. O. Hirschfelder, *J. Chem. Phys.* **32**, 1323 (1960).
5. J. O. Hirschfelder and W. J. Meath, *Advan. Chem. Phys.* **12**, 3 (1967).
6. R. T. Pack and J. O. Hirschfelder, *J. Chem. Phys.* **49**, 4009 (1968); A. Dalgarno and R. McCarroll, *Proc. Roy. Soc. Ser. A* **239**, 413 (1957).

7. M. Born and J. R. Oppenheimer, *Ann. Phys.* (*Leipzig*) **84**, 457 (1927); M. Born and K. Huang, "Dynamical Theory of Crystal Lattices," Chapter IV and Appendix VII. Oxford Univ. Press, London and New York, 1954.
8. A. D. Liehr, *Ann. Phys.* (*New York*) **1**, 221 (1957).
9. M. Moshinsky and C. Kittel, *Proc. Nat. Acad. Sci. U.S.* **60**, 1110 (1968).
10. H. A. Kramers, "Quantum Mechanics." North-Holland Publ., Amsterdam, 1958.
11. S. Epstein, *J. Chem. Phys.* **44**, 836 (1966).
12. M. Born, *Nachr. Akad. Wiss. Goettingen* **1951**, 1; M. Born and K. Huang, "Dynamical Theory of Crystal Lattices," Appendix VIII. Oxford Univ. Press, London and New York, 1954.
13. S. T. Epstein, *J. Chem. Phys.* **49**, 1436 (1968).
13a. H. C. Longuet-Higgins, *Advan. Spectrosc.* **2**, 429 (1961).
14. T. F. O'Malley, *Phys. Rev.* **162**, 98 (1967); *J. Chem. Phys.* **51**, 322 (1969); S. H. Lin, *Theor. Chimica Acta* **8**, 1 (1967).
15. J. von Neumann and E. Wigner, *Phys. Z.* **30**, 467 (1929).
16. M. Kotani, K. Ohno, and K. Kayama, *in* "Encyclopedia of Physics" (S. Flügge, ed.), Vol. XXXVII, pt. 2, p. 58. Springer–Verlag, Berlin and New York.
17. J. G. Valatin, *Proc. Phys. Soc. London* **58**, 695 (1946).
18. J. O. Hirschfelder and W. J. Meath, *Advan. Chem. Phys.* **12**, 14 (1967).
19. W. Kołos and L. Wolniewicz, *Rev. Mod. Phys.* **35**, 473 (1963).
20. G. Herzberg, "Spectra of Diatomic Molecules," Chapter 5. Van Nostrand-Reinhold, Princeton, New Jersey, 1950.
21. P. R. Fontana, *Phys. Rev.* **125**, 220 (1962); J. H. Van Vleck, *Ibid.* **33**, 467 (1929); *J. Chem. Phys.* **4**, 327 (1936).
22. W. Kołos, *Int. J. Quantum Chem.* **2**, 471 (1968).
23. G. Hunter, B. F. Gray, and H. O. Pritchard, *J. Chem. Phys.* **45**, 3806 (1966); G. Hunter and H. O. Pritchard, *Ibid.* **46**, 2146, 2153 (1967).
24. G. Hunter, *J. Chem. Phys.* **45**, 3022 (1966).
25. L. Wolniewicz, *J. Chem. Phys.* **45**, 515 (1966).
26. W. D. Hobey, *J. Chem. Phys.* **43**, 2187 (1965).
27. F. London, *Z. Phys.* **74**, 143 (1932).
28. R. Seiler, *Int. J. Quantum Chem.* **3**, 25 (1969).
29. W. Kołos and L. Wolniewicz, *J. Chem. Phys.* **41**, 3674 (1964).

C. Measurement of Potential Energy Curves

We now mention the available methods for obtaining information about the functions $U(R)$ from experimental measurement. In general, this means observation of the motion of the nuclei under the influence of $U(R)$ (or, in some cases, of several such functions). A very useful and complete review of the subject has been given by Mason and Monchick [1]. Others will be mentioned in due course and referred to often.

If the internuclear potential is negative for some R (assuming it is defined so as to approach zero for very large R), there is the possibility of formation of a bound molecule. The radial motion of the nuclei is then a vibration of R about an equilibrium value. This motion is quantized, and transitions between the energy levels may be induced and observed by spectroscopy. The more observed transitions, the more information is gained about the shape of the potential energy curve.

Where the interaction is repulsive, or where the energy of radial motion is too large for the nuclei to be bound by an attractive potential, the motion is not quantized. We have a scattering problem: Phase shifts and scattering cross sections are of interest. Like vibrational energy levels, these reflect $U(R)$, so that their measurement can be considered a measurement on the potential.

Intermediate between these two situations is the case where the potential goes through a maximum. While this is uncommon and has so far not been found for a ground state interaction (cases exist for excited states), the centrifugal term in Eq. (61) superposed on an attractive potential, can give a maximum. For energy below the top of the maximum, there are metastable states. Where they are very long-lived, spectroscopic measurements may be possible. Otherwise, they appear as scattering resonances.

In a gas, the individual scattering events are responsible for transport and other properties. Measurements of these can then also give information about the interaction potential $U(R)$. The macroscopic equation of state can, in principle, also be derived from the properties of individual molecules and their interactions. Thus, measurements of equilibrium properties of matter (both for gases and condensed phases) can be used to measure $U(R)$.

In what follows, we discuss first spectroscopic investigations, then simple scattering, and finally other methods.

1. *Spectroscopy*

We are concerned principally with rotation–vibration spectroscopy. After having measured and identified the spectral lines, we may construct the scheme of rotational and vibrational energy levels. The vibrational spacing depends on the shape of the potential energy curve; in particular, the energy of a given level is determined by the portion of the curve corresponding to values of R for which the vibrational wave function is appreciable, which amounts to the region between the classical turning points. The rotational spacing depends on the moment of inertia, which in turn depends on the average internuclear distance, and on its variation with vibrational state.

The monograph of Herzberg [2] is the preeminent reference on the theory of the spectroscopy of diatomic molecules. A compendium of information about the spectroscopic properties of diatomic molecules has recently appeared [3]. Mason and Monchick's review article [1, Chapter III] discusses the extraction of information on $U(R)$ from spectroscopic measurements.

Suppose $U(R)$ goes through a minimum at $R = R_e$. The vibrational wave functions will be largest near R_e, so it is convenient to expand

$$U(R) = U(R_e) + \tfrac{1}{2}U''(R_e)(R - R_e)^2 + \tfrac{1}{6}U'''(R_e)(R - R_e)^3 + \cdots \quad (66)$$

Primes are derivatives with respect to R; the first derivative vanishes at the minimum. If third and higher derivatives may be neglected (this is a better approximation for the lowest state wave functions, which are large only very near R_e) we have to solve

$$\left\{\frac{-\hbar^2}{2\mu R^2}\left[\frac{\partial}{\partial R}\left(R^2\frac{\partial}{\partial R}\right)\right] + U(R_e) + \tfrac{1}{2}U''(R_e)(R - R_e)^2 + \frac{J(J+1)\hbar^2}{2\mu R^2}\right\}\chi_R$$

$$= E_R\chi_R \quad\quad\quad (67)$$

For $J = 0$, the potential is that of a harmonic oscillator, and the energies are (PILAR, Sect. 4–5, KAUZMANN, Chapt. II-E)

$$E_n = U(R_e) + (n + \tfrac{1}{2})\hbar\omega_e \quad\quad (68)$$

where n is an integer and $\omega_e = [U''(R_e)/\mu]^{1/2}$. Where the harmonic oscillator approximation is valid, the spacing between the levels gives the curvature of the potential curve at the minimum.

The term due to rotation may crudely be taken as a perturbation, and its contribution estimated as its average value over a vibrational wavefunction. The wavefunctions being centered about $R = R_e$, a term $J(J+1)\hbar^2/2\mu R_e^2$ is added to (68). Thus the pure rotation spectrum gives the equilibrium internuclear distance. More accurately, one may expand [4] R^{-2} in a power series in $(R - R_e)/R_e$, stopping with the second power, and transform the resulting equation to a harmonic oscillator equation in the variable $R - R_e + a$, where a is a constant. The energy levels, after some simplification, are given by

$$E_n = (n + \tfrac{1}{2})\hbar\omega_e + U(R_e) + \frac{J(J+1)\hbar^2}{2\mu R_e^2} - \frac{J^2(J+1)^2\hbar^4}{2\omega_e^2(\mu R_e^2)^3} \quad (69)$$

the last term corresponding to centrifugal stretching. In fact, the energy levels of a vibrating rotator are given by a formula like (69), with the addition of a term in $(n + \frac{1}{2})^2$ and the replacement of the coefficients of $J(J + 1)$ and $J^2(J + 1)^2$ by similar quantities corresponding to averages over vibrational state. The former coefficient is the rotational constant B_n, an average of R^{-2} over the vibrational state n. Extrapolation of B_n to $n = -\frac{1}{2}$ gives R_e.

We note that $U(R_e)$, which gives the depth of the well relative to the energy for infinite R, enters only as a constant term, canceling out of the expression for a rotation–vibration transition frequency. It can be obtained from the dissociation energy by adding the zero-point vibrational energy, which is calculable when the vibrational frequency is known. The dissociation energy is the sum of all vibrational quanta, so it is, in principle, measurable spectroscopically as well as thermally. An alternative to measuring many energy differences between adjacent states is the observation of the transition corresponding to a final state where the molecule just dissociates. This appears as a convergence limit of a series of vibrational lines. Extrapolation is often necessary, the limit not being directly observed. Sometimes, one may observe the onset of the continuum via the long-wavelength limit of a continuous spectrum. For a fuller discussion, see Herzberg [2, Sect. VII.2].

Irregularities in the rotational structure of a vibration–rotation band may be evidence of predissociation [2, p. 405 et seq.; 1, p. 354; 5], which is due to a crossing of potential energy curves. If $U(R)$ is accurately known for some of these, data on predissociation can give information on others [1, p. 354; 5]. Predissociation due to tunneling through a centrifugal barrier can also give information on $U(R)$ [1, p. 354; 5]. In particular, the dissociation energy of the rotationless state may be obtained by extrapolation ("limiting curve of dissociation") [6–8].

It is clear that the harmonic oscillator is a poor representation of an actual attractive potential except very near the minimum, since the true U rises sharply for $R < R_e$ and goes up to a finite limiting value for $R \to \infty$. The first neglected term in (66), $U'''(R_e)$, provides the asymmetry. The eigenfunctions and eigenvalues of the anharmonic oscillator are discussed in Herzberg [2, Sect. III.2]. Observation of the spectra then gives $U'''(R_e)$. More and more terms of the expansion (66) are needed to describe adequately higher vibrational levels, since their wave functions sample regions of R progressively further from R_e. No finite power series, however, can describe the entire range of R, since a power series cannot go infinite except for infinite values of the argument, whereas an acceptable $U(R)$ must be infinite for $R = 0$ and finite for all $R > 0$. Therefore, we must be cautious in interpreting results according to a representation of U as a finite power-series expansion.

The technique for obtaining rotation–vibration energy levels from a potential given as a power series was developed by Dunham [9, 10]. It is convenient to replace R by the dimensionless variable

$$\xi = (R - R_e)/R_e \tag{70}$$

and put $\chi(R) = P(\xi)/R$. The equation for P is

$$-\frac{\hbar^2}{2\mu R_e^2} \frac{d^2P}{dR^2} + \left[V(\xi) + \frac{J(J+1)\hbar^2}{2\mu R_e^2(1+\xi)^2} - E \right] P = 0 \tag{71}$$

where $V(\xi) = U(R)$. We imagine the potential to be expanded as

$$V(\xi) = a_0\xi^2(1 + a_1\xi + a_2\xi^2 + a_3\xi^3 + \cdots) \tag{72}$$

where a_0 is in energy units and proportional to $U''(R_e)$; a_i $(i > 0)$ is dimensionless. The centrifugal term may also be put into the form of a power series:

$$(1 + \xi)^{-2} = 1 - 2\xi + 3\xi^2 - 4\xi^3 + \cdots$$

Then the effective potential is

$$J(J+1)B_e - 2J(J+1)B_e\xi + \left(a_0 + 3J(J+1)B_e \right)\xi^2$$
$$+ \left(a_0 a_1 - 4J(J+1)B_e \right)\xi^3 + \cdots$$

with $B_e = \hbar^2/2I_e$ and $I_e = \mu R_e^2$. The Schrödinger equation is solved by the WKB method. More accurate results are obtained in this semiclassical procedure when $J(J+1)$ is replaced by $(J + \frac{1}{2})^2$. The resulting eigenvalue may be written in the form

$$E_{vJ} = \sum_{l,j} Y_{lj}(v + \tfrac{1}{2})^l [J(J+1)]^j \tag{73}$$

where l, j, and v are integers, and the Y_{lj} are known functions of the a_i and B_e: each Y_j is a power series in B_e/a_0. In particular, $Y_{10} \sim \hbar\omega_e$, $Y_{01} \sim B_e$, $Y_{11} \sim 6(B_e^2/\hbar\omega_e)(1 + a_1)$. Since

$$a_0 = (\partial^2 V/\partial \xi^2)_{\xi=0} = R_e^2 U''(R_e) = \hbar^2\omega_e^2/2B_e$$

B_e/a_0 is the square of the ratio of a typical rotational energy to a typical vibrational energy. Referring to our discussion of the Born–Oppenheimer separation, we have $B_e/a_0 \sim \eta^4 \sim 10^{-4}$. Thus we would not go past terms linear in B_e/a_0 without considering the adequacy of the Born–Oppenheimer approximation itself.

If the observed spectral lines are fitted to differences of terms of the form (73), the Y_{lj} and hence the potential constants may be determined. From (72), $V(\xi)$ may be constructed, and we may speak of a measurement of the region of the potential curve sampled by the vibrational wave functions involved. In general, there will be more Y_{lj} than potential constants, so one may check for internal consistency. It was suggested [9, 10] that a lack of consistency indicates difficulties arising from the Born–Oppenheimer separation, but other sources are possible.

Only recently [11] has the effect of the breakdown of the Born–Oppenheimer approximation on the expressions for the Y_{ij} been treated completely. First, the adiabatic corrections [see Eq. (53)] give additional terms in the effective potential energy which may be written (again using an expansion in powers of ξ):

$$B_e(k_0 + k_1\xi + k_2\xi^2 + \cdots)$$

The effect of these terms may be ascertained by perturbation theory. Additional corrections to the energy levels arise from nonadiabatic theory. For H_2 and related molecules, where all the terms entering the corrections have been calculated, the proper Y_{ij} to give the energy levels according to (73) may be found. They agree with experimentally determined values. The corrections turn out to be as important as the corrections for anharmonicity. If accurate values of R_e, ω_e, and other constants are to be computed from spectral term values, these corrections must be considered.

There is also the problem of where to truncate the series for $V(\xi)$. The a_i depend strongly on the length of the series employed. The series may not converge rapidly, especially for a potential whose form is more complicated than expected. Then inconsistencies may be traced to an insufficient expansion length in (72). It also turns out that the Y_{lj} do not necessarily converge to unique values when additional spectral information is used in the fitting. In (73) as in (72) the values of the coefficients depend importantly on the expansion length. Cashion [12] has suggested that the Y_{lj} be regarded more as a set of parameters which can conveniently represent the observed term values than as physically meaningful quantities. An earlier paper [13] already emphasized some of the difficulties with the Dunham and related approaches, the suggestion being made that one deal only with the energy levels and not the "spectroscopic constants" which relate to them by various approximate formulas.

However, much earlier work uses a Dunham-type approach, and it is still often employed. Often, a numerical potential curve is fit to a power series which, via the Dunham treatment, is then used to produce a predicted spectrum to compare with experiment. The truncation problem for (72) in this case has been discussed by Beckel and Sattler [14]. For a numerical potential, it seems better to determine the vibrational energy levels directly, using a computer, by numerical integration of the radial Schrödinger equation [11]. The numerical methods employed are given in an article by Cashion [12]. If derivatives of the potential curve with respect to R are of particular interest, they may then be determined by numerical methods [11].

Where only rough or sparse experimental results are available, we may want to use a potential of simple functional form, with a few parameters, which may be adjusted to give a sufficiently good fit to the data available. Some of these are discussed in Chapter II, Section D. One source of error is removed in the Dunham treatment for a potential of given analytical form, in that the a_i are known exactly. But even here, the procedure of integration of the Schrödinger equation is probably more accurate. Furthermore, the use of a trial potential is meaningful only insofar as the form is in fact capable of representing the actual situation well, which one cannot know in general. There is also the danger that different-looking curves may predict essentially the same spectra.

We really would like a treatment which produces the potential energy curve from the measured term values instead of proceeding in the reverse direction. The Rydberg–Klein–Rees (RKR) procedure [15–18] does this, using the semiclassical approximation. According to this, the eigenvalue E_v obeys

$$\oint [2\mu(E_v - U')]^{1/2} \, dR = h(v + \tfrac{1}{2}) \tag{74}$$

where U' is the potential including the centrifugal energy

$$U' = U + J(J + 1)\hbar^2/2\mu R^2, \tag{75}$$

μ is the reduced mass, and v is an integer. The integral is over the classically allowed period of R, from one zero of the integrand to the other and back. One may start with an approximation to U and adjust it so that (74) is satisfied for all levels for which v and E_v are known. (74) does not suffice to define $U(R)$, since a translation along the R axis would still be permitted; one can demand, in addition, agreement with the observed rotational constants B_v, which are proportional to the average values of R^{-2} over the vibrational states. By semiclassical arguments,

$$\langle R^{-2}\rangle_v = \omega_v\left(\frac{\mu}{2}\right)^{1/2} \oint \frac{dR}{R^2(E_v - U)^{1/2}} \qquad (76)$$

with $\omega_v = (2\hbar)^{-1}(E_{v+1} - E_{v-1})$. Starting with (74) and (76), spectroscopic information can be used to deduce, for each energy level, the classical extremes of the motion $R_{\min}$ and $R_{\max}$, which are two points on the potential energy curve. The errors connected with the use of semiclassical formulas here do not appear to be serious, even for the lower vibrational levels.

The original procedure of Rydberg [15] employed graphical integration in (74) and (76), which led to difficulties because of the singular behavior of the integrands at the endpoints. Klein's modification [16] employed an auxiliary function in terms of which $R_{\min}$ and $R_{\max}$ may be expressed, and made it possible to use numerical integration. It later turned out to be possible [17, 18] to introduce analytical formulas for parts of the calculation. The relation between the RKR method and the method of Dunham was discussed by Jarmain [19]; they are equivalent to a first approximation. Simplifications to the RKR method have been proposed [20].

Zhirnov and Vasilevskii [21] have recently given a generalization of the RKR method for diatomic potential curves. In their method, which involves going to the next power of $\hbar^2$ in the semiclassical expansion, a second approximation is obtained, permitting estimation of the accuracy of the first approximation, which is closely equivalent to the RKR. New methods for estimation of errors in the turning points derived by the RKR have been developed [22].

On the other hand, the RKR method may also be used [23] to test spectroscopic constants obtained in other ways. The spectroscopic constants are used to calculate term values, from which an RKR curve is derived. This curve is examined to see if it obeys various necessary conditions on $U(R)$ (e.g., that dU/dR be negative for $R < R_e$, and that d^2U/dR^2 be positive for $R < R_e$).

The information available from the analysis of electronic spectra is the same as that from vibration–rotation spectra, if the resolution is sufficient. By looking at the lines of a single progression, one can study one electronic state at a time. A single progression can be isolated experimentally by resonance fluorescence, or theoretically by analysis of the spectrum. Often, however, there is not sufficient resolution in an electronic spectrum to measure individual rotational lines. One also may have the problem of identifying the state being studied. The use of intensities as a check on a potential curve here is more complicated, and usually is by way of Franck–Condon factors [24]. It is usually necessary to have extensive knowledge of the potential curve for one of the states involved [5].

The possibility of obtaining information from continuous electronic spectra has been discussed. Bandlike structure is observed [25, 26] in transitions between an attractive and a repulsive state. If the potential curve for one state is available, information about the other may be obtained [25–27].

The broadening of atomic spectral lines due to other atoms may be analyzed in terms of the potential curves, attractive and repulsive, for the states of the diatomic system which can be formed. Thus, Lewis et al. [28] studied the broadening of the Sodium D-line by atomic hydrogen in terms of calculated $U(R)$ for the states of NaH. Similarly, the hyperfine pressure shifts of H atoms in a rare gas atmosphere depend on the interatomic potentials as well as on the electronic wave functions [29]. Pressure effects (due to collisions) on spectral lines in the microwave and other regions may also be useful in studying $U(R)$ [1, p. 354; 5; 29].

For nonbinding (repulsive) states, the van der Waals attraction may provide a deep enough well to be studied by spectroscopic techniques [30]. The Mg_2 van der Waals molecule ($D_0^0 = 400 \text{ cm}^{-1}$ or $\frac{1}{20}$ eV) was the first for which an accurate RKR potential was derived [31], by studying transitions between the chemically nonbinding $^1\Sigma_g^+$ ground state and the stable $A^1\Sigma_u^+$ excited state. The vibrational constants for the van der Waals molecule Ar_2 have also been obtained spectroscopically [32]; D_0^0 for Ar_2 is only 77 cm^{-1}, and six vibrational levels were observed [42].

2. Scattering

The theory of atom–atom scattering is found in the treatise of Mott and Massey [33] and in the review by Massey [34] in the *Handbuch der Physik*. A wealth of important information, especially on molecular beam scattering, is found in the review by Pauly and Toennies [35], on the use of molecular beams to study intermolecular potentials. Other reviews discussing scattering measurement of $U(R)$ are those of Mason and Monchick [1, p. 354] and Bernstein and Muckerman [36].

The problem of an elastic collision between two particles reduces to the problem of scattering of a single particle in a potential field $U(R)$ when the motion of the center of mass is separated. Here $U(R)$ is the interaction potential. To solve the quantum mechanical scattering problem, one requires a solution to the differential equation

$$\nabla^2\psi + (2\mu/\hbar^2)(E - U(R))\psi = 0 \tag{77}$$

with boundary conditions corresponding to a plane wave plus a scattered spherical wave. From ψ, we obtain the scattering cross section and the probabilities for scattering through various angles, which is what we measure. It must be noted that the scattering angles in the center-of-mass system are not the same as in the laboratory system and that the cross section must correspondingly be converted [35, Sect. I-D; 37]. In some experiments, averaging over a velocity distribution may be necessary. For identical particles, the wave function must reflect their indistinguishability, which gives rise to interference terms in the scattering cross section.

The total scattering cross section is the integral of the (differential) cross section over all scattering angles. From measurements of its velocity dependence, one may obtain information about the interaction potential as follows [38]: At high energy, the cross section decreases in a way depending on the form of the repulsive part of $U(R)$ (assuming a long-range attraction and a short-range repulsion). The velocity dependence of the cross section at low impact energies gives information about the attractive part, including the bound states of the diatomic system. For example, a semiclassical treatment shows that, at low velocities, an attractive potential going as CR^{-S} ($S \geq 3$) leads to a total cross section Q going as $k^{-2/(S-1)}$, where $\hbar k/m$ is the velocity. The $k^{-2/5}$ behavior due to the R^{-6} van der Waals attraction has been observed, but reliable absolute values of Q are needed to determine C; and these values are difficult to obtain [39, 40]. Oscillations in the integral cross section as a function of energy are related to the presence of quantum mechanical bound states in the well. Bernstein and O'Brien [41] have shown how the spacing of the oscillations as a function of $1/k$ may be used to obtain information on the shape of the well. For a fuller discussion, see Pauly and Toennies [35, Sect. IV-B.

Expansion of ψ in spherical harmonics yields a set of one-dimensional equations for different l:

$$\frac{d^2\psi_l}{dR^2} + \frac{2\mu}{\hbar^2}\left(E - U(R) - \frac{l(l+1)}{R^2}\right)\psi_l = 0. \tag{78}$$

From the form of ψ_l for large R one extracts the phase shift η_l. The differential cross section is expressible as a sum of contributions of various l, which converges more rapidly for potentials of shorter range and for lower velocities. Often, an extremely large number of partial waves must be included, making the expansion method too difficult. It is manageable when one of the particles is light. Positive phase shifts are due to an attractive potential, negative phase shifts to a repulsive potential. Higher l corresponds to a larger distance of the particle from the scattering center. For the usual case, η_l should be negative for small l, go to 0 when attractive and repulsive forces balance, become positive due to attraction, then eventually die to 0. This parallels the discussion of ϑ versus b for the classical problem. There seems to be a correspondence between $U(R)$ and the phase shifts plotted as a function of l [41a, 42]. Determination of $U(R)$ from phase shifts (and of phase shifts from cross sections) has been discussed [43].

For large angles, a classical treatment of the scattering often suffices. A discussion of the classical mechanical treatment of a collision is found in the treatise of Hirschfelder *et al.* [44, Sect. 1.5]. Here, one calculates the scattering angle as a function of initial relative velocity and impact parameter. In the classical limit, the differential cross section for elastic scattering is given by

$$\sigma(\vartheta) = \frac{b}{\sin \vartheta} \frac{db}{d\vartheta} \tag{79}$$

with b the impact parameter (distance of closest approach in the absence of interaction) and ϑ the scattering angle. The dependence of ϑ on b [the "deflection function" $\vartheta(b)$] for a given U is [44]:

$$\vartheta = \pi - 2b \int_{R_\mathrm{m}}^{\infty} \frac{dR R^{-2}}{[1 - (2U/\mu g^2) - b^2/R^2]^{1/2}}$$

Here, g is the initial relative velocity and R_m the actual distance of closest approach, calculable by setting the square root equal to 0. The well-defined trajectory makes it possible to identify the region of R in $U(R)$ which causes the scattering through each angle. However, positive and negative ϑ are observed together in a typical experiment (see Fig. 2), and the inverse of $\vartheta(b)$ is generally not single valued.

Suppose $U(R)$ is attractive for large R and repulsive for small R. Scattering through large positive angles is due to the repulsive interaction, acting on particles with small impact parameters. Increasing b decreases ϑ, generally through zero (attractive and repulsive forces balance) to negative angles (attractive forces, for large b, dominate). Thus the small-angle scattering gives information on the attractive part of the potential (and also [45] on the curvature of $U(R)$ near R_o); the large-angle scattering depends mostly on the repulsive forces. In some cases, ϑ may become negatively infinite, which corresponds to orbiting. Since $\sin \vartheta$ vanishes for $\vartheta = 0$,

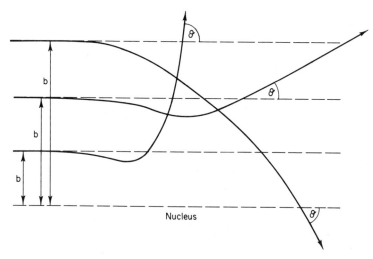

Figure 2. Classical trajectories, showing impact parameters and deflection angles.

the classical scattering cross section actually diverges for small angles. Quantum mechanical corrections remove the singularities in σ, but it still becomes very large. For this reason, integrated scattering cross sections from some small angle α out to π are sometimes considered.

One has $\sin \vartheta = 0$ when attractive and repulsive forces balance, and for $\vartheta = n\pi$, making $\sigma(\vartheta)$ singular for certain angles. These are called "glory effects." (An exception is for $n = 1$ in elastic collisions, where $b = 0$ as well.) Other angles for which $\sigma(\vartheta)$ is singular are called rainbow angles. For angles near a rainbow angle, the classical theory is incorrect and quantum mechanical corrections must be introduced [46]. The behavior of the rainbow angle as a function of energy gives information about the well depth. An analysis by Mason *et al.* [46] shows that, for a potential with a single minimum, the attractive part determines the extrema in $\sigma(\vartheta)$.

The problem of going from a trial potential to scattering cross sections for comparison with experiment can be difficult [46]. Usually, one assumes a potential of simple analytical form, calculates scattering angles or phase shifts, and adjusts parameters in the potential to fit experimental data. For high-energy scattering, a pure repulsive potential suffices [47], such as an exponential or an inverse power. For certain molecular beam experiments, one can assume an R^{-6} attraction and determine the van der Waals coefficient [48]. For a given form of potential, it should be possible to determine the parameters from a small number of phase shifts, but these are not directly measurable and must be derived from scattering cross sections. An analysis has been given [43] which simplifies this procedure. Again, one must note that the fact that a potential function may be parametrized to match a set of experimental results is no guarantee that the function itself represents $U(R)$ accurately for all R.

Often, calculated numerical potential curves are available, and we have to solve Eq. (77) or (78). For a discussion of possible methods and their validities see Pauly and Toennies [35]. It is possible to integrate (78) numerically to obtain the phase shifts [49], or a WKB procedure may be used [41a]. The accuracy of the latter vis-á-vis the former is generally less for smaller values of $(h/\sigma)(me)^{1/2}$ ($\sigma =$ cross section), representing the importance of quantum effects [42]. There is also a problem in using the WKB method in the case of a classically inaccessible region within the centrifugal barrier (tunneling), and a semiclassical treatment is invalid when the size of the scatterer is very small compared with the wavelength of the incoming particle and to the uncertainty in its position. Such procedures, however, have the advantage that, with certain additional approximations, one gets analytic closed-form expressions for the phase shifts.

The inversion problem in the WKB approximation has been discussed by Miller [50]. Given, for one energy E, the phase shift for all l values [which is equivalent to having the phase shift as a function of impact parameter $b = (l + \frac{1}{2})/k$], we can construct the classical deflection function $\vartheta(b)$. From $\vartheta(b)$ we may calculate an auxiliary function $T(u)$ (a transform of ϑ) such that $R = u \exp[T(u)]$ and $U = E\{1 - \exp[-2T(u)]\}$, provided that E is too great to permit orbiting. This is a parametric representation of the potential $U(R)$. Furthermore, $U(R)$ can also be generated from the knowledge of any one phase shift as a function of energy. The method is related to the RKR formula (74). The construction of $\vartheta(b)$ from the differential cross section for fixed energy is also discussed [50]. It is necessary to make certain approximations here. If the high-energy WKB approximation, in which the phase shift for a given b goes inversely as k, is valid, one may derive $U(R)$ from the total scattering cross section as a function of energy.

The complete inversion of molecular beam scattering data to yield an interatomic potential (for Na–Hg) was described by Buck and Pauly [51]. For angles larger than the rainbow angle, the large-angle scattering cross sections are used to construct $\vartheta(b)$. The other data needed are the positions of the rainbow extrema. The potential derived is compared to a Lennard-Jones potential: It has a broader well and goes to zero faster for large R.

Luoma and Mueller [45] discuss an iterative approach to the inversion problem. One starts by estimating a potential and calculating phase shifts, which are fit to a function of nine parameters. The resulting differential cross section is expanded in a Taylor series about the initial values of these parameters, whose values are then rechosen to minimize the deviation from that measured. The process is iterated. The potential is then expanded in a Taylor series, and a similar procedure is used to go from the phase shifts to the potential.

The Born approximation is often used for (77). It gives the cross section as a Fourier transform of the potential, and is useful for weak potentials or fast collisions. A Born approximation may also be used for the higher l phase shifts instead of integrating (78) numerically. There exist also integral equations for the phase shifts, and variational principles for the phase shifts and the scattering amplitude. Some approximation formulas are discussed by Bernstein and Kramer [52].

The centrifugal term in (78) leads to the existence of quasibound states ("resonances") which may be studied by numerical techniques applicable to bound states [53].

Where a change in internal state is possible (reactive scattering), the wave function at large R may be written as a sum of terms, each involving a factor for the radial motion and a factor giving the electronic wave function for the system of separated atoms [54]. The radial wave function $\psi_0(R)$ corresponding to the initial electronic state 0 will include a plane wave and a scattered wave, while the other radial wave functions will be scattered waves only, their size being related to the probability of reaction. If one expands in the separated atom states at all R, the radial wave functions $\psi_n(R)$ are determined by coupled equations of the form

$$(\nabla^2 + k_n{}^2)\psi_n(R) = \sum_m U_{nm}(R)\psi_m(R) \qquad (80)$$

U_{nm} is essentially a matrix element of the interatomic interaction between the separated atom states m and n.

It is often a good approximation to suppose that only direct reaction from 0 to n is important (no intermediate states). Then (80) reduces to

$$(\nabla^2 + k_n{}^2 - U_{nn})\psi_n = U_{n0}\psi_0 \qquad (81a)$$

$$(\nabla^2 + k_0{}^2 - U_{00})\psi_0 = U_{0n}\psi_n \qquad (81b)$$

It is consistent with the above approximations to suppose U_{0n} and ψ_n small, and thus put the right hand side of (81b) equal to zero. Then the elastic scattering still depends only on U_{00}, while the inelastic part is given by a Born-type approximation, with the replacement of plane waves by waves distorted by an average scattering potential. The diagonal elements U_{00} and U_{nn} may be considered zeroth-order approximations to the adiabatic potential curves. They are calculated supposing that the atoms maintain their identity at all R. The element U_{0n} is then responsible for the transition from 0 to n.

Alternatively, one notes that U_{0n}, the interaction between zero-order states, prevents their crossing (see Section B.4). A better approximation to the adiabatic potential curves, which do not intersect, is obtained by writing the wave function at each R as a linear combination of terms corresponding to 0 and n. Now, reaction is interpreted in terms of the breakdown of the adiabatic approximation. For the case of slow collisions, the fixed-nucleus molecular wave functions may thenselves be used to expand the wavefunctions for the scattering problem instead of the atomic wave functions for the scattering problem instead of the atomic wave functions [54]. The resulting equations resemble (80), except that V^2 is replaced by an operator including the effect of derivatives with respect to R on the electronic wave function, and U_{00} and U_{nn} are the adiabatic potential curves. The assumptions leading to (81) may then be invoked.

Measurements of reaction probabilities can thus be interpreted at least partly in terms of fixed-nucleus potential curves. However, for a quantitative treatment, one must be able to calculate reactive cross sections for comparison with experiment. This is extremely difficult for any real system. Furthermore, the wavefunctions and energies of more than one state are involved simultaneously. In favorable cases, chemical reactions can often be treated by considering only two potential curves. The same holds for electron detachment reactions, one of the curves being for the electrically charged diatomic system.

3. *Other Measurements*

In a gas of atoms, the collisions of all kinds, which are constantly taking place, are responsible for the approach to thermal equilibrium. In principle, we can average over velocity distributions, knowing the scattering cross section as a function of velocity and angle, to predict transport properties, such as conductivities, diffusion coefficients, and viscosities. Then a potential of assumed form may be parametrized to give agreement with observed transport coefficients, thus obtaining information on the potential. Much useful information on potential curves and their relation to transport properties is found in the review by Buckingham [55] and the book of

Cottrell and McCoubrey [56]. Extensive discussion of the theory involved, and of many aspects of the relation between transport coefficients and inter-atomic interactions, is presented by Hirschfelder *et al.* [44].

For dilute gases, one may use the Chapman–Enskog theory, which makes a knowledge of the temperature-dependence of a transport process equivalent to knowledge of the interatomic potential [44]. (It is possible [46] to check the validity of the Chapman–Enskog theory without making any assumptions about the form of $U(R)$, by consistency checks among certain transport properties and their composition and temperature dependence.) The transport coefficients depend on the collision integrals, which in turn depend on the differential cross sections appropriate to the transport properties. These cross sections, like the usual scattering cross sections, may be calculated from the phase shifts η_l obtained from Eq. (78) (cf. Hirschfelder *et al.* [42, Chapter 7]). Such calculations for a Buckingham potential are described by Mason [57]. If the collisions are treated classically, the deflection angle is needed as a function of relative velocity and impact parameter. From this we obtain temperature-dependent collision integrals in terms of which the transport coefficients are expressed [44, Sect. 7.4]. Usually, classical mechanics and Boltzmann statistics suffice.

Hirschfelder *et al.* [44, Chapter 8] give various formulas for the coefficients of viscosity, diffusion, thermal conductivity, thermal diffusion, etc., and also indicate how one goes about calculating transport coefficients from some simple potentials, i.e., rigid spheres, Sutherland, power series repulsions, and Lennard-Jones. Mueller and Brackett [42] discuss the sensitivity of measured transport coefficients, as well as scattering experiments, to the intermolecular potential. Considering Lennard-Jones potentials with various choices of the constants, they evaluate the cross sections for scattering, diffusion, and viscosity. Their conclusion is that transport properties are more sensitive to the repulsive part of the potential than low-energy scattering measurements. To obtain good information on the attractive part from diffusion measurements or viscosity measurements, low temperatures must be used. At low temperatures, the transport properties of a dilute gas are almost entirely due to the effect of the long-range potential. By extrapolation to absolute zero, one can deduce the coefficient of the R^{-6} attraction [46, 58].

The use of measured transport coefficients, as well as second virial and Joule–Thompson coefficients (see page 46), to determine the interatomic potential function has been discussed by Klein and Hanley [59]. They studied a model system such that the interaction was exactly given by the Lennard-Jones formula, and attempted to use other potential functions to fit the transport and other properties derived for the system. It appears that there is a temperature range in which the properties are insensitive to the function, and that three-parameter functions are not sufficiently flexible to describe properties both above and below this range.

It is possible to derive the interaction potential for rare gas atoms, without assuming an analytical form for the potential, from dilute gas transport coefficient data [60]. It is shown [60] that, so far as the collision integrals are concerned, the repulsive interaction may be replaced by a hard sphere potential in which the radius of the sphere depends on the strength of the interaction, on the temperature, and on the particular transport property being considered. This radius is the value of R for which the energy is a particular multiple of the temperature (the multiplicative constant is given). Thus a series of measurements at different temperatures maps out energy as a function of R.

The interaction between molecules of a gas is also responsible for non-ideality, i.e., deviations from the ideal gas law. It was recognized early that a long-range attraction and a short-range repulsion could be inferred from the measurements, as expressed, for instance, in the van der Waals equation of state. A convenient way of expressing nonideality is the virial expansion,

$$p\tilde{V}/RT = 1 + B(T)/\tilde{V} + C(T)/\tilde{V}^2 + D(T)/\tilde{V}^3 + \cdots \qquad (82)$$

We follow Hirschfelder *et al.* [44] by using the tilde to denote molar quantities. In general, positive virial coefficients are associated with repulsion. At low temperatures, colliding molecules are slowed by the attraction, and occupy less volume, so B is negative; for higher temperatures, the attractive well is not seen, but only the repulsion, so the pressure increases.

Statistical mechanical analysis shows that the first and leading term in the deviation from ideality, $B(T)$, is due to bimolecular collisions. The next coefficient $C(T)$ arises from configurations in which three molecules interact simultaneously, and so on. Thus information about diatomic interactions is contained in $B(T)$ and, if one can assume pairwise additivity, in higher coefficients as well. However, accurate measurements of the higher coefficients are hard to come by. Analysis of data for gas mixtures can give information about interactions between all the different pairs of species present, so that, provided one has accurate information about the homonuclear pairs, one obtains information about heteronuclear interactions.

When the behavior of $p\tilde{V}/RT$ is expressed as a finite polynomial in $\tilde{V}^{-1}$, the coefficients will depend on the number of terms taken and on the pressures for which measurements were made. Even if, say, the formula $1 + a/\tilde{V}$ suffices to fit the data, one cannot equate a with the second virial coefficient unless very low pressures were used. Rewriting (82), one sees that B may be obtained as

$$B = \lim_{p \to 0}(p\tilde{V}/RT - 1)\tilde{V}.$$

In general, we must assume an interatomic potential with embedded parameters, calculate virial coefficients, and adjust parameters for agreement with experiment. As we always expect, when different functions are parametrized to give best agreement with experiment, the resulting curves can look quite different. The calculation may proceed by way of evaluation of cluster integrals (integrals of Boltzmann factors over the coordinates of interacting particles). Details of the theory are given by Hirschfelder *et al.* [44, Chap. 3]. The expressions, assuming pairwise additivity and spherically symmetric particles, are integrals over the functions

$$f_{ij}(r_{ij}) = [\exp(-U_{ij}/kT)] - 1$$

where U_{ij} is the energy of interaction between particles i and j. For instance,

$$B(T) = -\tilde{N}/2V \int\int f_{12}\, \overrightarrow{dr_1}\, \overrightarrow{dr_2} = -2\pi\tilde{N}\int_0^\infty \{\exp[-U(R)/kT] - 1\}R^2\, dR$$

For small T, $B(T)$ is determined by the behavior of $U(R)$ near the minimum in the curve. For large T, high-energy collisions are important, and B measures the repulsive part of U. In the latter case, we may take $U(R)$ as monotonic in R. Then we can use U as the variable of integration. With Keller and Zumino [61], we introduce $\beta = (kT)^{-1}$ and $v = R^3$. Then, putting $b(\beta) = B/2\pi\tilde{N}$, we have

$$b'(\beta) = \tfrac{1}{3}(d/d\beta)\int_0^\infty (1 - e^{-\beta U})\, dv = \tfrac{1}{3}\int_0^\infty Ue^{-\beta U}\, dv = \tfrac{1}{3}\int_\infty^0 Ue^{-\beta U}(dU/dv)\, dU$$

assuming that $U \to \infty$ for $v \to 0$ and $U \to 0$ for $v \to \infty$. The last integral is minus the Laplace transform of $U(dU/dv)^{-1}$. This function, and hence U, is then determined when $b(\beta)$ is known. For a nonmonotonic potential with a single well Keller and Zumino [61] show that the virial coefficient determines only the positive part of $U(R)$ and the width of the well as a function of depth.

Classical statistical mechanics is employed in all derivations; it is rarely necessary to consider quantum effects. But one can also formulate the theory for the second virial coefficient in quantum statistical mechanics. The result [44, Chap. 6, especially Sect. 6.4] is an expression for the second virial coefficient in terms of the energies of bound states (if any) of the two-particle system and the phase shifts for collisions with positive energy. Since the phase shifts can be determined analytically only for square-well potentials, numerical calculation is necessary. By employing the WKB expansion, or by other methods, we may obtain the quantum mechanical virial coefficient as a series in h^2/m, of which the first term is the classical result and the rest quantum mechanical corrections. Quantum corrections to second virial coefficients for various forms of $U(R)$ have been evaluated [62].

The gas imperfections, and hence the interatomic interactions, are also reflected in the Joule–Thompson coefficient $\mu = (\partial T/\partial p)_{\mathrm{H}}$, which is zero for a perfect gas. The value of μ, extrapolated to zero pressure, may be expressed in terms of the second virial coefficient [63]. Similar considerations obtain for the coefficient of free expansion $(\partial T/\partial p)_{\mathrm{E}} = \eta$. Thus $U(R)$ may be determined from B, μ_0, or η_0 (subscript 0 means extrapolation to zero pressure). At high temperatures, η_0, but not B and μ_0, is sensitive to the attractive part of the potential. The quantum mechanical corrections to the formula arise mainly from molecule formation which gives discrete vibrational energy levels [63]. This correction may be isolated by considering isotopic molecules.

It is also possible to measure the radial distribution function of molecules in a gas by X-ray scattering [64]. If the density is small enough so that terms corresponding to nonadditivity of pair interactions may be neglected, $U(R)$ may be determined from the radial distribution function. Results have been given for the Ar–Ar system [64].

The gas imperfections are eventually responsible for condensation to a liquid. In principle, the thermodynamic properties of the liquid are predictable in terms of the interatomic interactions. Calculations of this kind have been done for rare gases [65]. Assuming that only pair interactions enter, one can average various functions of particle coordinates with a Boltzmann weight derived from a given potential. This gives the internal energy, pressure, and other thermodynamic properties of the liquid. (On the other hand, the entropy cannot be obtained by averaging over a representative number of particles.) Then parameters in the assumed potential may be adjusted to give thermodynamic properties in accord with experiment. The averaging may be performed by sampling a large number of configurations of the particles in the liquid, using Monte Carlo methods with probabilities designed to reproduce the canonical ensemble [65, 66].

An alternative to the average over configurations is a time average, where the motion of an isolated system is followed by stepwise integration of the equations of motion [67], the "molecular dynamics" technique. It appears from some of these calculations that either many-body effects are not important or, more likely, that one is dealing with an "effective" two-body interaction.

In a recent review [68] of liquid state theories, it is noted that the 6–12 potential, in which the parameters are chosen to fit higher temperature virial coefficient data, works well for the prediction of the properties of liquid argon (from computer calculations), but that this procedure does not work for other inert gases. Thus, it is accidental that the 6–12 potential works for Ar. In addition to the computer calculations, the equation of state and the intermolecular potential may be related [68] by various integral equations, for which, however, one needs the pair distribution function $g(R)$, i.e., the probability that, given a molecule at a point P, a second will be found a distance R away. The Percus–Yevick theory relates g and the intermolecular potential by another integral equation. Apparently this theory works well for dense gases, but not for liquids. One is more often in the position of testing the theory for a model system with a known potential function than using the theory to derive the function. The theory of quantum fluids is only now beginning to be sufficiently understood to lead to information on interatomic potentials. Data on solubility of gases in liquids, with a suitable theory of solubility, may also give information on interatomic interactions [69].

One may imagine further condensation to a solid, where the regular structure simplifies many calculations, and relate measured properties to interatomic interactions, assuming the nature of the atoms is maintained. If only pair interactions are important, the equilibrium interparticle distance in the solid can be calculated from the attractive and repulsive interactions, and if we can measure the change in atomic spacing with temperature, we have information about the shape of the potential curve. In general, thermodynamic measurements on condensed phases emphasize the short-range repulsive forces more than measurements on dilute gases, where interactions are due mainly to long-range attractions [70]. Solid state properties related to the lattice vibrations (like interatomic spacing and specific heat) are sensitive to the shape of the potential curve near the minimum. The heat of sublimation depends on the attractive portion (from the minimum on out) [71].

Considering a solid formed from atoms, we may assume some interatomic interaction potential and parametrize it to give agreement with the experimentally measured interatomic distance and sublimation energy. Guggenheim and McGlashan [72] have given the theoretical analysis necessary, and applied it to Ar–Ar interactions. Other properties which are available are heat capacity at zero pressure, dependence of volume on pressure, vapor pressure, and sound velocities, although not all are equally reliable [72].

Here, we must use the experimentally determined structure of the solid. We may also attempt to predict the structure. At absolute zero of temperature and zero pressure, the stability of different structures depends on the energy, which includes the static potential energy and the zero point vibrational energy. The latter is usually neglected. If the interatomic potential is taken as

$$U(R) = -\varepsilon_0 \frac{nm}{n-m} \left\{ -\frac{1}{m} \left(\frac{R_0}{R} \right)^m + \frac{1}{n} \left(\frac{R_0}{R} \right)^n \right\}, \qquad n > m$$

with R_0 the position of the minimum and ε_0 the depth of the well, the static energy of a crystal formed from N atoms (N is large) is [73]

$$E_L = \frac{N}{2} (-\varepsilon_0) \frac{nm}{n-m} \left\{ -\frac{A_m}{m} \left(\frac{r_0}{a} \right)^m + \frac{A_n}{n} \left(\frac{r_0}{a} \right)^n \right\}. \qquad (83)$$

In (83), a is the nearest neighbor distance and A_m and A_n are geometric constants depending on the type of lattice. The values of E_L computed for different types are compared, the lowest energy supposedly indicating the most stable. For rare gases, such procedures predict hexagonal close packing in contradiction to experiment (cubic close packing), but these differ only with reference to third nearest neighbors, so this test is a severe one. The question of the importance of three- and many-body interactions arises [70, 73–75]. If they enter, this would cast doubt on much of the interpretation of crystal properties in terms of potentials and mean such properties could not be used for the measurement of $U(R)$. For the rare gas crystals, the lattice molecules are essentially interacting with van der Waals forces. Generally, nonadditivity is connected with the repulsive interaction, rather than the van der Waals attraction [76]. Calculations of the size of nonadditivity effects have been made [71].

It is now believed [1, p. 354; 5] that three- and many-body forces are important, but their effects tend to cancel and be subsumed in effective pair potentials. To isolate many-body effects, the true two-body potentials are needed. These would be obtained from scattering or from measurements on dilute gases.

Having calculated the lattice energy as a function of interparticle distance (assuming the crystal form is maintained), we can derive an equation of state for the crystal [77]. This has been done for rare gases with various potentials parametrized according to theory, virial coefficient data, other measurements on the crystal, etc. ($U(R)$ from crystal structure has also been used to predict virial coefficients [78]). Similarly, one may derive force constants for vibration from the potential curve and attempt to predict the frequency spectrum and specific heat [76]. It is also possible to substitute, for the comparison of calculated equations of state (in particular, isotherms) with experiment, the calculation of more readily observable melting curves from assumed potentials [79]. Here, we require a theory that relates melting point to other properties of the solid.

In treating molecular solids like N_2, the intermolecular potential can be built up from atom–atom potentials [80]. These may be parametrized to reproduce thermodynamic, structural, or spectroscopic (lattice vibrations) properties of the crystal. Of course, the potential between atoms in molecules is not necessarily the same as that when the atoms interact directly.

Other problems related to extraction of information from crystal properties have been mentioned [81]: The data are not always accurate and there are a number of approximations and corrections that must be made in each case.

In general, only a fairly flexible potential curve can fit data obtained from a variety of measurements, simply because different measurements give information about different regions of the curve. The more realistic the potential function, the better agreement between the constants in it as determined from different measurements. In this way, one can begin to choose between various trial functions (see Chapter II, Section D).

It appears that no simple potential can give very good fits to measurements of all kinds. One can take care of all transport properties with a simple potential, but it will not do for crystal measurements. There seems to be a distinction between potentials derived from equilibrium properties and those derived from transport properties [71]. This is just another reason not to put too much weight in the interpretation of parameters deduced by assuming a form for $U(R)$ and fitting experimental results.

Rowlinson [82] has discussed to what extent information on intermolecular forces can be derived from macroscopic measurements without the assumption of a specific form for $U(R)$, emphasizing the case of argon atoms. Here $U(R)$ may be determined uniquely from the second virial coefficient as a function of T if $U(R)$ is monotonic, as already mentioned. The low-temperature limit of the transport coefficient depends on the rate of falloff of U at large distances.

REFERENCES

1. E. A. Mason and L. Monchick, *Advan. Chem. Phys.* **12**, 329 (1967).
2. G. Herzberg, "Spectra of Diatomic Molecules." Van Nostrand-Reinhold, Princeton, New Jersey, 1950.
3. B. Rosen, ed., "Spectroscopic Data Relative to Diatomic Molecules" (Tables Int. Constantes Selectionées, **17**). Pergamon, Oxford, 1970.
4. L. Pauling and E. B. Wilson, Jr., "Introduction to Quantum Mechanics, Sect. X-35. McGraw-Hill, New York, 1935.
5. M. S. Child, *J. Mol. Spectrosc.* **33**, 487 (1970).
6. R. Schmid and L. Gerö, *Z. Phys.* **96**, 198 (1935); **104**, 724 (1937).
7. R. B. Bernstein, *Phys. Rev. Lett.* **16**, 4 (1966).
8. T. Y. Chang, *Rev. Mod. Phys.* **39**, 911 (1967).
9. J. S. Dunham, *Phys. Rev.* **41**, 721 (1932).
10. C. L. Pekeris, *Phys. Rev.* **45**, 98 (1934); I. Sandeman, *Proc. Roy. Soc. Edinburgh* **60**, 210 (1940).
11. P. R. Bunker, *J. Mol. Spectrosc.* **28** 422 (1968); **35**, 306 (1970).
12. J. K. Cashion, *J. Chem. Phys.* **39**, 1872 (1963); J. W. Cooley, *Math. Comput.* **15**, 363 (1961).
13. A. S. Coolidge, H. M. James, and E. L. Vernon, *Phys. Rev.* **54**, 726 (1938).
14. C. L. Beckel, and P. J. Slatter, *J. Mol. Spectrosc.* **20**, 153 (1966).
15. R. Rydberg, *Z. Phys.* **73**, 376 (1932); **80**, 514 (1933).
16. O. Klein, *Z. Phys.* **76**, 226 (1932).
17. A. L. G. Rees, *Proc. Phys. Soc. London* **59**, 998 (1947).
18. J. T. Vanderslice, E. A. Mason, and W. G. Maisch, *J. Mol. Spectrosc.* **3**, 17 (1959); **5**, 83 (1960).
19. W. R. Jarmain, *Can. J. Phys.* **38**, 217 (1960).
20. F. R. Gilmore, Mem. RM-4034-1-PR. *Rand Corp.*, Santa Monica, California, 1966.
21. N. I. Zhirnov and A. S. Vasilevskii, *Opt. Spectrosc.* (*USSR*) **25**, 13 (1968); **26**, 387 (1969); *Opt. Spektrosk.* **25**, 28 (1968); **26**, 704 (1969).
22. R. J. LeRoy, *J. Chem. Phys.* **52**, 2683 (1970).
23. R. J. LeRoy and G. Burns, *J. Mol. Spectrosc.* **25**, 77 (1968).
24. R. N. Zare, *J. Chem. Phys.* **40**, 1934 (1964).
    R. P. Marchi and C. R. Mueller, *J. Chem. Phys.* **36**, 1100 (1962); **38**, 740, 745 (1963).
25. F. H. Mies and A. L. Smith, *J. Chem. Phys.* **45**, 994 (1966).
26. A. L. Smith, *Ibid.* **49**, 4813, 4817 (1968).
27. K.-W. Chow and A. L. Smith, *J. Chem. Phys.* **54**, 1556 (1971).
28. E. L. Lewis, L. F. McNamara, and H. H. Michels, *Phys. Rev. A* **3**, 1939 (1971).
29. G. Birnbaum, *Advan. Chem. Phys.* **12**, 487 (1967).
30. D. E. Beck, *Mol. Phys.* **14**, 311 (1968); *J. Chem. Phys.* **50**, 541 (1969).
31. W. J. Balfour and A. F. Douglas, *Can. J. Phys.* **48**, 901 (1970); W. C. Stwalley, *Chem. Phys. Lett.* **7**, 600 (1970).
32. Y. Tanaka and K. Yoshino, *J. Chem. Phys.* **53**, 2012 (1970); L. W. Bruch and I. J. McGee, *Ibid.* **53**, 4711 (1970).
33. N. F. Mott and H. S. W. Massey, "The Theory of Atomic Collisions." Oxford Univ. Press, London and New York, 1965.
34. H. S. W. Massey, *in* "Encyclopedia of Physics" (S. Flügge, ed.), Vol. XXXVI. Springer-Verlag, Berlin and New York.

35. H. Pauly and J. P. Toennies, *Advan. At. Mol. Phys.* **1**, 195 (1965).
36. R. B. Bernstein and J. T. Muckerman, *Advan. Chem. Phys.* **12**, 389 (1967).
37. R. K. B. Helbing, *J. Chem. Phys.* **48**, 472 (1968).
38. E. W. Rothe, P. K. Rol, and R. B. Bernstein, *Phys. Rev.* **130**, 2333 (1963).
39. E. W. Rothe and R. H. Neynaber, *J. Chem. Phys.* **42**, 3306 (1965).
40. E. W. Rothe and R. H. Neynaber, *J. Chem. Phys.* **43**, 4177 (1965).
41. R. B. Bernstein and T. J. P. O'Brien, *Discuss. Faraday Soc.* **40**, 35 (1965).
41a. R. P. Marchi and C. R. Mueller, *J. Chem. Phys.* **36**, 1100 (1962); **38**, 740, 745 (1963).
42. C. R. Mueller and J. W. Brackett, *J. Chem. Phys.* **40**, 654 (1964).
43. J. W. Brackett, C. R. Mueller, and W. A. Sanders, *J. Chem. Phys.* **39**, 2564 (1963); W. A. Sanders and C. R. Mueller, *Ibid.* **39**, 2572 (1963); T. J. P. O'Brien and R. B. Bernstein, *Ibid.* **51**, 5112 (1969).
44. J. O. Hirschfelder, C. F. Curtiss, and R. B. Bird, "Molecular Theory of Gases and Liquids." Wiley, New York, 1954.
45. J. Luoma and C. R. Mueller, *Discuss. Faraday Soc.* **40**, 45 (1965).
46. E. A. Mason, R. J. Munn, and F. J. Smith, *Discuss. Faraday Soc.* **40**, 27 (1965).
47. I. Amdur and J. Pearlman, *J. Chem. Phys.* **9**, 503 (1941).
48. R. Helbing and H. Pauly, *Z. Phys.* **179**, 16 (1964).
49. R. B. Bernstein, *J. Chem. Phys.* **33**, 795 (1960); R. B. Bernstein, C. F. Curtiss, S. Imam-Rahajoe, and H. T. Wood, *Ibid.* **44**, 4072 (1966).
50. W. H. Miller, *J. Chem. Phys.* **51**, 3631 (1969).
51. U. Buck and H. Pauly, *J. Chem. Phys.* **51**, 1662 (1969).
52. R. B. Bernstein and K. H. Kramer, *J. Chem. Phys.* **38**, 2507 (1963).
53. T. G. Waech and R. B. Bernstein, *J. Chem. Phys.* **46**, 4905 (1967).
54. D. R. W. Bates, H. S. W. Massey, and A. L. Stewart, *Proc. Roy. Soc. Ser. A* **216**, 437 (1953).
55. R. A. Buckingham, *Planet. Space Sci.* **3**, 205 (1961).
56. T. L. Cottrell and J. C. McCoubrey, "Molecular Energy Transfer in Gases." Butterworth, London, 1961.
57. E. A. Mason, *J. Chem. Phys.* **22**, 169 (1954).
58. R. J. Munn, *J. Chem. Phys.* **42**, 3032 (1965).
59. M. Klein and H. J. M. Hanley, *Trans. Faraday Soc.* **64**, 2927 (1968); H. J. M. Hanley and M. Klein, *J. Chem. Phys.* **50**, 4765 (1969).
60. J. M. Dymond, *J. Chem. Phys.* **49**, 3673 (1968).
61. J. B. Keller and B. Zumino, *J. Chem. Phys.* **30**, 1351 (1959); H. L. Frisch and E. Helfand, *Ibid.* **32**, 269 (1960).
62. R. D. Weir, *Mol. Phys.* **11**, 97 (1966).
63. J. O. Hirschfelder, R. B. Ewell, and J. R. Roebuck, *J. Chem. Phys.* **6**, 205 (1938).
64. P. G. Mikolaj and C. J. Pings, *Phys. Rev. Lett.* **16**, 4 (1966).
65. I. R. McDonald and K. Singer, *J. Chem. Phys.* **50**, 2308 (1969).
66. W. W. Wood and F. R. Parker, *J. Chem. Phys.* **27**, 720 (1957).
67. L. Verlet, *Phys. Rev.* **159**, 98 (1967).
68. D. Henderson, J. A. Barker, and S. Kim, *Int. J. Quantum Chem. Symp.* **3**, 265 (1969)
69. E. Wilhelm and R. Battino, *J. Chem. Phys.* **55**, 4012 (1971).
70. T. L. Cottrell, *Discuss. Faraday Soc.* **22**, 10 (1956).
71. G. L. Pollack, *Rev. Mod. Phys.* **36**, 748 (1964).
72. E. A. Guggenheim and M. L. McGlashan, *Proc. Roy. Soc. Ser. A* **255**, 456 (1960); M. L. McGlashan, *Discuss. Faraday Soc.* **40**, 59 (1965).

73. T. H. K. Barron and C. Domb, *Proc. Roy. Soc. Ser. A* **227**, 447 (1955).
74. T. H. K. Barron, *Discuss. Faraday Soc.* **40**, 69 (1965).
75. L. Jansen and E. Lombardi, *Discuss. Faraday Soc.* **40**, 117 (1965).
76. E. R. Dobbs and G. O. Jones, *Rept. Progr. Phys.* **20**, 516 (1957).
77. I. J. Zucker, *Nuovo Cimento B* **54**, 177 (1968).
78. D. D. Konowalow and J. O. Hirschfelder, *Phys. Fluids* **4**, 629 (1961).
79. I. J. Zucker, *Int. J. Quantum Chem. Symp.* **2**, 225 (1968).
80. T. S. Kuan, A. Warshel, and O. Schnepp, *J. Chem. Phys.* **52**, 3012 (1970).
81. General discussion, *Discuss. Faraday Soc.* **40**, 117 (1965).
82. J. S. Rowlinson, *Discuss. Faraday Soc.* **40**, 19, 53 (1965).

SUPPLEMENTARY BIBLIOGRAPHY

Below are listed, alphabetically by author, articles relevant to the subject of this chapter which came to our attention too late to be included in the manuscript. In most cases, we have given the title, sometimes adding in parentheses clarifying information. The letter at the end of each entry gives the section, A, B, or C, to which the article is relevant.

Amdur, Jordan, Chien, Fung, Hance, Hulpke, and Johnson, Scattering of Fast Potassium Ions by Helium, Neon, and Argon (to determine repulsive potentials). *J. Chem. Phys.* **57**, 2117 (1972). **C**

C. D. Andriesse and E. Legrand, Pair Potential for Argon from Neutron Diffraction at Low Intensity. *Physica* **57**, 191 (1972). **C**

V. Aquilanti, G. Liuti, F. Vecchio-Cattivi and G. G. Volpi, Absolute Total Cross Sections for Scattering of Hydrogen Atoms by Argon, Krypton and Xenon. *Chem. Phys. Lett.* **15**, 305 (1972). **C**

J. A. Barker, Interatomic Potential for Neon (experimentally determined, "superseding other potentials"). *Chem. Phys. Lett.* **14**, 292 (1972). **C**

R. F. Barrow, D. F. Broyd, L. B. Pederson and K. K. Yee, The Dissociation Energies of Gaseous Br_2 and I_2 (by spectroscopy). *Chem. Phys. Lett.* **18**, 357 (1973). **C**

E. Bar-ziv and S. Weiss, Information on Intermolecular Potentials Obtained from the Temperature Dependence of the IR Absorption Spectrum of Rare Gas Mixtures. *J. Chem. Phys.* **57**, 29 (1972). **C**

C. Battali-Cosmavici and K. W. Michel, Dissociation Energy of SrO from a Crossed-Beam Experiment. *Chem. Phys. Lett.* **16**, 77 (1972). **C**

H. G. Bennewitz, H. Busse, H. D. Dohmann, D. E. Oates, and W. Schrader, Evidence for Different Interaction Potentials for He^4–He^4 and He^3–He^3 from Scattering Cross-Section Measurements. *Phys. Rev. Lett.* **29**, 533 (1972). **B**

H. G. Bennewitz, H. Busse, H. D. Dohmann, D. E. Oates and W. Schrader, He^4–He^4 Interaction Potential from Low Energy Total Cross Section Measurements. *Zeits. Phys.* **253**, 435 (1972). **C**

J. Berkowitz, Experimental Potential Energy Curves for $X\,^2\Pi$ and $^2\Sigma^+$ states of HF^+. *Chem. Phys. Lett.* **11**, 21 (1971). **C**

U. Buck, M. Kick, and H. Pauly, High Resolution Molecular Beam Scattering Measurements and Potentials for K–Hg and Cs–Hg. *J. Chem. Phys.* **56**, 3391 (1972). **C**

P. R. Bunker, On the Breakdown of the Born–Oppenheimer Approximation for a Diatomic Molecule (consequences for determination of $U(R)$ from spectra). *J. Mol. Spec.* **42**, 478 (1972). **B**

P. Contini, M. G. Dondi, G. Scales, and F. Torello, Low-Energy Repulsive Interaction Potential for Helium (from scattering). *J. Chem. Phys.* **56**, 1946 (1972). **C**

D. Daumont, A. Jenouvrier, B. Pascat, and H. Guenebaut, Spectres Électroniques de la Molécule NSe... Analyses Vibrationnelles et Rotationnelles... *J. Chim. Phys. Physicochim. Biol.* **69**, 218 (1972). **C**

R. L. Fulton, Vibronic Interaction. The Adiabatic Approximation. *J. Chem. Phys.* **56**, 1210 (1972). **B**

M. L. Ginter and C. M. Brown, Dissociation Energies of X $^2\Sigma_u{}^+$He$_2{}^+$ and A $^1\Sigma_u{}^+$He$_2$ (experimental and theoretical considerations). *J. Chem. Phys.* **56**, 672 (1972). **C**

O. Goscinski and O. Tapia, Predissociation by Rotation and Long-Range Interactions. *Mol. Phys.* **24**, 641 (1972). **C**

O. Goscinski, Outer Vibrational Turning Points Near Dissociation in the B state of Br$_2$ and Cl$_2$. *Mol. Phys.* **24**, 655 (1972). **C**

D. W. Gough, G. C. Maitland, and E. B. Smith, The Direct Determination of Intermolecular Potential Energy Functions from Gas Viscosity Measurements. *Mol. Phys.* **24**, 151 (1972). **C**

E. F. Greene and E. A. Mason, Physical Interpretation of Glory Undulations in Scattering Cross Sections. *J. Chem. Phys.* **57**, 2065 (1972). **C**

Hanley, Barker, Parson, Lee, and Klein, Some Comments on the Interatomic Potential for Argon (treatment of experimental data from different sources), *Mol. Phys.* **24**, 11 (1972). **C**

D. L. Hildebrand, Thermochemistry of the Molecules CS and CS$^+$. *Chem. Phys. Lett.* **15**, 379 (1972). **C**

W. G. Hoover, D. Young, and R. Grover, Inverse Power Potentials and the Close-Packed to Body-Centered Cubic Transition. *J. Chem. Phys.* **56**, 2207 (1972), **C**

H. Inouye and S. Kita, Experimental Determination (by elastic scattering) of the Repulsive Potentials between K$^+$ Ions and Rare-Gas Atoms. *J. Chem. Phys.* **56**, 4877 (1972). **C**

H. Inouye and S. Kita, Elastic Scattering of Li$^+$ Ions in Helium (to get repulsive potential). *J. Chem. Phys.* **57**, 1301 (1972). **C**

S. E. Johnson, D. Connell, J. Lunacek and H. P. Broida, New Molecular Constants (from spectroscopy) for the Ground Electronic State of Pb$_2$. *J. Chem. Phys.* **56**, 5723 (1972). **C**

M. L. Klein, Comments on the (experimentally determined) Interatomic Potential of Ne$_2$. *Chem. Phys. Lett.* **18**, 203 (1973). **C**

R. Klingbeil, Determination of Interatomic Potentials by the Inversion of Elastic Differential Cross Section Data. I. An Inversion Procedure. II. ArH$^+$ at 5 eV. *J. Chem. Phys.* **56**, 132, **57**, 1066 (1972). **C**

R. J. LeRoy, Dependence of the Diatomic Rotational Constant B_v on the Long-Range Internuclear Potential. *Can. J. Phys.* **50**, 953 (1972). **C**

R. J. LeRoy, Improved Spectroscopic Dissociation Energy for Ground State Ar$_2$. *J. Chem. Phys.* **57**, 573 (1972). **C**

G. C. Maitland and E. B. Smith, The Direct Determination of Potential Energy Functions from Second Virial Coefficients. *Mol. Phys.* **24**, 1185 (1972). **C**

R. D. Murphy, Use of Ground State Energy Calculations (for liquid He) in Determining the He–He Interaction: *Phys. Rev.* **A5**, 331 (1972).   **C**

K. R. Naqvi and W. Byers Brown, The Non-Crossing Rule in Molecular Quantum Mechanics. *Int. J. Quantum Chem.* **6**, 271 (1972).   **B**

R. A. Nelson and H. R. Leribaux, A Further Test of Pair-Potential Theory (pairwise additivity of interactions) in Sodium Vapor. *Chem. Phys. Lett.* **11**, 623 (1971).   **C**

J. M. Parson, P. E. Siska, and Y. T. Lee, Intermolecular Potentials from Crossed-Beam Differential Elastic Scattering Measurements. IV. Ar + Ar. *J. Chem. Phys.* **56**, 1511 (1972).   **C**

V. Piacente and A. Desideri, Mass Spectrometric Determination of the Dissociation Energy of the GaBi Molecule. *J. Chem. Phys.* **57**, 2213 (1972).   **C**

M. Picot and R. D. Fink, H–Xe Potential Energy Function Determined from High Energy Elastic Scattering. *J. Chem. Phys.* **56**, 4241 (1972).   **C**

T. R. Powers and R. J. Cross Jr., Molecular-Beam Determination of Mercury Rare-Gas Intermolecular Potentials. *J. Chem. Phys.* **56**, 3181 (1972).   **C**

T. R. Powers and R. J. Cross Jr., Feasibility of the Experimental Detection of Long Range Retarded Dipole Forces. *Chem. Phys. Lett.* **15**, 293 (1972).   **C**

T. R. Powers and R. J. Cross, Jr., Molecular Beam Determination of Alkali Ion-Rare Gas Potentials. *J. Chem. Phys.* **58**, 626 (1973).   **C**

D. E. Pritchard, Finding the Potential Directly from Scattering Data. *J. Chem. Phys.* **56**, 4206 (1972).   **C**

K. M. Sando, Potential Curves from Continuous Spectra: He_2 X $^1\Sigma_g^+$ and A $^1\Sigma_u^+$. *Mol. Phys.* **23**, 413 (1972).   **C**

M. Shafi, C. L. Beckel, and R. Engelke, (a compilation of) Diatomic Molecule Ground State Dissociation Energies (an updating of *J. Mol. Spec.* **40**, 519 (1971).). *J. Mol. Spec.* **42**, 578 (1972).   **C**

W. C. Stwalley, RKR Potential Curves Directly from a Single Resonance Fluorescence Doublet Series. *J. Chem. Phys.* **56**, 2485 (1972).   **C**

B. T. Ulrich, L. Ford, and J. C. Browne, Translational Absorption of HeH. *J. Chem. Phys.* **57**, 2906 (1972).   **C**

W. M. Welch and M. Mizushima, Molecular Properties of the O_2 Molecule (from spectra). *Phys. Rev.* **A5**, 2692 (1972).

Chapter II   Qualitative Discussion of Potential Energy Curves

A. Very Large Internuclear Distances

In this chapter, we consider, without going into detail, the effects to be expected when two atomic systems are allowed to interact at large separation. More detailed discussions will be given subsequently. First, it is supposed that the internuclear distance is so large that the charge distributions do not overlap. Then the atoms maintain their identity, and there is no interference between their wave functions, so the problem reduces to the electrostatic interaction of two charge distributions. Each atom is characterized by its permanent moments (net charge, dipole moment, quadrupole moment, etc.) and polarizabilities, and the electrostatic interaction between two such charge distributions is calculated.

It is clear that, if each atom bears a net charge, the interaction will include the simple Coulomb attraction or repulsion. Similarly, there will be an electrostatic interaction between each permanent moment of atom A with each moment of atom B. If one atom is charged, its field can induce a moment in the other and this induced moment will interact with the field. Indeed, it suffices for either to have any permanent moment for an inductive interaction to occur.

The permanent moments are defined, following Hirschfelder *et al.* [1], in terms of the quantities

$$Q_l^m = \iiint r^l P_l^m(\cos\theta)e^{im\varphi}\varrho(r,\theta,\varphi)r^2\,dr\,\sin\theta\,d\theta\,d\varphi \tag{1}$$

$$l = 0, \ldots, \infty; \quad m = -l, \ldots, l$$

where ϱ is the charge density, the origin is at the nucleus for an atom, and $\theta = 0$ is conveniently taken along the line connecting the nuclei. Q_0^0 is the total charge, Q_1^m ($m = -1, 0, 1$) the components of the dipole moment in a space-fixed coordinate system, and so on. The nuclear charge density contributes only to Q_0^0, so as long as $l > 0$; ϱ in (1) is the electronic charge density, i.e.

$$\varrho = N \int \cdots \int |\Psi|^2\,d\tau_2 \cdots d\tau_N \tag{2}$$

for an N-electron system, if the wave function Ψ is normalized to unity.

The multipole moments usually used are Cartesian components rather than components of spherical tensors like the Q_l^m (a discussion of the tensor properties of the Cartesian components has recently been given [2]), and defined with respect to axes fixed in the charge distributions. Thus, for the dipole moment, we may choose axes such that the moment lies along the z axis. Then $\mu_x = \mu_y = 0$, $\mu_z = \mu$, the magnitude of the moment. For charge distribution A with dipole moment μ_A, let θ_A, φ_A give the orientation of these axes relative to the space-fixed system. Q_1^0 is the component of the dipole moment along the space-fixed polar axis, so $Q_1^0 = \mu_A \cos\theta_A$. Also, $\frac{1}{2}(Q_1^{-1} + Q_1^1)$ is the component along the space-fixed x axis, $\mu_A \sin\theta_A \cos\varphi_A$, and $(Q_1^1 - Q_1^{-1})/2i$ is the component along the space-fixed y axis, $\mu_A \sin\theta_A \sin\varphi_A$. Thus $Q_1^{-1} = \mu_A \sin\theta_A \exp(-i\varphi_A)$ and $Q_1^1 = \mu_A \sin\theta_A \exp(-i\varphi_A)$.

The interaction energy due to the permanent moments is [1]

$$W_{el} = \sum_{l_A=0}^{\infty} \sum_{l_B=0}^{\infty} \sum_{m=-l_<}^{l_<} \frac{(-1)^{l_B+m}(l_A + l_B)!}{(l_A + |m|)!(l_B + |m|)!}$$

$$\times (Q_{l_A}^m)^*(Q_{l_B}^m)R^{-(l_A+l_B+1)} \tag{3}$$

where $l_<$ is the lesser of l_A and l_B; $Q_{l_A}^m$ refers to moments of the charge distribution on A. For the dipole–dipole interaction ($l_A = l_B = 1$) we have a contribution

$$W_{dd} = (2/R^3)[\tfrac{1}{4}(Q_{A1}^{-1})^*Q_{B1}^{-1} - Q_{A1}^0 Q_{B1}^0 + \tfrac{1}{4}(Q_{A1}^1)^*Q_{B1}^1]$$

$$= (\mu_A\mu_B/R^3)[-2\cos\theta_A\cos\theta_B + \sin\theta_A\sin\theta_B\cos(\varphi_B - \varphi_A)].$$

This may also be written

$$W_{dd} = R^{-3}[\mathbf{\mu}_A \cdot \mathbf{\mu}_B - 3(\mathbf{\mu}_A \cdot \hat{R})(\mathbf{\mu}_B \cdot \hat{R})] \tag{4}$$

where $\hat{R}$ is a unit vector lying along the internuclear axis. Since the energy of a dipole $\mathbf{\mu}$ in a field $\mathbf{F}$ is $-\mathbf{\mu} \cdot \mathbf{F}$, we see that the field due to $\mathbf{\mu}_A$ at the position of B is

$$\mathbf{F}_{AB} = R^{-3}[-\mathbf{\mu}_A + 3(\mathbf{\mu}_A \cdot \hat{R}_{AB})\hat{R}_{AB}].$$

It should be noted that, for real dipoles, the energy is a series in R^{-1} of which (4) is just the first term. Moreover, for any charge distribution there will be a number of nonvanishing terms in (3). Usually, it is the leading term (lowest power of R^{-1}) which is of interest.

We turn now to the energy of induction. The polarizabilities may be defined as Cartesian tensors [3] by inspection of the expression for the energy of a charge distribution in an electric field, written as a power series in components of the field, the field gradients, and so on. On removing terms in the permanent moments, one has

$$W_{in} = -\tfrac{1}{2}\alpha_{ij}F_iF_j - \tfrac{1}{6}\beta_{ijk}F_iF_jF_k + \cdots$$
$$- \tfrac{1}{3}A_{ijk}F_iF_{jk} - \tfrac{1}{6}B_{ijkl}F_{ij}F_jF_{kl} + \cdots \tag{5}$$

Repeated indices are supposed to be summed over. Here F_i is a component of the electric field, F_{ij} of the field gradient. The ith component of the induced dipole moment would be

$$-\partial W_{in}/\partial F_i = \alpha_{ij}F_j + \tfrac{1}{2}\beta_{ijk}F_jF_k + \cdots$$
$$+ \tfrac{1}{3}A_{ijk}F_{jk} + \tfrac{1}{3}B_{ijkl}F_jF_{kl} + \cdots \tag{6}$$

The tensor α is the usual polarizability; the other terms refer to hyper-polarizabilities. Now if, for instance, one atom has permanent moments, the field, field gradients, etc. produced by these moments at the other atom may be calculated and used in (5). The field at B due to a charge Ze at A is $Ze\hat{R}_{AB}/R_{AB}^2$. With z axis ($i = 3$) along the internuclear line, the leading term in (5) is $-\tfrac{1}{2}\alpha_{33}(Ze)^2R^{-4}$. The field of a dipole was already given. The moment induced is proportional to R^{-3} and the contribution to the energy to R^{-6}.

It is found that, even for neutral atoms possessing complete spherical symmetry, there is an attractive force, as is evidenced by the term a/V^2 in the van der Waals equation of state for a gas

$$(p + a/V^2)(V - b) = RT.$$

We refer to these interactions as van der Waals or dispersive forces. In searching for the origin of a force between classical systems having no permanent moments, one is led to postulate the existence of fluctuating moments, such as dipole moments with randomly changing direction. If we take this as our model for the two atoms A and B, we see that at any instant the dipole on A, μ_A, will induce a moment in B. The size of the induced moment will be proportional to μ_A/R^3, the field due to μ_A at B, and to α_B, the polarizability of B. The energy lowering will then go as the induced moment times the field, that is as $\mu_A\alpha_B/R^6$. By similar reasoning there will be an energy lowering $\mu_B\alpha_A/R^6$ due to the moment μ_B. The average of μ_A vanishes, but the induced moment always has the same orientation to the moment inducing it, so the contribution to the energy does not vanish. The moment on B at any time consists of the randomly orienting μ_B and an induced moment whose orientation follows that of A. These are uncorrelated, and each averages to zero. We have assumed R is small enough so the fields are transmitted instantaneously and relativistic effects do not enter.

To make this definite, consider two classical charged harmonic oscillators centered at points A and B, separated by R. (This treatment is due to London [4].) Each consists of a negative charge $-e$ associated with a mass m, bound to a stationary charge $+e$ by a harmonic force with a force constant k. We write x_A, y_A, z_A for the Cartesian coordinates of the negative charge bound to A relative to an origin at A and correspondingly for B. In the absence of interaction, the energy is

$$E^0 = \tfrac{1}{2}m(\dot{x}_A{}^2 + \dot{y}_A{}^2 + \dot{z}_A{}^2) + \tfrac{1}{2}m(\dot{x}_B{}^2 + \dot{y}_B{}^2 + \dot{z}_B{}^2)$$
$$+ \tfrac{1}{2}k(x_A{}^2 + y_A{}^2 + z_A{}^2) + \tfrac{1}{2}k(x_B{}^2 + y_B{}^2 + z_B{}^2). \qquad (7)$$

Define $r_A{}^2 = x_A{}^2 + y_A{}^2 + z_A{}^2$ and similarly for r_B. Also, let r_I be the distance from the charge bound to B to A, r_{II} the distance from B to the charge bound to A, and r_{III} the distance between the negative charges. The interaction is

$$V = e^2/R + e^2/r_{III} - e^2/r_I - e^2/r_{II}. \qquad (8)$$

We are interested in values of R large compared to x_A, y_A, etc. Expanding r_I^{-1}, r_{II}^{-1}, and r_{III}^{-1} in powers of R^{-1}, we find the leading term in V,

$$V \approx (e^2/R^3)(x_A x_B + y_A y_B - 2z_A z_B),$$

the dipole–dipole interaction (4), with $\mu_A = er_A$ and $\mu_B = er_B$. The energy $E = E^0 + V$ is no longer a sum of terms for separate atoms, meaning that the motions of the oscillators are correlated.

However, if we introduce the new coordinates

$$x_\alpha = 2^{-1/2}(x_A + x_B), \quad y_\alpha = 2^{-1/2}(y_A + y_B), \quad z_\alpha = 2^{-1/2}(z_A + z_B)$$
$$x_\beta = 2^{-1/2}(x_A - x_B), \quad y_\beta = 2^{-1/2}(y_A - y_B), \quad z_\beta = 2^{-1/2}(z_A - z_B)$$
(9)

the energy may be written

$$E = \tfrac{1}{2}m(\dot{x}_\alpha^2 + \dot{y}_\alpha^2 + \dot{z}_\alpha^2 + \dot{x}_\beta^2 + \dot{y}_\beta^2 + \dot{z}_\beta^2)$$
$$+ \tfrac{1}{2}k(x_\alpha^2 + y_\alpha^2 + z_\alpha^2 + x_\beta^2 + y_\beta^2 + z_\beta^2)$$
$$+ (e^2/2R^3)(x_\alpha^2 - x_\beta^2 + y_\alpha^2 - y_\beta^2 - 2z_\alpha^2 + 2z_\beta^2) \quad (10)$$

This is a sum of energies for six, one-dimensional harmonic oscillators, with force constants $k + e^2/R^3$ (twice), $k + 2e^2/R^3$, $k - e^2/R^3$ (twice), and $k - 2e^2/R^3$. These represent the normal modes of the system; in each normal mode, the motions of A and B are coupled, which gives us our picture of coupled oscillating dipoles. Taking the oscillators to be quantum mechanical, the zero-point energy, which was $6(\hbar/2)(k/m)^{1/2}$ before interaction, is now

$$\tfrac{1}{2}\hbar m^{-1/2}[2(k + e^2/R^3)^{1/2}$$
$$+ (k + 2e^2/R^3)^{1/2} + 2(k - e^2/R^3)^{1/2} + (k - 2e^2/R^3)^{1/2}]$$

The energy difference is obtained in a convenient form on expanding the square roots in the small quantities e^2/kR^3. The leading term is

$$-\tfrac{3}{4}(e^2/kR^3)^2\hbar\omega, \quad \omega = (k/m)^{1/2}.$$

We now introduce the polarizabilities into the formula. Suppose an oscillator of the kind we are discussing were subjected to an electric field F in the x direction. A term $-eEx$ would be added to the energy, which one could now write as

$$E^E = \tfrac{1}{2}m(\dot{x}^2 + \dot{y}^2 + \dot{z}^2) + \tfrac{1}{2}k(X^2 + y^2 + z^2) - \tfrac{1}{2}(e^2F^2/k) \quad (11)$$

where $X = x - eF/k$. This is again a harmonic oscillator, but with the equilibrium position of the negative charge shifted to eF/k. The field has thus induced a dipole moment equal to e^2F/k. Thus the polarizability α is e^2/k, and it is of course independent of the direction of the field. The last term in (11) is the energy lowering of $-\tfrac{1}{2}\alpha F^2$ [cf. (5)]. The energy lowering due to the interaction of the two oscillators may now be written

$$\Delta E = -\tfrac{3}{4}\hbar\omega(e^2/kR^3)^2 = -\tfrac{3}{4}\hbar\omega\alpha^2/R^6. \quad (12)$$

A typical atomic polarizability is an atomic volume, say $10\,a_0^3$ if the atom is a sphere of radius 1–$2\,a_0$. To estimate $\hbar\omega$, we note that the zero point energy of $3\hbar\omega$ is twice the energy of an electron. If we take this as the ionization potential, say 10 eV, we can put $\hbar\omega \sim 7$ eV. Then the coefficient of $(R/a_0)^{-6}$ in (12) would be 5×10^2 eV. Actual values are an order of magnitude or so on either side of this.

We have already mentioned how the permanent moments are related to the quantum mechanical description of the charge distribution [Eq. (2)]. The ground-state wave function is required for their calculation. The polarizabilities needed for the inductive interactions may be calculated quantum mechanically by considering the change in energy of an atom due to electric field, field gradients etc. The van der Waals interaction is obtained by taking V of Eq. (8) as a perturbation on the system of two isolated atoms. For all three types of interactions it is necessary to be able to consider an isolated atom, that is, to ignore the possibility of exchange between the electrons assigned to the atom and any others.

Let $\psi_A(1\cdots k\cdots n_A)$ be a wave function for atom A, containing n_A electrons, and let $\phi(m)$ $(m > n_A)$ be a wave function for some other electron. Can the product wave function be used for the $(n_A + 1)$ electron system? The antisymmetrization of the overall wave function would give terms like $\psi_A(1, 2, \ldots, k - 1, m, k, \ldots, n_A')\phi(k)$. They may be ignored if products like

$$[\psi_A(1, 2, \ldots, n_A)\phi(m)]^*\psi_A(1, \ldots, k - 1, m, k + 1, \ldots, n_A)\phi(k)$$

$$(1 \leq k \leq n_A) \qquad (13)$$

vanish, since then the exchange will give no additional contribution to expectation values. (Note that a product like

$$|\,\psi_A(1, \ldots, k - 1, m, k + 1, \ldots, n_A)\phi(k)\,|^2$$

where the same electron appears in ϕ in both factors, gives the same contribution to an expectation value as $|\,\psi_A(1, \ldots, n_A)\phi(m)\,|^2$, since all electrons enter equivalently in physical operators.) The vanishing of (13) requires $\phi(k)$ to be zero where $\psi_A(1, \ldots, n_A)$ is not and vice versa, which is possible if these functions (for electron k) are essentially localized or confined to different regions of space. This is referred to as the case of zero differential overlap (ZDO), and all the preceding discussion presupposes its validity. Wave functions generally die off exponentially. Thus complete localization is impossible, but the overlaps go to zero faster than any inverse power of R, so the ZDO assumption may be considered valid for very large R. One need not antisymmetrize the overall wave function with respect to interchanges of electrons assigned to wave functions between which ZDO holds.

REFERENCES

1. J. O. Hirschfelder, C. F. Curtis and R. B. Bird, "Molecular Theory of Gases and Liquids." Wiley, New York, 1954.
2. J. A. R. Coope, R. F. Snider, and R. F. McCourt, *J. Chem. Phys.* **43**, 2269 (1965).
3. A. D. Buckingham, *Advan. Chem. Phys.* **12**, 107 (1967).
4. F. London, *Z. Phys. Chem. Abt. B* **11**, 222 (1930).

B. Large Internuclear Distances

If R is allowed to decrease, the assumption of zero differential overlap will become invalid, and it will be necessary to antisymmetrize the wave function with respect to exchange of electrons belonging to different atoms. Since the valence electrons are more extended in space than the core electrons, it is the overlap between valence electrons on different atoms that must first be considered. (Where there are no valence electrons, we mean the electrons of highest principal quantum number.) We will obtain the interaction energy from the expectation value of a suitable Hamiltonian over a wave function formed from the atomic orbitals of the valence electrons. The overlap between atomic orbitals of different atoms will play an important role here.

For simplicity, we suppose that there is only one valence electron on each atom. Only these two electrons are considered explicitly, the effect of the inner-shell electrons being simply to shield the nucleus and provide an effective potential in which the valence electrons move. This potential determines the form of the spin orbitals for the valence electrons. Denoting the spatial parts of these spin orbitals as ϕ_A and ϕ_B, we write wavefunctions for the two-electron system in the form

$$\Psi = (2\eta)^{-1/2}[\lambda_1(1)\lambda_2(2) - \lambda_2(1)\lambda_1(2)]. \tag{14}$$

Here, each of the spin orbitals λ_1 and λ_2 is ϕ_A or ϕ_B with α or β spin, and η is a normalizer of size unity. We have resolved the exchange degeneracy between $\lambda_1(1)\lambda_2(2)$ and $\lambda_2(1)\lambda_1(2)$ by taking the antisymmetric combination demanded by the Pauli principle. There is also a spin degeneracy: To a good approximation the energy does not depend on the spins of the two electrons (the Hamiltonian is spin-independent). We can choose our functions to be eigenfunctions of total electronic spin S as well as of its z component S_z. The matrix element of H between two functions of different spin vanishes.

Since there are four choices for λ_1 and λ_2 there are six functions of the form (14). In two of them, λ_1 and λ_2 have the same spatial part but differ in spin. These wave functions are clearly spin singlets:

$$\Psi_I = 2^{-1/2}\phi_A(1)\phi_A(2)[\alpha(1)\beta(2) - \beta(1)\alpha(2)] \tag{15a}$$

$$\Psi_{II} = 2^{-1/2}\phi_B(1)\phi_B(2)[\alpha(1)\beta(2) - \beta(1)\alpha(2)] \tag{15b}$$

In Ψ_I and Ψ_{II} both electrons are assigned to the same atom. At large R, it is energetically more advantageous to put one on each atom. We discuss wave functions like Ψ_I and Ψ_{II} in what follows (ionic bond). When λ_1 and λ_2 differ in space but not in spin, we have triplet wave functions:

$$\Psi_{III} = (2\eta)^{-1/2}[\phi_A(1)\phi_B(2) - \phi_B(1)\phi_A(2)]\alpha(1)\alpha(2) \tag{16a}$$

$$\Psi_{IV} = (2\eta)^{-1/2}[\phi_A(1)\phi_B(2) - \phi_B(1)\phi_A(2)]\beta(1)\beta(2) \tag{16b}$$

Ψ_{III} and Ψ_{IV} correspond to $S = 1$, $S_z = \pm 1$. Finally, if λ_1 and λ_2 differ in both space and spin, (14) is not itself an eigenfunction of total spin. But the linear combinations

$$\Psi_{V,VI} = 2\eta^{-1/2}[\phi_A(1)\phi_B(2) \pm \phi_B(1)\phi_A(2)][\alpha(1)\beta(2) \mp \beta(1)\alpha(2)] \tag{17}$$

are spin eigenfunctions, the upper signs corresponding to a singlet ($S = 0$, $S_z = 0$) and the lower signs to a triplet ($S = 1$, $S_z = 0$). The triplet Ψ_{VI} differs from Ψ_{III} and Ψ_{IV} in spin alone, so leads to the same energy.

Let V_A and V_B be the potentials due to the nuclei and core electrons of atoms A and B, respectively. The additional terms in the Hamiltonian due to the mutual approach of the two atoms are the interaction of the cores (approximately Coulombic), the repulsion between the two valence electrons, and the attraction of an electron in ϕ_A for the core of B and the electron in ϕ_B for the core of A. Because of the electronic exchange we cannot associate a particular electron with a particular orbital, so the latter two interactions cannot be isolated. We have to write $\sum_i [V_A(i) + V_B(i)]$, which includes interactions already present in the atoms, and subtract out the effect of these later. Note that if V_A and V_B are Coulombic, our "perturbing Hamiltonian" is just V, Eq. (8) of Chapter II, Section A.

We now calculate the expectation value of the two-electron Hamiltonian over the diatomic wave function and subtract off the energy at large R (atomic energies). Denoting the kinetic energy operator for electron i by $T(i)$ we have

$$\Delta E = \frac{\langle \Psi \mid e^2 R^{-1} + \sum_{i=1}^{2} [T(i) - eV_A(i) - eV_B(i)] + e^2/r_{12} \mid \Psi \rangle}{\langle \Psi \mid \Psi \rangle} - E_\infty \tag{18}$$

where $e^2 R^{-1}$ is the core–core repulsion. We consider the cases where $\Psi = \Psi_{\text{III}}$, Ψ_{IV}, Ψ_{V}, or Ψ_{VI}. The integration over spins may be carried out easily. It may also be seen that some of the contributions to ΔE differ only in the labeling of electrons, so we may simplify to

$$\Delta E = \frac{\langle \phi_A(1)\phi_B(2) \pm \phi_B(1)\phi_A(2) \mid T(1) - eV_A(1) + T(2) - eV_B(2)}{\langle \phi_A(1)\phi_B(2) \pm \phi_B(1)\phi_A(2) \mid \phi_A(1)\phi_B(2) \rangle} - E_\infty$$

The angular brackets now signify integration over spatial variables only. The atomic orbital ϕ_A is an eigenfunction of $T + V_A$ with eigenvalue ε_A while ϕ_B is an eigenfunction of $T + V_B$ with eigenvalue ε_B. Furthermore, E_∞, the energy when R is large enough for the diatomic system to be considered as separate atoms, is $\varepsilon_A + \varepsilon_B$. Then ΔE is just

$$\Delta E = \frac{e^2}{R}$$

$$+ \frac{\langle \phi_A(1)\phi_B(2) \pm \phi_B(1)\phi_A(2) \mid -eV_A(2) - eV_B(1) + e^2/r_{12} \mid \phi_A(1)\phi_B(2) \rangle}{\langle \phi_A(1)\phi_B(2) \pm \phi_B(1)\phi_A(2) \mid \phi_A(1)\phi_B(2) \rangle}$$

where we have used the fact that $e^2 R^{-1}$ is independent of electronic coordinates. The upper signs are for the singlet, the lower signs for the triplet. The expression for ΔE may also be written as

$$\Delta E = (C \pm D)/(1 \pm S_{AB}^2) \tag{19}$$

where

$$C = \langle \phi_A(1)\phi_B(2) \mid e^2 R^{-1} - eV_A(2) - eV_B(1) + e^2/r_{12} \mid \phi_A(1)\phi_B(2) \rangle$$

is called a Coulomb integral or Coulomb term,

$$D = \langle \phi_A(1)\phi_B(2) \mid e^2 R^{-1} - eV_A(2) - eV_B(1) + e^2/r_{12} \mid \phi_B(1)\phi_A(2) \rangle$$

is called an exchange integral or exchange term, and

$$S_{AB} = \langle \phi_A(1) \mid \phi_B(1) \rangle.$$

It usually turns out that C and D are both negative but that D is of larger magnitude, so that, if we construct a wave function from the orbitals of the interacting atoms, terms arising from exchange or antisymmetry are dominant in the interaction. For H_2, there is no core, and $V_A(2) = -1/r_{A2}$, $V_B(1) = -1/r_{B1}$. At $R = 1.5a_0$ and with hydrogen atom orbitals, Slater [1] cites these values: $S_{AB} = 0.7252$, $C = -0.0104e^2/a_0$, $D = -0.1616e^2/a_0$. Then the upper sign in (19) gives a net energy lowering of 0.1128 a.u., the lower sign a net energy increase of 0.3190 a.u. The exchange term leads to stabilization if the resulting configuration is a singlet (paired spins), and this we may refer to as a chemical bond. The magnitude of the integrals in D, and hence ΔE, increases with the overlap of the orbitals ϕ_A and ϕ_B. Because of the exponential fall-off with distance of the wave functions for Coulombic systems, the R-dependence of the integrals in C and D is exponential, as is S_{AB}. Then ΔE of (19) is composed of exponentials, and could perhaps be approximated as a single decreasing exponential in R.

In order to generalize to atoms with several valence electrons, we rewrite the wavefunction of (17) (to within a normalizing constant) as a linear combination of determinants which are degenerate at large R:

$$\Psi = |\ \phi_A(1)\alpha(1)\phi_B(2)\beta(2)\ | \pm |\ \phi_A(1)\beta(1)\phi_B(2)\alpha(2)\ |. \qquad (20)$$

The positive and negative signs yield the triplet and singlet (i.e., repulsion and attraction) respectively. Putting $\alpha = \beta$ in (20) allows it to include the triplets with $S_z = \pm1$: The negative sign is not allowed in this case. The bond is represented by a linear combination of determinants differing by an exchange of opposite spins between the atomic orbitals (one on each atom) involved. Where there are n unpaired valence electrons on each atom, n bonds may be formed. We can write 2^n determinants differing in assignment of spins to the orbitals involved in bonds, if we require that, for each pair of such orbitals, one have α spin and one β spin. Then the wave function is a linear combination of these with coefficients ±1. The signs are chosen so that the *total* wave function changes sign on exchange of spins between atomic orbitals participating in a bond. The choice of which orbitals to pair with which is made to maximize the overlap $\phi_A\phi_B$ between orbitals bonded to each other. The reason is that the strength of the bond (energy lowering)—cf. Eq. (19)—depends on the size of the exchange integral, which in turn depends on the size of the overlap density $\phi_A(1)\phi_B(1)$.

The energy may be expected to involve sums of Coulomb and exchange integrals like (19). It is expected that the energy lowering will become larger in magnitude, the more bonds that are formed. It may, in fact, be shown [2] that the expectation value of the energy is approximately $(C + \sum A)/(1 + S)$, where C is the Coulomb integral, S the overlap, and $\sum A$ a sum of exchange terms for the various bonds. Of course, electrons involved in one bond are not available for another. This is the saturation property of chemical valence.

If an atomic orbital on one atom is doubly occupied, exchange of, say, an α spin with a β spin associated with an orbital on the other atom would lead to two electrons of β spin in the same atomic orbital. This is not allowed; indeed, the determinant would have two identical rows and vanish. Exchange of an α spin with an α spin is analogous to the triplet case mentioned earlier [$\alpha = \beta$ in (20)]. According to the present model, the interaction of a doubly occupied orbital on one atom with an unpaired spin on the other can only lead to an increase of energy over the separated atoms (repulsion), since it is impossible to construct a linear combination of determinants of atomic orbitals for which the exchange integrals will enter the interaction energy with positive signs. (Because of the overall antisymmetry demanded by the Pauli principle, the spin orientations limit the possible spatial symmetries of the wave function. The use of spins is a convenient way to keep track of the symmetry properties of the wave function; there is no physical repulsion between the spins that produces the energy with which we are concerned here.) Where there are *no* unpaired electrons, but only closed shells, only repulsive interactions can occur by the spin-exchange mechanism and there will be no chemical bond. Closed shells thus repel each other when they begin to overlap.

One can imagine exciting the atom to a state in which the spin multiplicity (number of unpaired spins) is greater than in the ground state, allowing more bonds to be formed. If enough energy is recovered by bond formation to offset the initial cost of promotion, this will represent the ground state of the system. It must be emphasized that this procedure is meant as a way of describing the wave function, not as an actual description of the physical process taking place. In a similar manner, we may introduce "hybridization." The differential overlap for the atomic orbitals involved in the bonds may be increased by putting the bonding electrons into strongly directional orbitals, pointed toward each other. For this purpose, we may want to utilize orbitals not populated in the ground state of the atoms. This requires some promotion energy, which is regained in the form of added bond strength.

We now consider the "ionic" functions Ψ_I and Ψ_{II} [Eq. (15)]. Here, one or more electrons which would be assigned to one atom when the atoms are noninteracting are put into empty orbitals on the other. At large R, this corresponds to ion formation, which generally requires additional energy, which may be calculated as the difference between the ionization potential of one atom and the electron affinity of the other (both taken as positive). At small enough R, this energy is compensated by the stabilization associated with the $1/R$ attraction between the ions. To write the interaction energy quantum mechanically, we return to the two-electron case and suppose that the orbital on A is of lower energy than that on B, so the structure A^-B^+ is preferred to A^+B^-. Using the wave functions Ψ_{I} the expectation value of the Hamiltonian is

$$\frac{\langle \phi_A(1)\phi_A(2) \mid T(1) + T(2) - eV_A(1) - eV_A(2)}{\langle \phi_A(1)\phi_A(2) \mid \phi_A(1)\phi_A(2)\rangle}$$
$$\frac{-eV_B(1) - eV_B(2) + e^2R^{-1} + e^2/r_{12} \mid \phi_A(1)\phi_A(2)\rangle}{}$$
$$= e^2R^{-1} + 2\varepsilon_A + \langle \phi_A(1)\phi_A(2) \mid V_B(1)$$
$$+ V_B(2) + e^2/r_{12} \mid \phi_A(1)\phi_A(2)\rangle$$

and the interaction energy, after subtraction of $E_\infty = \varepsilon_A + \varepsilon_B$, is

$$\Delta E = e^2R^{-1} + 2\langle \phi_A \mid V_B \mid \phi_A\rangle + \varepsilon_A - \varepsilon_B$$
$$+ \langle \phi_A(1)\phi_A(2) \mid e^2/r_{12} \mid \phi_A(1)\phi_A(2)\rangle. \tag{21}$$

The last three terms express the net energy gained by taking an electron out of orbital ϕ_B and putting it into orbital ϕ_A. Since there are now two electrons in ϕ_A, an extra interelectronic repulsion integral will arise: This is the last term. Then the electron affinity of A is given by

$$-\varepsilon_A - \langle \phi_A(1)\phi_A(2) \mid e^2/r_{12} \mid \phi_A(1)\phi_A(2)\rangle$$

The ionization potential of B is approximated by $-\varepsilon_B$. The value of $\langle \phi_A \mid V_B \mid \phi_A\rangle$ is essentially $-e^2R^{-1}$, the attraction of an electron on A for the positive core of B.

It will be recognized that our treatment of the two-electron case is a generalization of the well-known valence bond theory of the hydrogen molecule (EYRING, Sect. 12a, 12b; KAUZMANN, Sect. 11-B.b). The emphasis on spin exchange, which is then used to discuss, qualitatively, more complex systems, is found in books like Heitler's [2]. Slater [1] gives a complete discussion of the valence bond treatment of H_2.

If there are no valence electrons, we expect the potential curve to exhibit the R^{-6} attraction due to the van der Waals interaction for large R, with this going over to a sharply rising repulsion at smaller R due to the closed shells (and eventually the Coulombic repulsion of the nuclei). The interaction potential will thus show a minimum. This actually occurs at internuclear distances of about $10\,a_0$; the depths are of size 10^{-2} eV. In the case where the interaction is bonding, the valence attraction will outweigh the van der Waals. Eventually, the attraction will be outweighed by the closed shell repulsion (except for first row elements) so a minimum will occur. Here we might express $U(R)$ as a difference of two exponential terms. The resultant minimum occurs for R of size $1\,a_0$, and its depth is several electron volts. For an idealized ionic bond, we would have no attraction until the e^2/R energy lowering is sufficient to compensate the cost of ion creation. Since ionization potentials (I.P.) run from 3.9 eV (Cs) to 25 eV (He) and electron affinities (E.A.) from 0 to 3.6 eV (F), we may take a typical value for $\varDelta = \text{(I.P.)}_B - \text{(E.A.)}_A$ as 5 eV. This equals e^2/R for $R \sim 5\,a_0$. For smaller R, the interaction potential would be the sum of the $1/R$ attraction and the exponential closed-shell repulsion, which leads to a minimum at a distance of the order of Bohr radii and depth of the order of electron volts.

It is common to call the wave functions of Eqs. (16) and (17) "covalent," and, in contrast, those of Eqs. (15) "ionic." These descriptions may be meaningful at large internuclear distances, but only if ϕ_A and ϕ_B are actually the orbitals of atoms A and B. However, the "covalent" and "ionic" wave functions are far from mutually orthogonal, and hence far from mutually exclusive, at small values of R. Furthermore, the covalent wavefunction itself has "ionic" character, since it does not give a symmetrical distribution of electrons. It may be noted that the ionic function resembles a molecular orbital function (Chapter II, Section C), in which the bonding orbitals are so polar as to be one-center.

REFERENCES

1. J. C. Slater, "Quantum Theory of Molecules and Solids," Vol. I, Chapter 3. McGraw-Hill, New York, 1963.
2. W. Heitler, "Elementary Wave Mechanics" (Oxford Univ. Press, London and New York, 1956); H. Eyring, J. Walter, and G. E. Kimball, "Quantum Chemistry," Chapters XII, XIII. Wiley, New York, 1944.

C. Bond Region and Smaller R

We have already shown how an energy minimum may arise due to a balance between attractive and repulsive covalent and/or ionic effects. The closed shell repulsions of inner shell electrons will become more important for smaller R. Eventually, of course, it is the internuclear repulsion which dominates, since the distance between the point charges of the nuclei may decrease to zero, while the electrons are always diffuse. One can continue to invoke the mechanisms of exchange attraction and repulsion of the valence bond theory, but as more and more interatomic overlaps become important, the theory becomes very complicated. Since the identity of the individual atoms is being lost because of these overlaps, it is suggested that it would be more fruitful to treat the diatomic system as a single entity. Except that one now considers interacting electrons in the field of two nuclei, the procedures used for calculation of atomic wave functions may be used here. In particular, one may consider each electron to be described by some one-electron function or orbital, which will in general be bicentric (molecular orbital), becoming more monocentric in character as the nuclei approach each other. Each orbital represents the motion of the electron in the field of the nuclei and the other electrons. The electron density and other properties are built up in terms of these orbitals.

To determine the orbitals one would try to solve a one-electron Schrödinger equation where the Hamiltonian includes kinetic energy and the potential due to the nuclei and the other electrons. Even without doing this, certain qualitative features of the orbitals can be anticipated. First, the cylindrical symmetry means that each is characterized by a quantum number referring to angular momentum about the figure axis. Where this is nonzero, the orbital wave function must vanish on the axis. If the molecule is homonuclear, the wave function for each orbital must be symmetric or antisymmetric with respect to an inversion through the center. Close to a nucleus, these functions resemble the orbitals for hydrogen-like atoms, since here the potential of the nucleus dominates. Far from both nuclei, the bicentric character of the field is less important, and we again get atomic behavior: There is an exponential fall-off with distance from the center.

Because of the resemblance to atomic orbitals, a description of the molecular orbitals in terms of atomic orbitals is appropriate and convenient for a qualitative understanding. Let a molecular orbital be expressed as a linear combination of atomic orbitals on the two atoms. Orbitals with different symmetries with respect to rotation about the figure axis do not combine, since cross terms of the Hamiltonian vanish between them. It is also unfavorable to mix orbitals whose energies in the atom are very different. All this is seen from the secular equation which arises from minimization of the expectation value of the Hamiltonian over the molecular orbital. In the simplest case, we consider the molecular orbital ψ to be formed as a linear combination of only two normalized atomic orbitals, one from atom A and one from atom B:

$$\psi = C_A\phi_A + C_B\phi_B. \tag{22}$$

The secular equations, from minimization of $\langle \psi \mid H \mid \psi \rangle / \langle \psi \mid \psi \rangle$, are

$$(H_{AA} - E)C_A + (H_{AB} - ES_{AB})C_B = 0$$
$$(H_{BA} - ES_{BA})C_A + (H_{BB} - E)C_B = 0 \tag{23}$$

where H is an effective Hamiltonian for the electron under consideration. The secular determinant may be written as

$$f(E) = (H_{AA} - E)(H_{BB} - E) - (H_{AB} - ES_{AB})^2 = 0.$$

$f(E)$ is positive for $E \to \pm\infty$ and negative for $E = H_{AA}$ or H_{BB}. Thus one of the roots is greater than either H_{AA} or H_{BB}, the other less than either. According to (23), $(C_A/C_B)^2 = (H_{BB} - E)/(H_{AA} - E)$. The amount of mixing is small if $H_{AA} \ll H_{BB}$, since $(C_B/C_A)^2$ becomes $(H_{AA} - E)/(H_{BB} - E)$, and ($E$ is less than H_{AA}) this is small, i.e., $C_B \to 0$. Similarly, $H_{BB} \ll H_{AA}$ makes $C_A \to 0$. The energies of the atomic orbitals in the molecule are H_{AA} and H_{BB}. Thus there is no mixing when they are very different.

Between the nuclei, there is a region of lower potential energy, so the system may be stabilized by putting more electron density here. In addition to lowering the energy, this tends to hold nuclei together: The attraction of the electron density for the nuclei outweighs their mutual repulsion. (If there are inner shells, the "nuclei" are effective charges due to the nuclear charge screened by the inner shells.) Molecular orbitals which build up electron density in the internuclear region (relative to the superposed atomic densities) are referred to as bonding. For values of R which are not too small, the energy of an electron in such an orbital decreases as the nuclei draw closer together. This is to be contrasted with orbitals whose energy is insensitive to changes in R ("nonbonding" orbitals centered on one nucleus) and those whose energy increases with decreasing R because their formation leads to a decrease of electron density between the nuclei and hence an increase on either side ("antibonding" orbitals). The energy lowering in the case of a bonding orbital is accompanied by increased interelectronic repulsion and kinetic energy. Furthermore, the Pauli principle prevents electrons of the same spin from being in the same place, effectively making it impossible to pack as many electrons as one would like into a given region of space, so bonding orbitals become less effective when the nuclei get very close together.

Having found a set of molecular orbitals we fill them in order of increasing energy, since we want to minimize the total energy of the system. The question is then whether, compared to the superposition of atomic densities, there is a buildup of electron density between the nuclei, or whether there is a decrease, which leads to repulsion between the atoms. If the orbitals ϕ_A and ϕ_B are defined so that the overlapping parts have the same sign ($\phi_A\phi_B$ is positive over most of the region where it is nonzero), the bonding orbitals are those for which C_A and C_B have the same sign. This is seen from the expression for the density of the normalized molecular orbital ψ:

$$\frac{\psi^*\psi}{\int \psi^*\psi \, d\tau} = \frac{C_A^2\phi_A^2 + 2C_AC_B\phi_A\phi_B + C_B^2\phi_A^2}{C_A^2 + 2C_AC_BS_{AB} + C_B^2}.$$

We illustrate the buildup of the molecular orbitals for the homonuclear case where there is one valence electron on each atom. If ϕ_A and ϕ_B are S orbitals, there are two σ orbitals, σ meaning angular momentum zero about the figure axis. First,

$$\psi(\sigma_g) = (\phi_A + \phi_B)/(2 + 2S_{AB})^{1/2} \qquad (24)$$

The other linearly independent combination is antibonding:

$$\psi(\sigma_u) = (\phi_A - \phi_B)/(2 - 2S_{AB})^{1/2} \qquad (25)$$

The subscripts g and u refer to symmetric or antisymmetric behavior on inversion through the origin. Putting both electrons in $\psi(\sigma_g)$ lowers the energy and leads to binding. If the electrons are in p orbitals, there are six orbitals all together. One can form two π_g and π_u orbitals (differing in the directions of the 1 unit of angular momentum) plus a σ_g and a σ_u orbital. The σ_g and π_u orbitals lead to binding, but which lies lowest in energy is not always easy to decide.

For heteropolar cases, of course, $|C_A|$ and $|C_B|$ in (22) need not be equal. In fact, for some orbitals one coefficient may be so much larger than the other that the molecular orbital is essentially an atomic orbital and may be considered nonbonding. If the energy of such an orbital is sufficiently low, electrons from the other atom go into the orbital, and the molecule may be thought of as composed of ions. The molecular energy is lowered because the orbital energy is low and also because of Coulomb stabilization. It is clear, however, that one can go continuously from the covalent to the ionic case in the molecular orbital model. It should be emphasized that the use of atomic orbitals in an expansion, while an aid to thought, may be replaced by any convenient basis set for purposes of calculation.

In the molecular orbital picture, an attempt to bring the nuclei too close together decreases the volume in which electrons are effective in bonding. The density in this region is limited by Pauli principle and interelectronic repulsion considerations. The energy eventually starts to rise. Thus the energy goes through a minimum before increasing sharply. Of course, if the situation is initially antibonding (repulsive) we obtain a monotonic curve: The energy rises increasingly sharply with decreasing internuclear distance.

Eventually, the nuclei are close enough together so that only the inner shell electrons contribute to the density between them. The dominant term in the energy is now

$$V = \frac{Z_A^{\text{eff}} Z_B^{\text{eff}}}{R} e^2 \tag{26}$$

where R is the internuclear distance. Also, Z_A^{eff} and Z_B^{eff}, which are effective charges, will approach the nuclear charges Z_A and Z_B as R goes to zero. The field of two nuclei very close together resembles the field of a single point charge over most of space. The electron density goes over to the electron density for the atom whose nuclear charge is $Z_A + Z_B$. The molecular orbitals go over naturally to the atomic orbitals of this "united atom." The infinite internuclear repulsion $Z_A Z_B / R$ ($R \to 0$) is a constant for the electronic coordinates, so the electronic wave function is not affected by the fact that it is formally infinite. The total energy is the energy of the united atom plus $Z_A Z_B / R$.

One can, in fact, consider starting from the united atom and considering the separation of the nuclei as a perturbation. The internuclear repulsion is considered separately, since it does not affect the wave function. The perturbation is the change in the nuclear–electronic Hamiltonian between the molecule and the united atom.

$$H_1 = \sum_i \left(-\frac{Z_B e^2}{r_{Bi}} - \frac{Z_A e^2}{r_{Ai}} + \frac{(Z_A + Z_B)e^2}{r_i} \right). \tag{27}$$

The sum is over electrons. The effect of H_1 on the electronic energy may be calculated by perturbation theory, which leads to the energy being $Z_A Z_B e^2 / R$ plus a power series in R. Initially, the energy decreases with R because the nuclear repulsion dominates.

D. Forms for $U(R)$

In this section, we enumerate and review briefly most of the analytical forms which have been proposed for $U(R)$. More complete discussions, especially for use in interpreting experimental data, are found in the book of Hirschfelder et al. [1], and in articles by Pauly and Toennies [2], Varshni [3], Cottrell [4], Preuss [5], Steele et al. [6], Abrahamson [7]—for the rare gases; Konowalow [8], and others. The first article, in addition to general discussion, treats the relation of the potential curves to scattering problems; most of the others emphasize spectroscopic measurements. Kihara [9] has discussed some of the functions in relation to virial coefficient measurements.

An analytical formula for the interaction energy of two atomic systems as a function of R is clearly easier to deal with than the energies corresponding to a series of values of R, such as are provided by most calculations. Many applications in fact require such an analytical form, in which one or more parameters are left to be evaluated. In general, more parameters mean greater flexibility and the possibility of a closer representation of the true $U(R)$, but the choice of the functional form is of great importance.

In postulating a form for $U(R)$ one has first to bear in mind the general theoretical results outlined in the preceding sections: behavior for large and small R, etc. In addition, one would like to use as simple a function as possible. But defining simplicity is not easy. For infrared spectroscopy, a "simple" function might be one whose vibrational levels are easy to calculate [like the Morse potential, Eq. (43)]. If one starts with a simple formula for the vibrational levels and constructs the corresponding potential, arranging things so that it has the desired qualitative behavior [10], the functional form will be far from simple. The choice among different n-parameter functions can sometimes be made by using the functions, with parameters inserted, to predict new experimental data (see page 80). Alternatively, one can compare the values of the parameters derived using different sets of experimental data. If the function chosen represents the true $U(R)$ faithfully, the same values should result.

There are really several different kinds of potential curves to be dealt with. Where there is no valence attraction, $U(R)$ includes a short-range repulsion and a long-range, weak attraction of the van der Waals type, which leads to a shallow minimum at perhaps $8\, a_0$, of depth a few times 10^{-2} eV. Chemical bonding means the minimum in $U(R)$ will occur for smaller R, say $2\, a_0$, and be much deeper, perhaps several electron volts.

It might also be expected that ionic bonding could give curves of somewhat different shape than covalent bonding because of the more long-range nature of the force. In addition to curves where $U(R)$ decreases from $U(\infty)$ as R decreases and goes through a single minimum, one may also have to deal with more complicated situations [5]: U may go through a maximum before decreasing and going through a minimum; the value of U at the minimum may be above or below $U(\infty)$; there may be several minima separated by relative maxima; and so on.

1. *Functional Forms*

We first consider simple functions which represent only certain features of $U(R)$. The choice of which to use depends on the application being considered, and hence on the range of R of interest. If the parameters in a complicated function are to be determined from experimental measurements, the range of R which is interrogated by the experiment must be ascertained; one need hardly emphasize that parameters affecting $U(R)$ mainly in other regions will be unreliably determined, if at all.

Spectroscopic measurements on a stable molecule are generally concerned with the region of R close to R_e, the equilibrium internuclear distance. Then a truncated power series about R_e is a convenient form for $U(R)$, despite the fact that the behavior of such a series (which becomes infinite for $|R - R_e| \to \infty$) is incorrect at $R = 0$ (where $U(R)$ should become infinite) and at $R \to \infty$ (where $U(R)$ approaches a constant value). Whether a convergent power series exists for $U(R)$ (the interval of convergence certainly cannot be greater than $|R - R_e| = R_e$) is likewise irrelevant; one simply wants to represent $U(R)$ over a range $R_1 \leqq R \leqq R_2$. Commonly one uses

$$U(R) = U(R_e) + \tfrac{1}{2}k(R - R_e)^2 + b(R - R_e)^3 + c(R - R_e)^4 \quad (28)$$

or this with c put equal to zero. The first derivative $U'(R)$ vanishes at R_e, k is the harmonic force constant, and b a constant of anharmonicity. One recognizes, however, that the addition of higher powers of $(R - R_e)$ to the series may have an effect on the coefficients of preceding powers. Beckel and co-workers [11] have discussed this and other problems related to the derivation of potential constants from power series expansions of computed vibrational potentials.

Transport and other bulk properties of gases are often concerned with the long-range forces. If there are no interactions involving permanent moments, we expect an attraction going as R^{-6}. This will actually be the first term in a series in R^{-1}. It may be fruitful to use several terms, such as

$$U(R) = -a_1 R^{-6} - a_2 R^{-8} \tag{29}$$

or to attempt to approximate such a function, over a restricted range of R, as

$$U(R) = -aR^{-n} \tag{30}$$

where both a and n are parameters to be determined. The value of n is not of fundamental significance like the 6 or 8 in (29): it depends on the range of R sampled.

If we are primarily interested in internuclear distances where the valence repulsion dominates, we may write

$$U(R) = Ae^{-\xi R} \tag{31}$$

since we have argued that valence-type interactions should vary exponentially with R. Over a limited range of R, it may be just as accurate (and lead to simpler computations) to put

$$U(R) = AR^{-n} \tag{32}$$

If we put $A = \sigma^n$ in (32) and let n become infinitely large, this becomes an infinitely steep and short-range repulsion or "hard-sphere potential":

$$U(R) = \infty, \quad R > \sigma; \quad U(R) = 0, \quad R \leq \sigma. \tag{33}$$

By combining (31) with the van der Waals attraction, we have a potential which should be useful for interactions between rare gas atoms:

$$U(R) = Ae^{-\xi R} - BR^{-6} \tag{34}$$

This "exponential-six" potential is associated with the name of Buckingham [12], who used it to discuss the properties of rare gases. Sometimes a term in R^{-8} is added. The minimum in the curve occurs at a value of R determined by $e^{-\xi R} = (6B/A\xi)R^{-7}$. There is also a maximum for a smaller value of R, $U(R)$ going to $-\infty$ as R goes to zero. To use the potential, we generally cut it off at the value of R corresponding to the maximum by putting in an infinite repulsion (hard sphere potential) at this point.

In using a potential like (34), care must be taken not to identify B with the van der Waals coefficient automatically. Where the constants are chosen to fit experimental data from a variety of sources, we are fitting the correct potential with an approximate form over a wide range of R. Since there is not much flexibility in the first term, the second term must represent more than just the large-R behavior, and B should be considered just as a curve-fitting parameter. To derive the van der Waals coefficient, one must be sure that only the long-range part of $U(R)$ is being fit, or use a more flexible potential which includes an R^{-6} term.

The Lennard-Jones or "six–twelve" potential uses (32) instead of (31) in conjunction with the R^{-6} long-range attraction, and chooses the power n as 12, partly for aesthetic reasons and mathematical convenience. It may be written

$$U(R) = 4\varepsilon[(\sigma/R)^{12} - (\sigma/R)^6].\qquad(35)$$

The attractive and repulsive terms balance at $R = \sigma$, while the minimum occurs for $R = 2^{1/6}\sigma$, the well having depth ε. This two-parameter function goes properly to.zero as R becomes infinite and is infinite for $R = 0$. Since the use of the 12th power in the repulsive term is arbitrary, one may write instead [2]

$$U(R) = \frac{6\varepsilon}{n-6}\left[\left(\frac{R_m}{R}\right)^n - \frac{n}{6}\left(\frac{R_m}{R}\right)^6\right].\qquad(36)$$

Here, ε is still the well depth and the minimum is at $R = R_m$. (36) of course reduces to (35) for $n = 12$. Other powers of R may be used in the attractive part as well, giving

$$U(R) = -aR^{-m} + bR^{-n}.\qquad(37)$$

Such a form was proposed and discussed by Sutherland [13]. Sometimes a fifth parameter is introduced by putting $R - d$ for R in (37).

The values of m and n may be assumed on physical grounds or taken as additional parameters. The form (37) may be used to describe minima due to valence interactions as well as those due to the van der Waals interaction. It appears that, for series of related molecules, m and n can be held constant. Buckingham [12] took $m = 6$ and tried varying n to fit virial coefficient data for rare gas interactions but found values between 8 and 14 worked equally well. An interesting analysis was given of how we go about employing experimental results from different sources to determine parameters, and what the parameters signify.

Kratzer [14] considered the series of inverse powers of R,

$$U(R) = (-e^2/R_e)[\alpha(R/R_e)^{-1} + \tfrac{1}{2}\beta(R/R_e)^{-2} + \tfrac{1}{3}\gamma(R/R_e)^{-3} + \cdots]\qquad(38)$$

and proposed the simplified formula

$$U(R) = D_e[(R - R_e)/R]^2. \tag{39}$$

The Fues potential [15]

$$U(R) = K[-(R/R_e)^{-1} + b(R/R_e)^{-2}] \tag{40}$$

is also a special case of (38). Parr and White [16] showed that (13) works well in predicting relations between force constants. They gave a simple physical interpretation to the coefficients (see Vol. 2, Ch. III, Sect. B), and also a derivation of the inverse power series (11) by perturbation theory (see Chapter III, Section B4). Wooley [17] suggested a three-constant form for $U(R)$ which is an infinite power series like (38) with relations between the coefficients; it works well for many diatomics.

Just as (35) was generalized to (37), (34) may be generalized to

$$U(R) = -Ae^{-\xi R} + BR^{-n} \tag{41}$$

which is associated with the name of Linnett [18]. With A and B positive, this could represent a valence attraction and an electrostatic repulsion. On the other hand, Buckingham [12] and Margenau [19] used (41) for the rare gases with A and B negative.

As we have mentioned, the valence attraction and repulsion should vary exponentially with R, so that both attractive and repulsive parts of $U(R)$ could be represented by exponentials:

$$U(R) = ae^{-bR} + ce^{-dR}. \tag{42}$$

The Morse [20] potential, one of the most often used descriptions of a potential curve containing a minimum, is of this form. We may write it in a more usual way:

$$U(R) = D_e\{\exp[-2a(R - R_e)] - 2\exp[-a(R - R_e)]\}. \tag{43}$$

It has only three parameters, D_e being the depth of the well and R_e its location. The parameter a relates to the curvature at $R = R_e$. The Morse function approaches zero for large R and a very large value for $R = 0$.

Its frequent use is due to the fact that the vibrational levels may be exactly calculated, and, with reasonable values for the parameters D_e, a, and R_e, agree fairly well with observation for many molecules. The energy of the vibrational level with quantum number v ($v = 0, 1, 2, \ldots$) is a quadratic in $(v + \frac{1}{2})$, depending only on the two constants a and D_e. The value R_e is obtained from the spacing of rotational energy levels. Pekeris [21] solved the vibrational equation with a centrifugal term added. Actually, the curve is *not* a good representation of the true $U(R)$ over a wide range of R. If R_e is determined from the rotational constants and D_e and a from vibrational spectra (ω_e and $\omega_e x_e$), the dissociation energy may be as much as two times too high compared with an independently measured value. The Morse potential thus falls off too fast at large R. Konowalow [8] has suggested joining it smoothly to the van der Waals attraction, $U = -CR^{-6}$, which is correct for large R.

The hard-sphere repulsion may also be combined with attractive forces. For example, the Sutherland potential [1, p. 33],

$$U(R) = \infty, \quad R < \sigma; \qquad U(R) = -aR^{-m}, \quad R > \sigma \qquad (44)$$

is quite realistic for many cases while easy to handle mathematically. In the square-well potential, both attractive and repulsive forces are infinitely strong and of zero range:

$$U(R) = \infty, \ R < \sigma_1; \ \ U(R) = -\varepsilon, \ \sigma_1 < R < \sigma_2; \ \ U(R) = 0, \ R > \sigma_2 \quad (45)$$

The mathematical simplicity compensates the physical unreality. The three constants are the depth of the well ε, the radius of the repulsive core σ_1, and the extent of the attractive region $\sigma_2 - \sigma_1$.

The two-parameter Eckart potential [22],

$$U(R) = -abe^{-bR}(ae^{-bR} + 1)^{-2} \qquad (46)$$

which may represent pure attraction or go through a minimum, seems more complicated in form. However, it is of interest in scattering problems since the $l = 0$ phase shift can be given in closed form [23, 24].

With more and better experimental measurements, it becomes possible to use functions containing more parameters, and hence more flexible functions. Improved computing methods compensate for the inconveniences associated with their use. Often, the more flexible functions are derived by modifying simpler forms of $U(R)$. For instance, Hulburt and Hirschfelder [25] considered improving the Morse function (43) to take into account measured values for five spectroscopic constants. Various functions of three, four, and five parameters were considered by Hulburt and Hirschfelder before they decided on

$$U(R) = D[(1 - e^{-x})^2 + cx^3 e^{-2x(1+bx)}] \qquad (47a)$$

where

$$x = 2\beta(R - R_e)/R_e. \tag{47b}$$

To compare with the Morse potential, note that (43) may be written

$$U(R) = D(1 - e^{-x})^2 - D$$

and the last term shifts the zero of energy.

In a modification to the Morse curve suggested by Huggins [26], the repulsive term $\exp[-2a(R - R_e)]$ in (43) is replaced by $(c/D_e) \exp[-2a(R-R_{12})]$. The parameters a and R_{12} are chosen according to the inner-shell structure of the atoms involved, and the other parameters are determined from experiment. The resulting curve seems valid over a wider range of R than the unmodified Morse curve, but, like the Morse, it always gives small values for higher anharmonicities and thus fails for cases where they are not small. Arguments based on Huggins' curve can be used to justify Badger's Rule (cf. Chapter III, Section B of Volume 2).

Kihara [9] modified the Lennard-Jones potential [Eq. (35)] with addition of an additional parameter ϱ; ϱ may be considered to represent the strongly repulsive cores of the atoms. Kihara's potential is

$$U(R) = 4\varepsilon\left[\left(\frac{\sigma - 2^{-1/6}\varrho}{R - \varrho}\right)^{12} - \left(\frac{\sigma - 2^{-1/6}\varrho}{R - \varrho}\right)^{6}\right], \qquad R \geq \varrho \tag{48a}$$

$$U(R) = \infty, \qquad\qquad\qquad\qquad\qquad\qquad\qquad R \leq \varrho \tag{48b}$$

The potential becomes infinite at $R = \varrho$ instead of at $R = 0$, but the minimum and well depth are unchanged. Evidence has been marshalled [27] that the Kihara potential is the best three-parameter potential.

One can augment the function (41) with additional inverse powers of R on theoretical grounds (see Section A). Thus,

$$U(R) = Ae^{-R/\varrho} - e^2R^{-1} - e^2(\alpha_A + \alpha_B)R^{-4} - 2e^2\alpha_A\alpha_BR^{-7} - CR^{-6} \tag{49}$$

seems to be useful for ionic diatomic molecules [28]. The first and last terms are the inner-shell repulsion and van der Waals attraction, while the others are suggested by the successful model of polarizable ions (see Chapter III, Section B of Volume 2). Because of the physical interpretation, the coefficients of these terms are written in terms of the atomic polarizabilities α_A and α_B, so that their values may be derived from measurements on the atoms, leaving only three parameters to be derived from spectroscopic data on the molecule. But even (49) is not adequate: The use of a sum of three exponentials for the first term seems to be necessary if (49) is to fit a large number of molecules [28].

Sometimes, simplicity of form and accuracy can be obtained simultaneously by joining several functional forms, i.e., by using different potentials for different values of R. The Buckingham–Corner function [1, p. 32] is of this type. It derives from (34) (including an R^{-8} term) and has four parameters:

$$U(R) = be^{-dR/\sigma} - (cR^{-6} + c'R^{-8})f(R) \qquad (50)$$

The function $f(R)$ is unity for $R > \sigma$ and $\exp[-4(\sigma R^{-1} - 1)]^3$ for $R < \sigma$; σ is the value of R for which U has its minimum. Another modification of the exponential-six potential, given by Hirschfelder *et al.* [29] is

$$U(R) = A\left[\frac{6}{\alpha}\exp\left(\alpha - \frac{\alpha R}{\sigma}\right) - \left(\frac{\sigma}{R}\right)^6\right], \qquad R > \sigma \qquad (51a)$$

$$U(R) = \infty, \qquad\qquad\qquad\qquad\qquad\qquad R < \sigma \qquad (51b)$$

Here, σ is the value of R for which (51a) would have a maximum. This three-constant potential proved useful [29] for second virial coefficient calculations.

We may now ask about the possibility of choosing among several functions of comparable qualitative behavior and mathematical complexity, with the same number of parameters. This may be done by invoking more detailed results of theory. Second, we may investigate the ability of a function to predict new properties, not involved in the parametrization. An important example, for bound states, is prediction of relations between potential constants. A three-parameter $U(R)$ is defined when R_e, k_e, and D_e are specified; anharmonicities and other properties thus are calculable in terms of these quantities. Various authors have attempted to choose among functions in this way. Linnett [18] took as the criteria of goodness of a potential function (a) that it provide an explanation of the correlation of small values of R_e with large values of ω_e and (b) that it predict correctly relations between force constant, R_e, anharmonicity, and D_e (dissociation energy). On this basis he found the form (42) more satisfactory than (37) or (41).

A subsequent attempt at assessing the relative quality of potential functions is that of Varshni [30] for binding states of diatomic molecules. At the outset, he listed necessary conditions on $U(R)$:

(1) $U(R) \to$ const as $R \to \infty$;
(2) $U(R)$ goes through a minimum for some $R = R_e$;
(3) $U(R)$ becomes infinite, or at least very large, as $R \to 0$.

An additional desirable characteristic is that $U(R)$ be flexible enough to describe a function having a maximum as well as a minimum, since such cases exist and we do not want to exclude them arbitrarily. There were also conditions related to theoretical results which we discuss later. A critical review of the different potential curves that had been proposed, including some we have not yet mentioned, was given, as well as a discussion of the problems involved in the determination of the parameters from spectroscopic measurements. In the course of the article, seven new forms for $U(R)$ were proposed by Varshni.

Five spectroscopic parameters could be measured. Then, for a three-parameter function, three could be used to define $U(R)$ completely, and predictions of the other two could be compared with experiment. It was concluded that no three-parameter function is wholly adequate, and that one should not look for a simple and universal $U(R)$ which would describe many states of many molecules for different parametrizations. The three-parameter functions judged best for overall representation of the potential curves were [30]

$$U(R) = -D_e(1 + b[R - R_e]) \exp[-b(R - R_e)] \qquad (52)$$

(proposed by Rydberg),

$$U(R) = D_e\{1 - (R_e/R) \exp[-\beta(R^2 - R_e^2)]\}^2 \qquad (53)$$

and

$$U(R) = D_e\{1 - (R_e/R) \exp[-a(R - R_e)]\}^2 \qquad (54)$$

(proposed in Varshni's article).

Another function with a Gaussian R-dependence was proposed by Lippincott [31, 32]

$$U(R) = D_e(1 - \exp[-n(R - R_e)^2/2R])[1 + af(R)] \qquad (55)$$

Here, $f(R) \to \infty$ for $R \to 0$ and $f(R) \to 0$ for $R \to \infty$, but it may be neglected (i.e., a may be set equal to zero) for some applications. This function was considered in a further study of the potential curves by Steele et al. [32]. These authors reassessed the accuracy to which various forms of $U(R)$ allow the prediction of one potential constant from another. They also compared calculated curves, from parametrization of $U(R)$ in each case, with the curves derived via the RKR method from the experimentally measured energy levels. Both ground and excited states were considered. The better three-parameter functions can give accuracy of 2–3% for

$| U - U_{RKR}|/D_e$, while the five-parameter functions only improve this to 1–2%.

Hirschfelder et al. [1, Sects. 3.9, 8.5] discuss the parametrization of potentials of simple form to fit experimentally determined virial coefficients and transport coefficients, and hence the adequacy of these potentials to reflect reality. Jenč [33] has reviewed the tests on the Lennard-Jones and Buckingham potentials for the rare gases.

Most recently, Jain [34] has studied the Morse, Hulburt–Hirschfelder, and Lippincott functions by determining the vibrational energy levels and wavefunctions (required for determination of rotational constant) for a number of electronic states of some related diatomic molecules. Spectroscopic term values were used as input data, and the derived curves for various $U(R)$ were compared with the RKR curves derived from this data. This is the procedure of Steele et al. [32]. Due to a change in the method for construction of the curves, better results were obtained for the derived values of some spectroscopic constants. In contrast to what was found by previous workers [30–32] good correlations were found between the deviations of the potential functions from RKR curves and their ability to predict potential constants. Overall, the Hurlburt-Hirschfelder function again performed best. It is also interesting that in tests of a method for improving the RKR procedure [35] which requires a zero-order approximation to the curve in conjunction with spectroscopic data, the Hurlburt–Hirschfelder function was preferable to other empirical potentials as a starting point.

The use of reduced potential curves, which are discussed in Chapter III, Section B of Volume 2, was proposed [33] as a means to decide on the relative merits of different potential functions. The reduced potential curve is obtained when U/D_e is plotted as a function of ϱ, the reduced distance. ϱ is itself a function of R and of a parameter characteristic of the atoms in the molecule. A close coincidence was found between reduced potential curves for molecules in a group; for the ground states of the heavier molecules, the reduced potential curves were broadened and lowered in the region of $R > R_e$. Thus proposed potentials for Ar_2 and Xe_2 were deemed unacceptable because the reduced curves derived from them lay well above the He_2 reduced curve in this region.

When there is a large amount of accurate experimental data of different kinds, none of the simple forms of $U(R)$ will be adequate, and one must tailor-make a function with a large number of parameters to fit various parts of the curve. Thus, for He_2 a potential was proposed by Beck [36] to fit scattering, transport coefficient and virial coefficient data for the short-range part, quantum mechanical calculations for intermediate R, and the calculated van der Waals coefficients (R^{-6} and R^{-8} terms) for large R. The potential was

$$U(R) = A \exp(-\alpha R - \beta R^6) - B(R^2 + a^2)^{-3}[1 + (C + 3a^2)/(R^2 + a^2)]$$

But even this was not adequate for *all* the available data [36].

Guggenheim and McGlashan [37] considered the Ar–Ar interaction, about which much is known from experimental measurements of different types. They used virial coefficient data over a wide range of temperature, crystal properties, and viscosity, to discuss and evaluate various possibilities for $U(R)$. All the data could not be explained with a function involving fewer than six parameters.

Dymond and Alder [38] reviewed many of the analytical potentials which have been used for Ar–Ar, as well as the experimental results available, before concluding that a numerically tabulated potential would be necessary to fit all properties of the dilute gas. The advantage of the numerical potential is its flexibility: We can vary it for a particular region of R without affecting it elsewhere, which cannot be done rigorously for parametrized analytical functions. The potential they derived differed significantly from previously postulated potentials (analytic functions). Barker and Pompe [39], noting that several analytical potentials could fit everything except for viscosity and crystal structure data, proposed explanations for the discrepancies in these cases. Their conclusions are not shared by other workers [38–40]. They proposed an analytical potential with nine independent parameters, some of which were fixed from theoretical values of van der Waals coefficients. Recently obtained spectroscopic data [41] on the Ar_2 van der Waals molecule point out inaccuracies in the Barker–Pompe potential, but also in that of Dymond and Alder [38]. The former gives the vibrational levels correctly but not the dissociation energy, while the latter gives incorrect vibrational constants.

2. Theoretical Considerations

There have been attempts to combine theoretical knowledge, of a more detailed sort than what has been mentioned, with empirical parameters, to produce semiempirical forms for $U(R)$. Several authors have suggested consideration of the properties of the atom formed by letting the nuclei coalesce (see Chapter I, Section C of Volume 2). As we have mentioned, if we split $U(R)$ into the nuclear repulsion $Z_A Z_B / R$ and the electronic energy $U^e(R)$, $U^e(R)$ for small R will go smoothly into the electronic energy of the united atom. This is the atom whose nuclear charge is the sum of the nuclear charges of the molecule. It is also expected that dU^e/dR vanishes for $R \to 0$, since this is the electronic contribution to the force on the nucleus (see Chapter III, Section D), which must vanish for the spherically symmetric (atomic) charge distribution for $R = 0$. Then

$$U^e(R) = E_0 + E_2 R^2 + \cdots \tag{56}$$

where E_0 is the energy of the united atom. For large R, U^e must approach $-Z_A Z_B/R$, since eventually $U(R)$ will go as R^{-6} or something else which vanishes faster than $1/R$.

On these grounds, plus the requirement that $U(R)$ provide a minimum, Frost and Musulin [42] proposed

$$U(R) = e^{-aR}(R^{-1} - b) \tag{57a}$$

which means

$$U^e(R) = -R^{-1}(1 - e^{-aR}) - be^{-aR}. \tag{57b}$$

This is the simplest $U^e(R)$ having the desired qualities (except for vanishing slope at $R = 0$). The exponential fall-off of $U^e(R)$ is expected from quantum mechanical calculations. This two-parameter function was tested on the H_2^+ and H_2 ground states. The united atom energy could be taken as one of the experimental quantities to be predicted, a and b being chosen so (57) fits experimental values for two quantities. Alternatively, if the united atom energy is known, there is only one free parameter in (57), since $E_0 = -(a + b)$. To assure $dU^e/dR = 0$ at $R = 0$, one can take

$$U^e(R) = -R^{-1}[1 - e^{-aR}(1 + \tfrac{1}{2}aR)] - be^{-aR}(1 + aR). \tag{58}$$

Putting $E_0 = -(a + b)$ leaves one parameter to be determined, but this function was unsatisfactory, and had to be modified with introduction of more empirical parameters.

The function (57) was subsequently generalized [43] to

$$U(R) = e^{-aR}(cR^{-1} - b), \tag{59}$$

corresponding to the use of effective nuclear charges on the interacting atoms. The parameters a, b, and c were determined from the spectroscopic constants R_e, D_e, and k_e. For hydrides and homonuclear molecules, the predicted third and fourth derivatives of $U(R)$ at R_e were no better than what we obtain from a Morse function. For ionic molecules, the function was further modified by Frost and Woodson [44] to

$$U(R) = c'R^{-1} + e^{-aR}[(c - c')/R] + \bar{U}(R) \tag{60}$$

where c' is the product of charges on the ions, c is an available parameter, and $\bar{U}(R)$ is supposed to include exchange and dispersion energies.

For the interaction of two atoms in s states, Buckingham [45] considered

$$U(R) = (Z_A Z_B/R)p(R)e^{-aR} \tag{61a}$$

with $p(R)$ a polynomial in R:

$$p(R) = 1 + p_1 R + p_2 R^2 + p_3 R^3 + \cdots \tag{61b}$$

At small R, $U(R) \to Z_A Z_B/R$; for large R, the exponential dominates the behavior. The electronic energy $U^e(R)$ is a polynomial times an exponential, as we get from the valence bond treatment of the H–H interaction, and as we expect in general. The region of R where the van der Waals interaction dominates is not well represented. If one requires, for small R, that $U^e(R)$ approach $E_0 + E_2 R^2$, the coefficients p_i are not arbitrary. The first two are calculable from E_0, while to get p_3 we need E_2. It was emphasized [45] that the function, carrying terms through p_3, can to be parametrized to fit $U(R)$ as well as possible, so p_3 need not be computed from E_0 and E_2.

Preuss [5] considered the form $U(R)$ should take to satisfy the conditions mentioned earlier, as well as behavior as a series of inverse powers of R for $R \to \infty$. He also emphasized the need for a function which can describe purely repulsive interactions (no minimum), as well as cases where there is a single minimum, a minimum plus a maximum, etc. He proposed

$$U(R) = U(\infty) + Z_A Z_B R^{-1} + [E_0 - U(\infty)]S(R) \tag{62a}$$

where

$$S(R) = \left[\sum_{k=0}^{m-1} c_k R^k \right] \Big/ \left[\sum_{l=0}^{m} d_l R^l \right] \tag{62b}$$

with $c_0 = d_0 = 1$. If we know R_e, k_e, $D_e = U(\infty) - U(R_e)$, E_0, E_1, E_2, E_3 (from the united atom expansion of the electronic energy) and the first n coefficients in the large-R expansion (in powers of R^{-1}) one can fix $n + 6$ parameters in S. Then m in S is $\frac{1}{2}(n + 7)$. The procedure for determination of the c_k and d_k from the values of these parameters was discussed and used for H_2 ($n = 11$) where a highly accurate curve was derived [5].

Clinton [46], in a discussion of the virial theorem (Chapter III, Section C), derived the relation

$$R^2 \, d^2U/dR^2 + 4R \, dU/dR + 2U = 2S \tag{63}$$

where S is formally a sum of excited state contributions from perturbation theory. Supposing $S/U = \alpha$ may be considered a constant, one has a second-order differential equation with the general solution

$$U(R) = AR^{-K} + BR^{-L}. \tag{64}$$

K and L are the roots of

$$K(K + 1) - 4K + 2(1 - \alpha) = 0.$$

If $\alpha = -\frac{1}{8}$, the two roots become identical ($K = L = \frac{3}{2}$) and $U(R)$ has to be replaced by

$$U(R) = (A + B)R^{-K} + CR^{-K} \ln R. \tag{65}$$

Functions of the form (64) have already been mentioned above, while Clinton [46] introduced a form suggested by (65). Putting $A + B = \ln \lambda_2$, this is

$$U(R) = \lambda_1 R^{-K} \ln(\lambda_2 R). \tag{66}$$

For $\lambda_1 < 0$, $\lambda_2 > 0$, and $K > 0$, $U(R)$ has the proper behavior for $R \to 0$ and $R \to \infty$ and goes through a minimum. It is interesting to note that, even for $K = 1$, $U^e(R)$ is nonanalytic at $R = 0$.

Parr, Borkman, and co-workers [47] have recently used equation (63) to investigate shapes of potential curves. Explicitly, the solution to this equation may be written

$$\frac{U - U(\infty)}{D_e} = \frac{-2 + \int_1^r q(r') \, dr'}{r} + \frac{1 - \int_1^r r'q(r') \, dr'}{r^2} \tag{67}$$

where D_e is the dissociation energy,

$$D_e = U(\infty) - U(R_e), \tag{68a}$$

and scaled variables are being used:

$$q = 2[S - U(\infty)]/D_e \tag{68b}$$

$$r = R/R_e \tag{68c}$$

Here, R_e is the equilibrium internuclear distance as usual. The implication of certain forms for q was discussed [47] in terms of the form of the resulting U. In particular, if S or q is constant near $R = R_e$, U is a series in inverse powers of R in this region, as in Eq. (38):

$$U = S + a_1 R^{-1} + a_2 R^{-2}. \tag{69}$$

The equilibrium internuclear distance is $R_e = -2a_2/a_1$, and the quadratic and higher force constants are expressible in terms of a_1 and a_2. This means relations between them are predicted, and, as we discuss in Chapter III, Section B of Volume 2, these are verified. Explicitly, $2S$ is $R^{-1}(d/dR)(R^2\bar{T})$, where $\bar{T}$ is the kinetic energy expectation value. The constancy of this implies $\bar{T}$ can be written as a constant term plus a term in R^{-2}, which is characteristic of the behavior of noninteracting electrons in a box. Parr, Borkman, and co-workers [47] have developed a simple model for the bond which uses this picture. We discuss it in Chapter III, Section B, Volume 2.

REFERENCES

1. J. O. Hirschfelder, C. F. Curtiss, and R. B. Bird, "Molecular Theory of Gases and Liquids." Wiley, New York, 1954.
2. H. Pauly and J. P. Toennies, *Advan. At. Mol. Phys.* **1**, 195 (1965).
3. Y. P. Varshni, *Rev. Mod. Phys.* **29**, 664 (1937).
4. T. L. Cottrell, *Discuss. Faraday Soc.* **22**, 10 (1956).
5. H. Preuss, *Theor. Chim. Acta* **2**, 102 (1964).
6. D. Steele, E. R. Lippincott, and J. T. Vanderslice, *Rev. Mod. Phys.* **34**, 239 (1962).
7. A. A. Abrahamson, *Phys. Rev.* **130**, 693 (1963).
8. D. D. Konowalow, *J. Chem. Phys.* **50**, 12 (1969).
9. T. Kihara, *Rev. Mod. Phys.* **25**, 831 (1953).
10. A. Cohen and C. H. Blanchard, *J. Chem. Phys.* **36**, 1402 (1962).
11. C. L. Beckel, *J. Chem. Phys.* **33**, 1885 (1960); E. J. Finn and C. L. Beckel, *Ibid.* **33**, 1887 (1960); C. L. Beckel and J. P. Sattler, *Ibid.* **42**, 2620 (1965); *J. Mol. Spectrosc.* **20**, 153 (1966); C. L. Beckel and R. Engelke, *J. Chem. Phys.* **49**, 5199 (1968).
12. R. A. Buckingham, *Proc. Roy. Soc.* **168**, 264 (1938).
13. G. B. B. M. Sutherland, *J. Chem. Phys.* **8**, 161 (1940).
14. A. Kratzer, *Z. Phys.* **3**, 289 (1920).
15. E. Fues, *Ann. Phys.* **80**, 367 (1926).
16. R. G. Parr and R. J. White, *J. Chem. Phys.* **49**, 1059 (1968).
17. H. W. Wooley, *J. Chem. Phys.* **37**, 1307 (1962).
18. J. W. Linnett, *Trans. Faraday Soc.* **36**, 1123 (1940); **38**, 1 (1942).
19. H. Margenau, *Phys. Rev.* **56**, 1000 (1939).
20. P. M. Morse, *Phys. Rev.* **34**, 57 (1929).
21. L. Pekeris, *Phys. Rev.* **45**, 98 (1934).
22. C. Eckart, *Phys. Rev.* **35**, 1303 (1930).

23. V. Bargmann, *Rev. Mod. Phys.* **21**, 488 (1949).
24. T. J. P. O'Brien and R. B. Bernstein, *J. Chem. Phys.* **51**, 5112 (1969).
25. H. M. Hulburt and J. O. Hirschfelder, *J. Chem. Phys.* **9**, 61 (1941).
26. M. L. Huggins, *J. Chem. Phys.* **3**, 473 (1935); **4**, 308 (1936).
27. R. D. Weir, *Mol. Phys.* **11**, 97 (1966).
28. R. L. Redington, *J. Phys. Chem.* **74**, 181 (1970); Y. P. Varshni and R. C. Shukla, *J. Mol. Spectrosc.* **16**, 63 (1965).
29. W. E. Rice and J. O. Hirschfelder, *J. Chem. Phys.* **22**, 187 (1954).
30. Y. P. Varshni, *Rev. Mod. Phys.* **29**, 664 (1957).
31. E. R. Lippincott, *J. Chem. Phys.* **21**, 2070 (1953).
32. D. Steele, E. R. Lippincott, and J. T. Vanderslice, *Rev. Mod. Phys.* **34**, 239 (1962).
33. F. Jenč, *Collect. Czech. Chem. Commun.* **29**, 2881 (1964).
34. D. C. Jain, *Int. J. Quantum Chem.* **4**, 579 (1970).
35. N. I. Zhirnov and A. S. Vasilevskii, *Opt. Spectrosc. (USSR)* **25**, 13 (1968); **26**, 387 (1969); *Opt. Spektrosk.* **25**, 28 (1968); **26**, 704 (1969).
36. D. E. Beck, *Mol. Phys.* **14**, 311 (1968); W. L. Taylor and J. M. Keller, *J. Chem. Phys.* **54**, 647 (1971).
37. E. A. Guggenheim and M. L. McGlashan, *Proc. Roy. Soc.* **255**, 456 (1960).
38. J. H. Dymond and B. J. Alder, *J. Chem. Phys.* **51**, 309 (1969).
39. J. A. Barker and A. Pompe, *Aust. J. Chem.* **21**, 1683 (1968).
40. W. C. Stwalley, *J. Chem. Phys.* **55**, 170 (1971).
41. Y. Tanaka and K. Yoshino, *J. Chem. Phys.* **53**, 2012 (1970); L. W. Bruch and I. J. McGee, *Ibid.* **53**, 4711 (1970).
42. A. A. Frost and B. Musulin, *J. Chem. Phys.* **22**, 1017 (1954).
43. P. S. K. Chen, M. Geller, and A. A. Frost, *J. Phys. Chem.* **61**, 828 (1957).
44. A. A. Frost and J. H. Woodson, *J. Amer. Chem. Soc.* **80**, 2615 (1958).
45. R. A. Buckingham, *Trans. Faraday Soc.* **54**, 453 (1958).
46. W. L. Clinton, *J. Chem. Phys.* **36**, 555, 556 (1962).
47. R. G. Parr and R. F. Borkman, *J. Chem. Phys.* **46**, 3683 (1967); **48**, 1116; **49**, 1055 (1968); **50**, 58 (1969).

SUPPLEMENTARY BIBLIOGRAPHY

Below are given references to work which came to our attention too late to be included in the manuscript. The references are given alphabetically by author, and the Section to which they are most relevant is indicated by a letter at the end of each reference. We have given the title, with occasional clarifying information in parentheses.

R. E. Caligaris, A. L. Calvo, and J. C. Grangel, Potential Intermoléculaire Simple Dans la Théorie des Fluides Classiques en Équilibre. (modified square well). *J. Chim. Phys.* **69**, 649 (1972). **D**

R. E. Caligaris and A. E. Rodriguez, On the m-6-8 Potential Function. *J. Chem. Phys.* **56**, 4715 (1972). **D**

D. W. Calvin and T. M. Reed II, Relationship Between the n, 6 Potential and Exponential-6 Potential. *J. Chem. Phys.* **56**, 2484 (1972). **D**

B. Hutchinson and W. R. Conkie, Thermodynamically Self-Consistent Radial Distribution Functions for Inverse Power Potentials. *Mol. Phys.* **24**, 567 (1972),   **D**

D. D. Konowalow and D. S. Zakheim, Morse-6 Hybrid Potentials for Pair Interactions of Rare Gas Atoms. *J. Chem. Phys.* **57**, 4375 (1972).   **D**

C. F. Melius and W. A. Goddard III, The Theoretical Description of an Asymmetric, Nonresonant Charge Transfer Process. *Chem. Phys. Lett.* **15**, 524 (1972).   **B**

S. B. Rai, V. N. Sharma, and D. K. Rai, On the Validity of Tietz Potential Functions for Diatomic Molecules. *Can. J. Phys.* **50**, 428 (1972).   **D**

T. Tietz, Potential-Energy Functions for Diatomic Molecules (allowing exact solution for energy levels). *Can. J. Phys.* **49**, 1315 (1971).   **D**

D. G. Truhlar, S. Trajmar, and W. Williams, Electron Scattering by Molecules with and without Vibrational Excitation. IV and V. (to test potential functions). *J. Chem. Phys.* **57**, 3250, 3260 (1972).

S. I. Vetchinkin and V. L. Bachrach (gives Green's functions for several potentials, to be used for molecular vibration problems), *Int. J. Quantum Chem.* **6**, 143 (1972).

Chapter III Methods of Calculation

A. Variation Methods

1. General

Since the Schrödinger equation

$$H\Psi = E\Psi \tag{1}$$

is intractable by direct solution in all but the simplest cases, the associated variational principle is used as the basis for most calculations. Let Φ be a function satisfying the known boundary conditions for the particular problem. Then the expectation value of the Hamiltonian

$$\langle H \rangle_\Phi \equiv \langle \Phi \mid H \mid \Phi \rangle / \langle \Phi \mid \Phi \rangle \tag{2}$$

is stationary if and only if Φ obeys the eigenvalue equation (1) (EYRING, Sect. 7c, PILAR, Sect. 10–2, McWEENY, Sect. 2.1–2.3). The angular brackets here mean integration over space and spin coordinates.

For the ground state, the stationary principle is, in fact, a minimum principle. Let Ψ_0 be the normalized eigenfunction corresponding to the lowest eigenvalue E_0, and let Φ be $\Psi_0 + \delta\Psi$, with $\langle \delta\Psi \mid \Psi_0 \rangle = 0$. Then

$$\langle H \rangle_\Phi = \frac{E_0 \langle \Psi_0 \mid \Psi_0 \rangle + \langle \delta\Psi \mid H \mid \delta\Psi \rangle}{1 + \langle \delta\Psi \mid \delta\Psi \rangle} \tag{3}$$

89

which differs from E_0 by quantities of second and higher order in $\delta\Psi$. If $\delta\Psi$ is expanded in the orthonormal eigenfunctions Ψ_i, with expansion coefficients a_i, $\langle\delta\Psi\mid H\mid\delta\Psi\rangle$ becomes a weighted sum of energies, which is not less than $\sum_i\mid a_i\mid^2 E_0 = \langle\delta\Psi\mid\delta\Psi\rangle E_0$. Therefore

$$\langle H\rangle_\Phi \geqq E_0. \tag{4}$$

Now for a trial function Φ which we allow to vary only within certain limits (for example, the allowed variation may be in a limited number of parameters in Φ), one chooses the best Φ as that one which makes $\langle H\rangle_\Phi$ stationary within the subspace of Hilbert space defined by the allowed variations. It is certainly true that (a) for the ground state, minimizing the energy will always bring us as close as possible to the true energy, while guaranteeing an upper bound, and (b) as the subspace is enlarged, we eventually obtain the exact eigenfunction by this method. Let Φ_b be the "best" function in the subspace according to the variation method. Then

$$\langle\widetilde{\delta\Phi}\mid H\mid\Phi_b\rangle = 0 \tag{5}$$

where $\widetilde{\delta\Phi}$ is any function in the subspace and orthogonal to Φ ($\widetilde{\delta\Phi}$ is an allowed variation), but, since $\widetilde{\delta\Phi}$ is not perfectly general, this does not imply that Φ is an eigenfunction of H.

Clearly, H plays a central role in the variation method. While $\langle H\rangle_\Phi$ goes monotonically toward E as the subspace grows, there is no such necessary rule for the wave function itself, or for any properties other than the energy calculated with it. Thus it is perfectly possible that expectation values of operators other than H will not always become closer to the correct values on improvement of the wave function. It is also well known that errors in the energy are generally smaller than errors in other properties.

This is a consequence of the variational principle. Formally, write

$$\Phi = (1 + \varepsilon^2)^{-1/2}(\Psi + \varepsilon X) \tag{6}$$

with $\langle\Phi\mid\Phi\rangle = \langle X\mid X\rangle = 1$ and $\langle\Psi\mid X\rangle = 0$. If Φ approximates Ψ well, ε is small, and we take it as a "parameter of smallness." One has

$$\langle\Phi\mid H\mid\Phi\rangle = (1 + \varepsilon^2)^{-1}(E + \varepsilon^2\langle X\mid H\mid X\rangle) \tag{7}$$

where $\langle\Phi\mid H\mid\Phi\rangle$ is the approximate energy and E the exact energy. Putting $(1 + \varepsilon^2)^{-1} = 1 - \varepsilon^2 + \cdots$, it is seen that $\langle\Phi\mid H\mid\Phi\rangle$ and E differ in terms second order in ε. For an arbitrary operator Q, however,

$$\langle \Phi \mid Q \mid \Phi \rangle = (1 + \varepsilon^2)^{-1}[\langle \Psi \mid Q \mid \Psi \rangle + \varepsilon(\langle \Psi \mid Q \mid X \rangle$$
$$+ \langle X \mid Q \mid \Psi \rangle) + \varepsilon^2 \langle X \mid Q \mid X \rangle] \qquad (8)$$

with the terms in ε not necessarily disappearing. Thus, we should expect in general expectation values of operators other than the Hamiltonian to be less accurately calculated, with an approximate function, than the energy itself. Note also that the terms in ε may be positive or negative: we have no upper or lower bound on $\langle \Psi \mid Q \mid \Psi \rangle$.

There are thus several important sources of trouble in the use of the simple variation method, which turn out to be related. First, it gives only an upper bound on the energy, and so no estimate of the error. Second, there is neither an upper nor a lower bound on other expectation values, so we know even less about the exact value than in the case of the energy. This is aggravated by the fact that the errors are likely to be larger for a given expectation value than for the energy. While we can argue roughly [1] about the relative sizes in terms of the size of ε, this can give very poor estimates in some cases.

2. *Bounds*

The number ε is essentially the mean-square deviation of Φ from Ψ. For small ε,

$$\Phi(1 + \tfrac{1}{2}\varepsilon^2 + \cdots) = \Psi + \varepsilon X$$

and

$$\langle \Phi - \Psi \mid \Phi - \Psi \rangle \sim \langle \varepsilon X - \tfrac{1}{2}\varepsilon^2 \Phi \mid \varepsilon X - \tfrac{1}{2}\varepsilon^2 \Phi \rangle$$
$$\sim \varepsilon^2 - \varepsilon^3 \operatorname{Re}\langle X \mid \Phi \rangle + \tfrac{1}{4}\varepsilon^4.$$

Since $\langle X \mid \Phi \rangle \sim \varepsilon$, ε is approximately $\langle \Phi - \Psi \mid \Phi - \Psi \rangle^{1/2}$. As pointed out by Eckart [2], one may obtain a bound on this quantity if something is known about the eigenenergies. Since $\langle X \mid \Psi \rangle = 0$, X can be expanded in the higher eigenfunctions of H (Ψ is assumed nondegenerate). Then $\langle X \mid H \mid X \rangle$ must be at least as great as E_1. From (7), putting E_0 for E,

$$\langle \Phi \mid H \mid \Phi \rangle - E_0 = \varepsilon^2(\langle X \mid H \mid X \rangle - \langle \Phi \mid H \mid \Phi \rangle) \geq \varepsilon^2(E_1 - \langle \Phi \mid H \mid \Phi \rangle) \quad (9)$$

Within terms of second order, we may replace $\langle \Phi \mid H \mid \Phi \rangle$ on the right side by E_0 [Eq. (7)]. Rearranging, we obtain Eckart's criterion:

$$\varepsilon^2 \leq (\langle \Phi \mid H \mid \Phi \rangle - E_0)/(E_1 - E_0). \tag{10}$$

Without making the approximation $E_1 - \langle \Phi \mid H \mid \Phi \rangle \sim E_1 - E_0$ we obtain a lower bound for $a_0{}^2$, the component of Ψ_0 in Φ. Let Φ be expanded in the exact eigenfunctions:

$$\Phi = \sum_{i=0}^{\infty} a_i \Psi_i, \qquad \sum_{i=0}^{\infty} \mid a_i \mid^2 = 1. \tag{11}$$

In terms of these coefficients,

$$\langle \Phi \mid H \mid \Phi \rangle - E_0 = \sum_{i>0} a_i{}^2 (E_i - E_0) \geq \sum_{i>0} a_i{}^2 (E_1 - E_0).$$

Using the normalization condition of (11) and rearranging,

$$a_0{}^2 \geq 1 - [(\langle \Phi \mid H \mid \Phi \rangle - E_0)/(E_1 - E_0)]. \tag{12}$$

An upper bound to $a_0{}^2$ may be obtained [3] in terms of

$$U^2 = \langle (H - E_0)^2 \rangle = \sum_{i>0} a_i{}^2 (E_i - E_0)^2$$

which is related to W_Φ (see equation 13) but is less useful unless we have a good estimate of E_0.

Weinhold [4] has shown that, having found the Eckart bound on ε^2 for a wave function Φ' which is better than Φ, we can derive a bound on the overlap of Φ with the exact function, using the overlap of Φ and Φ'. This bound can be better than what we obtain from using (10) on Φ directly. Another extension of the Eckart formula is given by Weinberger [5].

We now turn to the problem of lower bounds on the energy. An early formula, due to Weinstein [6], requires the quantity

$$W_\Phi = \langle \Phi \mid H^2 \mid \Phi \rangle - \langle \Phi \mid H \mid \Phi \rangle^2 \ (\langle \Phi \mid \Phi \rangle = 1), \tag{13}$$

which we refer to as the width of Φ. Here, W_Φ is nonnegative, since it may be rewritten as

$$W_\Phi = \langle \Phi \mid (H - \langle H \rangle_\Phi)^2 \mid \Phi \rangle \geq 0$$

and is evidently zero if $H\Phi = E\Phi$. It is a measure of the quality of the approximate function, with the advantage that its value for the exact function is known to be zero. In fact, W_Φ may be thought of as the variance of the local energy $H\Phi/\Phi$, which is constant over configuration space for an exact eigenfunction [see Chapter III, Section E, Eqs. (7–8)].

Now suppose it can be assumed that $\langle\Phi\,|\,H\,|\,\Phi\rangle$ is closer to E_0 than to E_1, or

$$\langle\Phi\,|\,H\,|\,\Phi\rangle - E_0 \leq E_1 - \langle\Phi\,|\,H\,|\,\Phi\rangle < E_k - \langle\Phi\,|\,H\,|\,\Phi\rangle, \quad k > 1 \quad (14)$$

Now the width for the wave function written in the form (6) is

$$W_\Phi = (1 + \varepsilon^2)^{-1}\big(\langle\Psi\,|\,(H - \langle H\rangle_\Phi)^2\,|\,\Psi\rangle + \varepsilon^2\langle X\,|\,(H - \langle H\rangle_\Phi)^2\,|\,X\rangle\big)$$
$$\geq (1 + \varepsilon^2)^{-1}([E_0 - \langle H\rangle_\Phi]^2 + \varepsilon^2[E_0 - \langle H\rangle_\Phi]^2) \quad (15)$$

The last follows because, if $X = \sum_{i>0} b_i\Psi_i$ with $\sum_{i>0}|\,b_i\,|^2 = 1$,

$$\langle X\,|\,(H - \langle H\rangle_\Phi)^2\,|\,X\rangle = \sum_{i>0}|\,b_i\,|^2(E_i - \langle H\rangle_\Phi)^2 \geq \sum_{i>0}|\,b_i\,|^2(E_0 - \langle H\rangle_\Phi)^2.$$

From (15),

$$|\,W_\Phi^{1/2}\,| \geq |\,\langle H\rangle_\Phi - E_0\,|$$

that is

$$W_\Phi^{1/2} \geq \langle H\rangle_\Phi - E_0 \geq -W_\Phi^{1/2} \quad (16)$$

and a lower bound is given by

$$E_0 \geq \langle H\rangle_\Phi - W_\Phi^{1/2}. \quad (17)$$

This may be extended to excited states. We still need to know something about E_1 to ensure that $\langle\Phi\,|\,H\,|\,\Phi\rangle$ is closer to E_0 than to E_1, but it may be expected that the condition is fulfilled for a reasonably good approximate wave function.

The preceding argument holds if $\langle H\rangle_\Phi$ is replaced [7] by any number closer to E_0 than to E_1. Let such a number be a. Then (15) becomes

$$\langle(H - a)^2\rangle_\Phi \geq (E_0 - a)^2$$

and a lower bound is

$$E_0 \geq a - (\langle(H - a)^2\rangle_\Phi)^{1/2} \quad (18)$$

Stevenson [7] shows that the right side of (18) is monotonic with a. The best value is the highest value satisfying the condition, i.e., $a = \frac{1}{2}(E_0 + E_1)$.

This gives the formula given by Temple [8]. With this value of a

$$(E_0 - E_1)/2 \geq - (\langle H^2 \rangle - 2a\langle H \rangle + a^2)^{1/2}$$

Squaring (and changing the sign of the inequality since both sides are negative) and canceling,

$$-E_0 E_1 \leq \langle H^2 \rangle - (E_0 + E_1)\langle H \rangle.$$

Again rearranging,

$$E_0 \geq \langle H \rangle_\Phi - W_\Phi/(E_1 - \langle H \rangle_\Phi). \tag{19}$$

For the connection between these and other formula, cf. Goodisman and Secrest [9] and Walmsley [10]. Knowledge of E_1 may be replaced by the more easily accessible knowledge of $\Delta E = E_1 - E_0$ [11, 12].

Note that minimization of W_Φ requires accurate evaluation of the expectation value of the square of the Hamiltonian. This is so difficult that only for one- and two-electron molecules have the formulas been used [9, 13, 14]. For the lower bounds, a rough evaluation of $\langle H^2 \rangle$ often suffices.

Other alternatives to the variation method, requiring evaluation of expectation values of higher powers of H than the second, have also been proposed [15]. The difficulties in the evaluation of the necessary integrals are even worse here. Other lower bound procedures are derived from the partitioning method [16], which we discuss in Section 4 and in Chapter III, Section B, equations (37) and (38). At this writing, molecular systems with several electrons have not been treated by this method.

A third kind of lower-bound procedure is being developed by Bazley, Fox, and co-workers [11, 12]. Let the Hamiltonian be divided into two parts

$$H = H_0 + H' \tag{20}$$

If H' is positive definite, $\langle H' \rangle_\Phi \geq 0$ for all Φ, we may show that the nth eigenvalue of H is greater than the nth eigenvalue of H_0. Stated differently, one can find a lower bound to the nth eigenvalue of H by obtaining the nth eigenvalue of the Hamiltonian H_0, where

$$\langle H_0 \rangle_\Phi \leq \langle H \rangle_\Phi \qquad \text{for all} \quad \Phi.$$

A trivial, but not very useful, case is obtained on taking H' as the interelectronic repulsion, so H_0 represents noninteracting electrons. This is not very useful because of the large size of $\langle H' \rangle$: The lower bounds are too low by hundreds of atomic units for small molecules. To obtain useful results, one must construct "intermediate Hamiltonians," H_1, H_2, H_3, etc., such that

$$\langle H_0 \rangle_\Phi \leq \langle H_1 \rangle_\Phi \leq \langle H_2 \rangle_\Phi \leq \cdots \leq \langle H \rangle_\Phi \qquad \text{for all} \quad \Phi. \qquad (21)$$

The eigenvalues of H_i give successively better lower bounds as i increases.

Bazley and Fox have used projection operators to do this. For example, they define the projection operator P^k with reference to a set of k linearly independent functions p_i. For Φ arbitrary,

$$P^k \Phi = \sum_{i=1}^{k} a_i p_i. \qquad (22)$$

The numbers a_i are chosen such that for some positive definite H'

$$\langle p_j \mid H' \mid \Phi \rangle = \langle p_j \mid H' \mid P^k \Phi \rangle = \left\langle p_j \mid H' \mid \sum_{i=1}^{k} a_i p_i \right\rangle.$$

so they depend on Φ as well as on H'. It may be shown from the definition that $(P^k)^2$ and P^k are identical, so that P^k is idempotent as a true projection operator must be, and that for arbitrary ϕ and ψ

$$\langle \phi \mid H' P^k \mid \psi \rangle = \langle \psi \mid H' P^k \mid \phi \rangle. \qquad (23)$$

Then

$$\langle \Phi \mid H' P^k \mid \Phi \rangle = \langle \Phi \mid H' P^k P^k \mid \Phi \rangle = \langle P^k \Phi \mid H' \mid P^k \Phi \rangle \geq 0$$

where (23) was used with $\phi = P^k \Phi$ and $\psi = \Phi$. Thus $H' P^k$ is nonnegative definite, assuming that H' has this property. It can further be shown that

$$\langle \Phi \mid H' P^k \mid \Phi \rangle \leq \langle \Phi \mid H' P^{k+1} \mid \Phi \rangle \leq \langle \Phi \mid H' \mid \Phi \rangle.$$

Thus we have a set of intermediate Hamiltonians satisfying (21) when we take $H_i = H_0 + H' P^i$. As i becomes infinite and the set $\{p_i\}$ becomes complete, P^i approaches the unit operator and H_i approaches the true Hamiltonian $H_0 + H'$.

Another method [11, 12] of forming the intermediate Hamiltonians is by truncation of H_0. If the eigenfunctions Ψ_i^0 of H_0 are known together with their eigenvalues E_i^0, the truncated Hamiltonian $H^{l,0}$ is defined by

$$H^{l,0} \Phi = \sum_{i=1}^{l} \langle \Psi_i^0 \mid \Phi \rangle E_i^0 \Psi_i^0 + E_{l+1}^0 \left[\Phi - \sum_{i=1}^{l} \langle \Psi_i^0 \mid \Phi \rangle \Psi_i^0 \right] \qquad (24)$$

It is easy to see that

$$\langle \Phi \mid H^{l,0} \mid \Phi \rangle = \sum_{i=1}^{l} |\langle \Psi_i^0 \mid \Phi \rangle|^2 E_i^0 + \sum_{i=l+1}^{\infty} |\langle \Psi_i^0 \mid \Phi \rangle|^2 E_{l+1}$$

$$\leq \sum_{i=1}^{l+1} |\langle \Psi_i \mid \Phi \rangle|^2 E_i^0 + \sum_{i=l+2}^{\infty} |\langle \Psi_i \mid \Phi \rangle|^2 E_{l+2}^0 = \langle \Phi \mid H^{l+1,0} \mid \Phi \rangle,$$

which means a set of intermediate Hamiltonians may be constructed by putting $H_i = H^{i,0} + H'$. As i increases, H_i approaches $H_0 + H'$. A combination of the two methods is possible. Define $H^{l,k}$ such that

$$H^{l,k}\Phi = H^{l,0}\Phi + \sum_{i=1}^{k} a_i H' p_i$$

where the a_i are chosen so that

$$\sum_{i=1}^{k} a_i \langle p_j \mid H' \mid p_i \rangle = \langle p_j \mid H' \mid \Phi \rangle.$$

Such a method has been used for H_2^+ [17]. The lower bounds obtained were unsatisfactory due to the slow convergence of the intermediate Hamiltonians to the correct one as k or l grows larger.

A related method, for which very promising results have in fact been obtained [18] for H_2 and He_2^{++} at a series of internuclear distances, was given by Miller [19]. Dividing H as in (20) and truncating H_0 as in (24) we have

$$H^l = H^{l,0} + H' \leq H \tag{25}$$

whether or not H' is positive definite. To find $E^{(l)}$, the eigenvalue of H^l, consider the eigenvalue equation

$$H^l \Psi = E^{(l)} \Psi.$$

Using (24), this becomes

$$(E_{l+1}^0 - E^{(l)} + H')\Psi = \sum_{i=1}^{l} (E_{l+1}^0 - E_i^0)\langle \Psi_i^0 \mid \Psi \rangle \Psi_i^0 \tag{26}$$

so that, so long as H' is purely multiplicative, Ψ is a linear combination of the l functions:

$$\Phi_i = (E_{l+1}^0 - E^{(l)} + H')^{-1}\Psi_i^0$$

Writing Ψ as a linear combination of the Φ_i leads to a kind of secular equation from which $E^{(l)}$ can be determined. Then the l lowest roots are lower bounds to the l lowest eigenvalues of H. There is one restriction: The inverse operator which appears in the definition of the Φ_i must be well defined. This requires $E_{l+1}^0 - E^{(l)}$ to be positive. It is possible to generalize to the case where H' is any positive, but not necessarily multiplicative, operator. Unlike most other methods, this one seems to work better for higher excited states. The calculation [18] for He_2^{++} is interesting because it showed conclusively that the maximum in $U(R)$ for the ground state, found by variational calculations, will not disappear as better variational functions are used.

An article by Wilson [20] shows how this method appears as a special case of the method of intermediate Hamiltonians, and how the lower-bound formulas of Weinstein, Temple, and others can be derived from it. Recently, Epstein [21] has discussed the lower bound formulas as part of a general review of the variation method. The methods we have discussed, as well as others, are included, with a large number of references. The relations between different kinds of lower bound formulas are given.

The search for formulas giving bounds on expectation values has gone hand in hand with investigations of lower bound procedures for the energy. Indeed, the problem of bounding the error in the energy given by a variational calculation is exactly the lower bound problem for E. Thus, Gordon [15] used similar methods in developing formulas for both problems. Bazley and Fox [22] and Lowdin [23] have reviewed some of the earlier work on bounds for expectation values and given some new results.

Consider the difference in expectation values of the Hermitian operator F for the exact function Ψ and the approximate function Φ,

$$\Delta F = \langle \Phi \mid F \mid \Phi \rangle - \langle \Psi \mid F \mid \Psi \rangle. \tag{27}$$

Bazley and Fox [22] show that $\mid \Delta F \mid$ can be bounded if F itself is a bounded operator, i.e., there is a positive constant c such that

$$\mid \langle \Phi \mid F \mid \Phi \rangle \mid < c \langle \Phi \mid \Phi \rangle \tag{28}$$

for all wave functions Φ. The problem of unbounded operators is more complicated [22, 24]. Sometimes, very poor expectation values are obtained for such operators for fairly good wave functions.

Jennings and Wilson [25] could show, for arbitrary Hermitian F

$$\mid \Delta F \mid \leq [\Phi - \Psi \mid \Phi - \Psi]^{1/2} [\langle \Psi \mid F^2 \mid \Psi \rangle^{1/2} + \langle \Phi \mid F^2 \mid \Phi \rangle^{1/2}] \tag{29}$$

using the Schwarz inequality. Replacing $\langle \Psi \mid F^2 \mid \Psi \rangle$ by $\langle \Phi \mid F^2 \mid \Phi \rangle$ and obtaining a bound on the first factor of (29) from the Eckart criterion, one has a bound on $\mid \Delta F \mid$. The accuracy of putting $\langle \Psi \mid F^2 \mid \Psi \rangle \sim \langle \Phi \mid F^2 \mid \Phi \rangle$ may be checked by repeating (29) with F^2 for F. Also, ΔF is unchanged by putting $F \rightarrow F - a$ (a a constant), but not the upper bound. Therefore, one may minimize the bound with respect to a. Applications to two-electron diatomics are made. For the expection value of a one-electron operator over a closed-shell single determinant function (see Volume 2, Chapter II, Section A), Alexander [26] derived error bounds in terms of the overlap between approximate and exact functions [see Eqs. (9) *et seq.*]. This overlap could be estimated [26] if a limited configuration interaction function was also available.

Other formulas are beginning to appear [27], but many demand computation of expectation values of H^2 or similarly complicated quantities, which prohibits their use on any but the simplest (usually atomic) systems. Wang [27] reviews many of these.

3. *Correcting Expectation Values*

A fruitful analysis of the problem of errors in expectation values results from imagining the approximate function to be the eigenfunction of a Hamiltonian H_Φ. Write [28]

$$H_\Phi \Phi = E_\Phi \Phi \tag{30}$$

where $E_\Phi = \langle H \rangle_\Phi$. It will not be necessary to give H_Φ explicitly. Since Ψ and Φ differ by terms of order ε, H and H_Φ differ by terms of order ε, and E and E_Φ differ by terms of order ε^2. Write Ψ in the form (6) and note that $H - E$ annihilates Ψ. Then to first order in ε

$$(H - E)\Phi \cong (H - E)(\varepsilon X) = (H_\Phi - E_\Phi)\varepsilon X.$$

Again to first order, (8) gives

$$\langle \Psi \mid Q \mid \Psi \rangle - \langle \Phi \mid Q \mid \Phi \rangle = 2\langle \varepsilon X \mid Q \mid \Phi \rangle$$

(taking all wave functions as real). We can obtain the quantity on the right without knowing Ψ. Consider instead solving the equation

$$(Q - \langle \Phi \mid Q \mid \Phi \rangle)\Phi = (H_\Phi - E_\Phi)y \tag{31}$$

with $\langle y \mid \Phi \rangle = 0$. The correction to the expectation value can be calculated from y, since to first order in ε

$$\langle \varepsilon X \mid Q \mid \Phi \rangle = \langle \varepsilon X \mid H_\Phi - E_\Phi \mid y \rangle = \langle \Phi \mid H - E \mid y \rangle = \langle \Phi \mid H \mid y \rangle$$

and knowledge of H_Φ is unnecessary here. To remove H_Φ from (31), put $y = f\Phi$, and suppose that H_Φ has the usual kinetic energy operator and differs from H by a potential energy term only. (Note that H_Φ may then be displayed explicitly as $T - \Phi^{-1}T\Phi + E_\Phi$ with T the kinetic energy operator, although this is not necessary here.) Then

$$(H_\Phi - E_\Phi)f\Phi = f(H_\Phi - E_\Phi)\Phi + T(f\Phi) - fT\Phi = T(f\Phi) - fT\Phi$$

so (31) becomes an equation for f which involves only Φ and Q.

We now want to solve (31) for y, or rather to solve

$$Tf\Phi - f(T\Phi) = (Q - \langle \Phi \mid Q \mid \Phi \rangle)\Phi$$

for f. Then the leading correction to $\langle \Phi \mid Q \mid \Phi \rangle$ may be calculated from $\langle \Phi \mid H \mid f\Phi \rangle$. A variational principle equivalent to this equation is also given by Chen and Dalgarno [28]. Applications to the Helium atom are given. The formalism of double perturbation theory [29] can be applied to this problem [30].

The sometimes discouraging coupling of good energy with poor expectation values has suggested that by forcing the wave function to give the correct expectation values for one or more operators a better wave function can be obtained for other properties. This approach was suggested by the work of Mukherji and Karplus [31] and Bader and Jones [32] in connection with the electrostatic Hellmann–Feynman Theorem (see Chapter III, Section D). In the former case, it was shown that an approximate wave function could be corrected to give a significant improvement in several expectation values with only a small sacrifice in the energy. This is a consequence of the stationary property of the energy. Rasiel and Whitman [33] suggested a "constrained variation" method, in which the energy is minimized while requiring, as a constraint, that the expectation value of some operator M be equal to some predetermined value.

They took a wave function of linear variational form (see the next section) and demanded that $\langle H \rangle$ be stationary with the auxiliary condition that $\langle M \rangle$ be equal to a fixed value μ. For convenience, the basis functions were taken as the solutions to the unconstrained variational problem—these are simply linear combinations of the original basis functions. Letting the expansion coefficients be a_i, they had

$$E = \sum_{i,j} a_i a_j H_{ij} \Big/ \sum_{i,j} a_i a_j S_{ij}, \qquad \mu = \sum_{i,j} a_i a_j M_{ij} \Big/ \sum_{i,j} a_i a_j S_{ij} \qquad (32)$$

and $H_{ij} = E_i \delta_{ij}$, $S_{ij} = \delta_{ij}$. With the use of a Lagrangian multiplier λ the variational condition was

$$(\partial E/\partial a_k) - \lambda(\partial \mu/\partial a_k) = 0 = \sum_i a_i[(E_k - E)\,\delta_{ki} - \lambda(M_{ki} - \mu\delta_{ki})]. \qquad (33)$$

In the resulting secular determinant, λ may be thought of as a perturbation parameter. When $\lambda = 0$, we have the free variation problem. Let E_1 be the lowest energy obtained from the free variation. We seek a solution such that $\Delta E = E_1 - E$ is small. If the method is to be of any use, the corresponding wave function should not differ much from the eigensolution of the free variation problem; this is just $a_i = \delta_{i1}$. This implies that λ is small as well as ΔE, so the secular determinant was expanded in powers of λ and ΔE. It turned out [33] that ΔE was of order λ^2 while λ was of order $\Delta\mu$, the correction to the expectation value: $\Delta\mu = \mu - M_{11}$. Dropping terms of order greater than $(\Delta\mu)^2$, they found

$$\Delta E = \lambda(\mu - M_{11}) - \lambda^2 \sum_{i\neq 1} (M_{1i})^2(E_i - E_1)^{-1}. \qquad (34)$$

Substituting into (33) and involking normalization, the leading terms in a_1 and a_i $(i \neq 1)$ can be found, and finally

$$\lambda = \frac{1}{2}\,\Delta\mu \Big/ \Big[\sum_{i\neq 1} (M_{1i})^2(E_i - E_1)^{-1} \Big] \qquad (35)$$

The solution can be iterated. Other methods for treating the constrained variation exist [34, 35].

Rasiel and Whitman [33] demonstrated their method for a three-configuration wave function for LiH, which gave an energy of -219.40 eV (experimental -219.71 eV) while yielding a dipole moment off by 12%. The constrained variation, using the dipole moment operator for M, raised the energy only by 0.14 eV. At the same time, the predicted forces on the nuclei as well as other properties were significantly improved.

For further theoretical work on the method, including relation to the hypervirial theorem, see the work of Byers Brown and Chong [36]. Chong and co-workers have been particularly active in investigation and implementation of the constrained variation technique [37]. There has recently been considerable interest in formulation of the equations of constrained variation [38] in terms of density matrices. Studies are under way to establish the effects of different kinds of constraints and hopefully make the method into a systematic way of producing wavefunctions suited for particular applications. Other attempts to improve variational wavefunctions with respect to particular expectation values involve weighting factors [39] and the hypervirial theorem [40] (see Section C). Local energy methods (Section E) may also be useful here.

4. *Linear Variation*

In searching for a wave function Φ which makes $\langle H \rangle_\Phi$ as low as possible, it is usually extremely convenient to vary Φ by keeping its *form* fixed and allowing parameters in it to vary. Then $\langle H \rangle_\Phi$ is a function of these parameters, whose minimum (or minima) can be found by a search over the space of the parameters. This becomes rapidly more difficult as the number of parameters increases. One important exception is the case of linear variation, because H is a linear operator.

We write

$$\Phi = \sum_{i=1}^{N} c_i \Phi_i \tag{36}$$

with the Φ_i fixed and vary the c_i. The variation is in the linear subspace spanned by the Φ_i. Substitution of (36) into $\langle H \rangle_\Phi$ gives the familiar result

$$\langle H \rangle_\Phi = \sum_{i,j} c_i^* c_j H_{ij} \Big/ \sum_{i,j} c_i^* c_j S_{ij} \tag{37}$$

where

$$H_{ij} = \int \Phi_i^* H \Phi_j \, d\tau; \qquad S_{ij} = \int \Phi_i^* \Phi_j \, d\tau. \tag{38}$$

Minimizing (37) with respect to a coefficient leads to

$$\sum_{j=1}^{n} H_{ij} c_j = \langle H \rangle_\Phi \sum_{j=1}^{n} S_{ij} c_j. \tag{39}$$

and the familiar secular determinant,

$$| H_{ij} - \varepsilon S_{ij} | = 0. \tag{40}$$

An N-dimensional basis gives N roots ε_k; each of these, with equation (39), allows calculation of a set of coefficients, say $c_i^{(k)}$, within an overall constant of normalization.

Solutions $\Phi^{(k)} = \sum_i c_i^{(k)} \Phi_i$ corresponding to different ε_k are automatically orthogonal, and $\langle \Phi^{(k)} \mid H \mid \Phi^{(l)} \rangle$ vanishes for $k \neq l$ (McWEENY, Sect. 2.3; LEVINE, Sect. 8.5). If several solutions correspond to the same energy one can choose the wave functions to be orthogonal. The statement that $\langle \Phi^{(k)} \mid H \mid \Phi^{(l)} \rangle$ vanishes for $k \neq l$ is just the statement of (5) applied to variation in a linear subspace. The $\{\Phi^{(k)}\}$, which are obtained from the $\{\Phi_i\}$ by a nonsingular transformation, may be taken as basis functions. The allowed variation to $\Phi^{(k)}$ [which is Φ_b in (5)] is admixture of any function in the subspace orthogonal to it, i.e.,

$$\delta\Phi^{(k)} = \sum_{i \neq k} \delta_i \Phi^{(i)} \tag{41}$$

with the constants δ_i small and independent (like Monaco). Then (5) is equivalent to

$$\langle \Phi^{(i)} \mid H \mid \Phi^{(k)} \rangle = 0, \qquad i \neq k.$$

Formal similarities between results in this case and exact eigenfunctions are related to the fact that the subspace, like the full Hilbert space, is linear.

If the basis set is orthonormal, (40) simplifies to

$$\mid \mathbf{H} - \varepsilon\mathbf{1} \mid = 0,$$

so one is simply looking for the eigenvalues of $\mathbf{H}$. If the basis functions are not orthonormal, one would like to go to an orthonormal basis, in which $\mathbf{S} = \mathbf{1}$. Let $\mathbf{T}$ be a transformation that produces such a basis, i.e., in the basis formed by

$$\Psi_k = \sum_l T_{lk} \Phi_l \tag{42}$$

the elements of the overlap matrix are

$$S'_{ij} = \langle \Psi_i \mid \Psi_j \rangle = (\mathbf{T}^\dagger \mathbf{S} \mathbf{T})_{ij} = S_i \, \delta_{ij}.$$

One way of obtaining $\mathbf{T}$ is by the Schmidt orthogonalization process.

Another, which has certain advantages, is symmetric orthogonalization. We first construct $\mathbf{S}^{-1/2}$ as follows: Suppose the unitary matrix $\mathbf{U}$ diagonalized $\mathbf{S}$, so

$$\mathbf{U}^{-1}\mathbf{S}\mathbf{U} = \mathbf{D} \tag{43}$$

where $\mathbf{D}$ is a diagonal matrix with elements d_i, which are necessarily real. Form

$$\mathbf{T} = \mathbf{U}\mathbf{D}^{-1/2}\mathbf{U}^{-1} \tag{44}$$

where $\mathbf{D}^{-1/2}$ is the diagonal matrix with elements $d_i^{-1/2}$, so $(\mathbf{D}^{-1/2})^2\mathbf{D} = \mathbf{1}$. Now

$$\mathbf{TTS} = \mathbf{U}\mathbf{D}^{-1/2}(\mathbf{U}^{-1}\mathbf{U})\mathbf{D}^{-1/2}\mathbf{U}^{-1}\mathbf{S} = \mathbf{U}(\mathbf{U}^{-1}\mathbf{S}^{-1}\mathbf{U})\mathbf{U}^{-1}\mathbf{S}$$

using the inverse of (43). Thus $\mathbf{TTS} = \mathbf{1}$, and we can identify $\mathbf{T}$ as $\mathbf{S}^{-1/2}$. If this $\mathbf{T}$ is used in (42), the new overlap matrix $\mathbf{S}'$ will be

$$\mathbf{T}^\dagger\mathbf{S}\mathbf{T} = \mathbf{U}\mathbf{D}^{-1/2}\mathbf{U}^{-1}\mathbf{S}\mathbf{U}\mathbf{D}^{-1/2}\mathbf{U}^{-1} = \mathbf{1}$$

The new basis functions obtained from the $\{\Phi_l\}$ by (42) with $\mathbf{T}$ of (44) resemble the original functions as much as possible in the sense of minimizing

$$\sum_k \langle \Psi_k - \Phi_k \mid \Psi_k - \Phi_k \rangle = \sum_k \left\langle \sum_l T_{lk}\Phi_l - \Phi_k \mid \sum_m T_{mk}\Phi_m - \Phi_k \right\rangle$$

$$= \sum_{k,l,m} T_{lk}^* T_{mk} S_{lm} - \sum_{k,l} T_{lk}^* S_{lk} - \sum_{k,m} T_{mk} S_{km} + \sum_k S_{kk}$$

while producing an orthonormal set. Since $\mathbf{T}^\dagger\mathbf{S}\mathbf{T}$ is to be the unit matrix, the first term is the number of basis functions, while the last term does not depend on $\mathbf{T}$. We thus want to maximize

$$q = \sum_{k,l} (T_{lk}^* S_{lk} + T_{lk} S_{kl}) \tag{45}$$

with the constraints

$$r_{kj} = \sum_{l,m} T_{lk}^* S_{lm} T_{mj} = \delta_{kj}. \tag{46}$$

We introduce Lagrange multipliers λ_{kj} and make $q - \sum \lambda_{kj} r_{kj}$ stationary with respect to variations in the T_{lk}^* (which can be taken as independent of the T_{lk}). Setting the coefficient of δT_{pn}^* equal to zero yields

$$S_{pn} = \sum_{j,m} \lambda_{nj} S_{pm} T_{mj}$$

and, using (46), one can show

$$\lambda_{nj} = \sum_m T_{mj}^* S_{mn}.$$

When this is substituted into the equation preceding, one has the condition

$$S = STT^tS$$

which is satisfied by $T = S^{-1/2}$.

The transformation to an orthonormal basis is much to be preferred to the rewriting of (40) as

$$|S| |S^{-1}H - \varepsilon 1| = 0$$

and finding the eigenvalues of $S^{-1}H$. First, $S^{-1}H$ is not a symmetric matrix, so finding its eigenvalues is much more difficult. Second, finding S^{-1} leads to numerical instabilities for large basis sets because S is more nearly singular ($|S| = 0$) as the basis functions become closer to linearly dependent. To see this, imagine orthogonalizing the nth function to all the others by replacing Φ_n by

$$\Phi_n' = \Phi_n - \sum_{m \neq n} \Phi_m \langle \Phi_m | \Phi_n \rangle.$$

For a large basis set, $\langle \Phi_n' | \Phi_n' \rangle$ will become small and the overlap matrix in the Φ' basis (which is diagonal) will have small elements, so numerical errors will be very important in S^{-1}. Several other strategies are available [41] to solve (40) without obtaining S^{-1}, but the problem of how best to do it continues to attract attention.

The partitioning or bracketing method of Löwdin [16, 41a], already mentioned in connection with lower bounds that may be derived from it, represents a formal solution to Eq. (39). Writing M_{ij} for $H_{ij} - \langle H \rangle_\Phi \delta_{ij}$, Löwdin partitioned the basis set into subsets "a" and "b," so that the matrix equation $Mc = 0$ breaks down into

$$M_{aa}c_a + M_{ab}c_b = 0, \qquad M_{ba}c_a + M_{bb}c_b = 0.$$

The matrix M is partitioned into submatrices M_{aa}, M_{ab}, M_{ba}, and M_{bb} according to the partitioning of the basis set. Formally, the second equation gives

$$c_b = -M_{bb}^{-1}M_{ba}c_a$$

which may be substituted into the first equation to give

$$(M_{aa} - M_{ab}M_{bb}^{-1}M_{ba})c_a = 0 \qquad (47)$$

If the subspace "a" consists of one function, c_a may be chosen as unity and one has an implicit equation for the energy E,

$$E = f(E) = (S_{11})^{-1}[H_{11} - (H_{1b} - ES_{1b})(H_{bb} - ES_{bb})^{-1}(H_{b1} - ES_{b1})].$$

Löwdin [16] considered first- and second-order iterative processes for solution of this equation. Some of these led to lower and upper bounds for E.

5. Single-Determinant Functions

The computations involved in evaluating the expectation value of the Hamiltonian are simplified if the trial function is a product of functions (spin orbitals) for the individual electrons: The results involves only integrals over clusters of one and two electrons. With the computational simplification goes ease in physical interpretation: The particles may be considered independent, and each is described by a wave function. Restricting oneself to a wave function of product form, and allowing the form of the individual functions to vary, lead [42, 43, 43a] to the Hartree wave function. One obtains certain one-electron eigenvalue equations which the "best" orbitals must satisfy, and the operators in these equations can be interpreted as the effective Hamiltonians (kinetic energy plus potential energy due to the nucleus and the other electrons) in which an electron moves. Essentially the same Hamiltonian was derived by Hartree on physical grounds.

The product wave function does not satisfy the Pauli exclusion principle. To make it antisymmetric with respect to an exchange of electrons, it suffices to use the antisymmetrizing operator

$$\mathscr{A} = (N!)^{-1/2} \sum_P (\delta_P P) \tag{48}$$

Here, N is the number of electrons and the sum is over all permutation operators P of N electrons (meaning their space and spin coordinates simultaneously). The factor δ_P depends on the parity of P, being -1 or $+1$ according as P is equivalent to an odd or even number of interchanges. It is easy to show that

$$\mathscr{I}\mathscr{A}\Phi = -\mathscr{A}\Phi, \tag{49}$$

where $\mathscr{I}$ is an operator which interchanges two electrons, whatever the nature of Φ. The $(N!)^{-1/2}$ in (48) is a normalizing factor in the following sense: If $\langle \Phi \mid P\Phi \rangle = 0$ for all P except the identity, and Φ is normalized, $\mathscr{A}\Phi$ will be normalized. To show this, consider

$$\langle \mathscr{A}\Phi \mid \mathscr{A}\Phi \rangle = \sum_P \delta_P \left\langle P\Phi \, \middle| \, \sum_Q \delta_Q \, \middle| \, Q\Phi \right\rangle \Big/ N! \tag{50}$$

The integral may be rewritten as

$$\left\langle \Phi \,\middle|\, \sum_Q \delta_Q P^{-1} Q \Phi \right\rangle = \left\langle \Phi \,\middle|\, \sum_S \delta_S \delta_P S \Phi \right\rangle$$

where $S = P^{-1}Q$. We have operated on the integrand with the permutation P^{-1} (the inverse to P) which cannot change anything because the electronic coordinates are all integrated over, and we have put $S = P^{-1}Q$. Now (50) becomes

$$\langle \mathscr{A}\Phi \,|\, \mathscr{A}\Phi \rangle = \sum_P \left\langle \Phi \,\middle|\, \sum_S \delta_S S\Phi \right\rangle \Big/ N! \tag{51}$$

The sum over P simply yields $N!$ identical terms. By hypothesis, only one term survives from the sum over S, so

$$\langle \mathscr{A}\Phi \,|\, \mathscr{A}\Phi \rangle = \langle \Phi \,|\, \Phi \rangle.$$

Operating with $\mathscr{A}$ on the Hartree product function makes it into a determinant:

$$\Psi = \mathscr{A} \prod_{k=1}^{N} \lambda_k(k) = (N!)^{-1/2} \det[\mathbf{M}], \tag{52}$$

where the matrix $\mathbf{M}$ has elements

$$M_{ij} = \lambda_i(j). \tag{53}$$

The physical interpretation is that each one-electron state described by λ_k is occupied by one electron, but one cannot specify which one. The energy calculated with the antisymmetrized Hartree function is lower than that calculated with the Hartree function itself, because, for an antisymmetric function, there is zero probability of finding two electrons with the same spin at the same point in space. This keeps the electrons apart and decreases the interelectronic repulsion. An even lower energy can be obtained from a function of form (52) if the spin orbitals are chosen to minimize the expectation value of the Hamiltonian over Ψ. A determinant being a sum of products, the computations for a determinantal wave function are hardly more complicated than those for a product wave function. In fact, the Hartree function is hardly used for molecular calculations. The conditions on the spin orbitals which lead to a determinantal function of minimum energy are the Hartree–Fock equations.

   We derive these by calculating the expectation value of the Hamiltonian with the determinantal function and minimizing with respect to variations

A. Variation Methods

which maintain determinantal form. The allowed variations are thus changes in the spin orbitals, i.e., $\lambda_i \to \lambda_i + \delta\lambda_i$. This procedure was first carried out by Fock [43, 43a]. The determinantal wave function constructed from solutions to these equations will have energy stationary to the allowed variations, and (provided the energy is a true minimum) thus will be the best wave function, in the sense of the variation method, in the space of single-determinant functions. We will refer to it also as the unrestricted Hartree–Fock function (UHF) since any variation in the spin orbitals is permissible. Of course, it cannot be the exact wave function, as is seen immediately from the form of H. A detailed derivation of the important Hartree–Fock equations follows.

We write the trial function as

$$\Phi = \mathscr{A} \prod_{k=1}^{N} \lambda_k(k) \tag{54}$$

where the spin orbitals λ_k are functions of both space and spin coordinates. Using Lagrangian multipliers, we try to minimize $\langle \Phi \mid H \mid \Phi \rangle$ for Φ normalized. The Hamiltonian consists of operators which are constants (not involving electronic coordinates), one-electron [of the form $\sum_i h(i)$], and two-electron [of the form $\sum_{i<j} g(i, j)$]. The constant or zero-electron operators are the internuclear repulsions, the two-electron $g(i, j)$ will usually be the interelectronic repulsions e^2/r_{ij}, and the one-electron operator $h(i)$ includes electronic kinetic energy and electron–nuclear attraction. All are symmetrical with respect to relabelings of the electrons:

$$P \sum_{i=1}^{N} h(i) = \sum_{i=1}^{N} h(i) \tag{55}$$

$$P \sum_{i<j}^{(N)} g(i, j) = \sum_{i<j}^{N} g(i, j). \tag{56}$$

Therefore, we can use the arguments given in going from Eq. (50) to Eq. (51) to write

$$\langle \Phi \mid H \mid \Phi \rangle = \left\langle \prod_{k=1}^{N} \lambda_k(k) \,\middle|\, \sum_{i=1}^{N} h(i) + \sum_{i<j}^{(N)} g(i, j) \,\middle|\, \sum_{P} \delta_P P \left[\prod_{l=1}^{N} \lambda_l(l) \right] \right\rangle. \tag{57}$$

The properties of determinants allow a simplification. Let Ψ be written as in (52)–(53) and consider the matrix $\mathbf{N}$, where $\mathbf{N} = \mathbf{TM}$ and $\det[\mathbf{T}]$ is not zero. Since $\det[\mathbf{N}] = \det[\mathbf{T}] \det[\mathbf{M}]$, the many-electron functions $\det[\mathbf{M}]$ and $\det[\mathbf{N}]$ differ only by a multiplicative constant. Now

$$N_{ij} = \sum_{k} T_{ij} M_{kj} = \sum_{k} T_{ik} \lambda_k(j) \tag{58}$$

so we have shown that Φ is unchanged, except by multiplication by the constant $\det[\mathbf{T}]$, by the linear transformation of the occupied spin orbitals among themselves, given by (58). The constant cannot affect the minimization of $\langle H \rangle_\Phi$. This arbitrariness allows us to go to a new set of orbitals which are orthonormal, just as we transformed basis functions to an orthonormal set in the linear variation problem. Without loss of generality we therefore require

$$\langle \lambda_i \mid \lambda_j \rangle = \delta_{ij} \tag{59}$$

and this simplifies the evaluation of $\langle H \rangle_\Phi$. Note that (59) implies $\langle \Phi \mid P\Phi \rangle = 0$ for P not equal to the identity so Φ is normalized. There is still some arbitrariness in the spin orbitals, since further transformations which are unitary leave Φ and (59) invariant, and we use this later.

In the one-electron part of (57), we have

$$\sum_{i=1}^{N} \left\langle \prod_{k=1}^{N} \lambda_k(k) \mid h(i) \mid \sum_{P} \delta_p P\left[\prod_{l=1}^{N} \lambda_l(l) \right] \right\rangle$$

$$= \sum_{i=1}^{N} \left\langle \prod_{k=1}^{N} \lambda_k(k) \mid h(i) \mid \prod_{l=1}^{N} \lambda_l(l) \right\rangle$$

$$= \sum_{i=1}^{N} \langle \lambda_i(i) \mid h(i) \mid \lambda_i(i) \rangle \tag{60}$$

since permutations P other than the identity lead to nothing by virtue of (59). For the two-electron part, we get

$$\sum_{i<j} \left\langle \prod \lambda_k(k) \mid g(i,j) \mid (1 - \mathscr{T}_{ij})\left[\prod_{l=1}^{N} \lambda_l(l) \right] \right\rangle$$

$$= \sum_{i<j} \langle \lambda_i(i)\lambda_j(j) \mid g(i,j) \mid \lambda_i(i)\lambda_j(j) - \lambda_i(j)\lambda_j(i) \rangle \tag{61}$$

In this case the orthonormality of the $\{\lambda_i\}$ means that permutations other than the identity and $\mathscr{T}_{ij}$, the interchange of i and j, give nothing. Combining (60) and (61), we have

$$\langle \Phi \mid H \mid \Phi \rangle = \sum_{i=1}^{N} \langle \lambda_i \mid h \mid \lambda_i \rangle + \sum_{i<j=1}^{N} (\langle \lambda_i\lambda_j \mid g \mid \lambda_i\lambda_j \rangle - \langle \lambda_i\lambda_j \mid g \mid \lambda_j\lambda_i \rangle). \tag{62}$$

We minimize this subject to (59), i.e.,

$$\delta\left\{ \langle \Phi \mid H \mid \Phi \rangle - \sum_{i,j} \varepsilon_{ij} \langle \lambda_i \mid \lambda_j \rangle \right\} = 0 \tag{63}$$

where the ε_{ij} are a matrix of Lagrangian multipliers. We consider first-order variations in the $\{\lambda_k\}$, and take the variations $\delta\lambda_k$ and $\delta\lambda_k^*$ as independent.

Setting the coefficient of $\delta\lambda_k^*$ equal to zero, we have

$$h(2)\lambda_k(2) + \sum_{i \neq k} [\langle \lambda_i(1) | g(1, 2) | \lambda_i(1)\rangle \lambda_k(2)$$

$$- \langle \lambda_i(1) | g(1, 2) | \lambda_k(1)\rangle \lambda_i(2)] = \sum_j \varepsilon_{jk}\lambda_j(2) \qquad (64)$$

The restriction $i \neq k$ may be dropped now. In Eq. (64), the angle brackets mean integration over space and spin coordinates of electron 1. Equation (64) may be written as

$$h^{\mathrm{HF}}(2)\lambda_k(2) = \sum_i \varepsilon_{ik}\lambda_i(2) \qquad (65)$$

where the Hartree–Fock one-electron Hamiltonian h^{HF}, defined by equations (64) and (65), is nonlocal and involves the spin orbitals themselves. It is straightforward to give an explicit expression for ε_{lk}:

$$\varepsilon_{lk} = \langle \lambda_l(2) | h(2) | \lambda_k(2)\rangle + \sum_{i \neq k} [\langle \lambda_i(1)\lambda_l(2) | g(1, 2) | \lambda_i(1)\lambda_k(2)\rangle$$

$$- \langle \lambda_i(1)\lambda_l(2) | g(1, 2) | \lambda_k(1)\lambda_i(2)\rangle] \qquad (66)$$

Evidently, $\varepsilon_{lk} = \varepsilon_{kl}^*$, so these form a Hermitian matrix. Equations (64) may be simplified by taking advantage of the residual arbitrariness in the spin orbitals. Under the transformation

$$\lambda_i' = \sum_j t_{ji}\lambda_j$$

with t a unitary matrix, the matrix of Lagrange multipliers transforms according to the similarity transformation

$$\varepsilon_{jk}' = \sum_{m,n} t_{jm}^{-1}\varepsilon_{mn}t_{nk}.$$

Transforming (64) directly by multiplication by t_{kl} and summing over k,

$$h(2)\lambda_l'(2) + \sum_{i \neq k} (\langle \lambda_i(1) | g(1, 2) | \lambda_i(1)\rangle \lambda_l'(2) - \langle \lambda_i(1) | g(1, 2) | \lambda_l'(1)\rangle \lambda_i(2))$$

$$= \sum_{j,k} \lambda_j(2)\varepsilon_{jk}t_{kl} = \sum_{j,k,m} t_{mj}^{-1}\lambda_m'(2)\varepsilon_{jk}t_{kl} = \sum_m \lambda_m'(2)\varepsilon_{ml}.$$

In the summation on the left side, we may replace λ_i by λ_i', since

$$\sum_i \lambda_i^* \lambda_i = \sum_i \sum_k \sum_l (t_{kl}^{-1}\lambda_k')(t_{li}^{-1}\lambda_l')$$

$$= \sum_i \sum_k \sum_l \lambda_k'^* \lambda_l' t_{ik}' t_{li}^{-1} = \sum_l \lambda_l'^* \lambda_l' \qquad (67)$$

irrespective of the arguments of the spin orbitals. Thus the λ_i' satisfy the Hartree–Fock equations with ε_{ij} replaced by ε_{ij}'. Since ε_{mn} is Hermitian, a unitary matrix exists which makes ε_{jk}' diagonal.

With this transformation, the equations determining the orbitals are the canonical Hartree–Fock equations

$$h^{\mathrm{HF}}(i)\lambda_k(i) = \varepsilon_k \lambda_k(i) \qquad (68)$$

where

$$h^{\mathrm{HF}}(i) = h(i) + \sum_{l=1}^{N} \left(\langle \lambda_l(j) | g(i,j) | \lambda_l(j) \rangle - \langle \lambda_l(j) | g(i,j)\mathscr{I}_{ij} | \lambda_l(j) \rangle \right) \quad (69)$$

The integration represented by the triangular brackets is over the dummy coordinate j. The interchange operator $\mathscr{I}_{ij}$ means that the coordinates i and j are to be interchanged before integration. For instance,

$$\langle \lambda_l(j) | g(i,j)\mathscr{I}_{ij} | \lambda_l(j) \rangle f(i) = \langle \lambda_l(j) | g(i,j) | f(j) \rangle \lambda_l(i).$$

The operator h^{HF} is sometimes written $h + F$, with F referred to as the Fock operator. Since h includes the kinetic energy and electron–nucleus attraction energy, the two remaining terms represent the effect of the other electrons. The first, or Coulomb, operator is simply the e^2/r_{12} interaction averaged over the wave function. The second, or exchange, operator removes the interaction of an electron on itself [since $\langle \lambda_l(j) | g(i,j)\mathscr{I}_{ij} | \lambda_l(j) \rangle$ operating on $\lambda_l(i)$ for $l \leq N$ gives the same result as does the Coulomb operator] and gives an energy lowering—the exchange energy—having no classical analog. Thus, an electron in one of the occupied spin orbitals sees the repulsive field of $N - 1$ electrons. However, F operating on an electron in some other orbital effectively represents the effect of N electrons.

Equations (68) for different spin orbitals are coupled by the sum in the operator h^{HF}, but no longer by the Lagrangian multipliers. The eigenvalues ε_k in (68) are called "orbital energies." The form of the equations suggests an iterative approach to their solution. Guessing a set of $\{\lambda_i\}$ to construct h^{HF}, one solves for a new set of eigenfunctions and orbital energies. The new $\{\lambda_i\}$ are inserted in h^{HF} and the process is repeated until the $\{\lambda_i\}$ no longer change. Hence the name "self-consistent field" (SCF). While this simple scheme was used in the early work and continues to be used, it sometimes leads to convergence and other problems, as has been shown in detail [44] for the two-valence-electron diatomic. More powerful schemes are available [44–46]. For instance, it has been suggested that the minimization of $\langle H \rangle_\Phi$ be attacked directly. This is a problem in nonlinear variation. The orbitals obtained by minimization of $\langle H \rangle_\Phi$ of course automatically obey Eqs. (64).

Each spin orbital appearing in h^{HF} is an eigenfunction of this operator, but there are also other solutions to (68), called virtual orbitals, which are not occupied in the ground state. Eigenfunctions of h^{HF} corresponding to different eigenvalues are of course orthogonal. In solving the equations iteratively, it seems reasonable to use, at each stage, the λ_i corresponding to the lowest orbital energies, since one is trying to minimize the total energy. This seems to work, but is not rigorously justified because the total energy is not just the sum of orbital energies. Cases are known [47] where this procedure does not lead to the determinant of minimum energy.

The determinant as a whole is an eigenfunction of the many-electron Hamiltonian

$$H^{HF} = \sum_{i=1}^{N} h^{HF}(i) \tag{70}$$

with eigenvalue equal to the sum of orbital energies. The term "Hartree–Fock energy" is not generally applied to this, but to the expectation value of the Hamiltonian over the Hartree–Fock function. As is seen from Eq. (62),

$$E^{HF} = \langle \Phi^{HF} | H | \Phi^{HF} \rangle = \sum_{i=1}^{N} \varepsilon_i - \tfrac{1}{2} \sum_{i,j=1}^{N} (\langle \lambda_i \lambda_j | g | \lambda_i \lambda_j \rangle - \langle \lambda_i \lambda_j | g | \lambda_j \lambda_i \rangle)$$

$$= \sum_{i=1}^{N} (\tfrac{1}{2}\varepsilon_i + \tfrac{1}{2}\langle \lambda_i | h | \lambda_i \rangle) \tag{71}$$

since the interaction between each pair of electrons is counted twice in $\sum_i \varepsilon_i$. Koopmans [48] showed that the ε_i themselves can be interpreted as ionization energies (Koopmans' theorem).

Following Fock [49], the $(N + 1)$-electron problem was treated in terms of the N-electron problem by considering a determinantal wave function formed by occupation of the canonical Hartree–Fock orbitals for the N-electron problem, λ_k ($k = 1, \ldots, N$) plus one additional spin orbital λ_{N+1}. Here λ_{N+1} is chosen to minimize the energy while constrained to be normalized and orthogonal to the λ_k $(k = 1, \ldots, N)$. These being fixed, we have to make stationary [see Eqs. (62) and (63)]:

$$E_{N+1} - \sum_{i=1}^{N+1} \bar{\varepsilon}_i \langle \lambda_{N+1} \mid \lambda_i \rangle = \langle \lambda_{N+1} \mid h \mid \lambda_{N+1} \rangle + \sum_{i=1}^{N} [\langle \lambda_i \lambda_{N+1} \mid g \mid \lambda_i \lambda_{N+1} \rangle$$

$$- \langle \lambda_i \lambda_{N+1} \mid g \mid \lambda_{N+1} \lambda_i \rangle] - \sum_{i=1}^{N+1} \bar{\varepsilon}_i \langle \lambda_{N+1} \mid \lambda_i \rangle. \quad (72)$$

The $\bar{\varepsilon}_i$ are Lagrange multipliers corresponding to the constraints on λ_{N+1}. Setting the coefficient of $\delta \lambda_{N+1}^*$ equal to zero, we have

$$h\lambda_{N+1} + \sum_{i=1}^{N} [\langle \lambda_i(1) \mid g(1, 2) \mid \lambda_i(1) \rangle \lambda_{N+1}(2)$$

$$- \langle \lambda_i(1) \mid g(1, 2) \mid \lambda_{N+1}(1) \rangle \lambda_i(2)] = \sum_{i=1}^{N+1} \bar{\varepsilon}_i \lambda_i. \quad (73)$$

Multiplication by λ_k^* $(k \leq N)$ and integration, plus use of the adjoints of (68) and (69), yields

$$\varepsilon_k \langle \lambda_k \mid \lambda_{N+1} \rangle = \bar{\varepsilon}_k$$

so $\bar{\varepsilon}_k = 0$. In (73), the sum over i on the left can be extended to include $i = N + 1$. Then this shows that λ_{N+1} obeys the canonical Hartree–Fock equations—(68) and (69) for the $(N + 1)$-electron problem—and $\bar{\varepsilon}_{N+1}$ is an orbital energy ε_{N+1}. Now multiplying (73) by λ_{N+1} and integrating, we see that $\bar{\varepsilon}_{N+1}$ is identical to E_{N+1} defined by Eq. (72). To the extent that the addition of the $(N + 1)$th electron does not affect the other N orbitals, the difference between the energies of the best determinantal wave functions for the N- and $(N + 1)$-electron problems is the orbital energy ε_{N+1}. This should then approximate an ionization potential.

Koopmans also considered the determinantal wave function for $N - 1$ electrons formed by occupying $N - 1$ spin orbitals formed as linear combinations of the N Hartree–Fock orbitals for the N-electron problem. Energy minimization then shows that the orbital which should be removed is one of the canonical spin orbitals. The energy difference is again the orbital energy. It is expected that if the $N - 1$ functions are now allowed to vary to minimize the energy, they will not change much, so that the energy will be close to that calculated. The applicability of Koopmans' theorem to open-shell states (see Volume 2, Chapter II, Section B) has recently been discussed [50].

The sum of orbital energies may fruitfully be considered as the zero-order energy of a perturbation scheme, with H^{HF} as zero order Hamiltonian. The sum of zero plus first-order energies is then the expectation value

of the full Hamiltonian over the zero order wave function, or E^{HF} of (71).
It should be noted that the Hartree–Fock determinant does not determine
uniquely the Hamiltonian of which it is an eigenfunction, and it is possible
[51] to construct other Hamiltonians which possess Φ^{HF} as lowest eigenfunction with $\sum_i \varepsilon_i$ as eigenvalue. These differ in higher eigenfunctions.

A perturbation approach was used by Møller and Plesset [52] to prove
a very important theorem. They computed the first-order correction to the
Hartree–Fock wave function, where V, the perturbation, is the difference
between the exact and Hartree–Fock Hamiltonians. It was assumed that
the antisymmetric eigenfunctions of the Hermitian operator H^{HF} form a
complete set. These eigenfunctions are determinants formed from the $\{\lambda_k\}$
which satisfy the eigenvalue equation (68), i.e., the filled and virtual orbitals.
We may imagine them to be numbered such that $\lambda_1 \cdots \lambda_N$ are the filled
or occupied orbitals and $\lambda_{N+1} \cdots$ are the virtual orbitals. One can now
expand the first-order wave function of perturbation theory in the eigenfunctions of H^{HF} other than the ground state (which is the zero-order
function). Let D be such a determinant. It may differ from the ground state
by one, two, or more orbitals. In the last case, it cannot contribute to the
first-order wave function, since the perturbation V has only one-- and two-
electron parts, so $\langle \Phi^{\text{HF}} | V | D \rangle$ vanishes if D is orthogonal to Φ^{HF} in
more than two electronic coordinates. The remarkable point is that
$\langle \Phi^{\text{HF}} | V | D \rangle$ also vanishes if D differs from Φ^{HF} by only one orbital, as
we show later.

The consequence of this is that the expectation values of one-electron
operators, i.e., those of the form $Q = \sum_{i=1}^{N} q(i)$, are calculated to the second
order of accuracy by Hartree–Fock functions. In Eq. (8), the first-order
term is $\langle X | Q | \Psi \rangle$ plus its complex conjugate. Keeping only leading terms
in ε,

$$\langle X | Q | \Psi \rangle \cong \langle X | Q | \Phi \rangle$$

and X is the first-order correction to Φ^{HF}, so is a sum of determinants
differing from Φ^{HF} in two electrons. We may write the contribution of
one of these to $\langle X | Q | \Phi^{\text{HF}} \rangle$ as

$$\sum_i \left\langle \prod_{j=1}^{N} \lambda_j(j) \, \frac{\lambda_{\bar{m}}(m)\lambda_{\bar{n}}(n)}{\lambda_m(m)\lambda_n(n)} \,\middle|\, q(i) \,\middle|\, \sum_P \delta_P P \left[\prod_{k=1}^{N} \lambda_k(k) \right] \right\rangle$$

where the determinant differs from Φ^{HF} in the replacement of λ_m and λ_n
by $\lambda_{\bar{m}}$ and $\lambda_{\bar{n}}$ ($m, n \leq N$; $\bar{m}, \bar{n} > N$). No permutation P can give anything
other than zero for any choice of i because $\lambda_{\bar{m}}$ and $\lambda_{\bar{n}}$ are orthogonal to

all λ_k for $k \leq N$. Thus $\langle X \mid Q \mid \Phi^{\mathrm{HF}} \rangle$ vanishes and $\langle \Phi^{\mathrm{HF}} \mid Q \mid \Phi^{\mathrm{HF}} \rangle$ differs from $\langle \Psi \mid Q \mid \Psi \rangle$ in second order, just as $\langle \Phi^{\mathrm{HF}} \mid H \mid \Phi^{\mathrm{HF}} \rangle$ does from $\langle \Psi \mid H \mid \Psi \rangle$. Explicitly,

$$\langle \Phi^{\mathrm{HF}} \mid Q \mid \Phi^{\mathrm{HF}} \rangle = (1 + \varepsilon^2)^{-1} (\langle \Psi \mid Q \mid \Psi \rangle + \varepsilon^2 \langle X \mid Q \mid X \rangle),$$

so that the error is, to terms in ε^4,

$$\langle \Phi^{\mathrm{HF}} \mid Q \mid \Phi^{\mathrm{HF}} \rangle - \langle \Psi \mid Q \mid \Psi \rangle = \varepsilon^2 (\langle X \mid Q \mid X \rangle - \langle \Phi^{\mathrm{HF}} \mid Q \mid \Phi^{\mathrm{HF}} \rangle).$$

It has been argued [25] that the two terms tend to cancel in many cases, yielding relative errors even smaller than ε^2.

The theorem that X may be written as a sum of determinants doubly orthogonal to Φ is often referred to as the Møller–Plesset theorem, but more often as Brillouin's theorem. Brillouin [53] gave a more extensive discussion of its meaning, including special cases. In fact, the theorem is seen to be an immediate consequence of the variational condition leading to the Hartree–Fock equations. The function Φ^{HF} minimizes the energy within the subspace of single determinants. Therefore $\langle \Phi^{\mathrm{HF}} \mid H \mid \delta\Phi^{\mathrm{HF}} \rangle$ vanishes when $\delta\Phi^{\mathrm{HF}}$ is an allowed variation of Φ^{HF}. The variation $\delta\Phi^{\mathrm{HF}}$ is obtained by changing one orbital in Φ^{HF} slightly (changing two would be a second-order variation), so $\delta\Phi^{\mathrm{HF}}$ is a determinant differing from Φ^{HF} in one orbital. Now $H = H^{\mathrm{HF}} + V$. Also, Φ^{HF} is an eigenfunction of H^{HF}, so $\langle \Phi^{\mathrm{HF}} \mid H^{\mathrm{HF}} \mid \delta\Phi^{\mathrm{HF}} \rangle$ is a multiple of $\langle \Phi^{\mathrm{HF}} \mid \delta\Phi^{\mathrm{HF}} \rangle$ which vanishes. Therefore $\langle \Phi^{\mathrm{HF}} \mid V \mid \delta\Phi^{\mathrm{HF}} \rangle = 0$ which is Brillouin's theorem. If restrictions on Φ other than determinantal form are imposed in the variations, the theorem holds only in a modified form [53].

The argument can be reversed [54]. If we demand that the determinant give an energy which is stationary to small variations in the occupied orbitals which preserve normalization, we have the requirement

$$\langle \Phi_{i \to a} \mid H \mid \Phi^{\mathrm{HF}} \rangle = 0$$

where $\Phi_{i \to a}$ is a determinant differing from Φ^{HF} by the replacement of λ_i by λ_a, where λ_a is orthogonal to all the orbitals of Φ^{HF}. By arguments like those employed earlier we can show that

$$\langle \Phi_{i \to a} \mid H \mid \Phi^{\mathrm{HF}} \rangle = \langle \lambda_a(i) \mid h_{\mathrm{HF}}(i) \mid \lambda_i(i) \rangle.$$

If this is to vanish for all λ_a so long as λ_a is orthogonal to all the occupied orbitals, h_{HF} must be a linear combination of these orbitals. Thus we have derived (65) from Brillouin's theorem. The theorem can, in fact, be made the basis of an iterative scheme for finding the self-consistent field function [55, 56].

The Hartree–Fock function is the determinantal function which gives the best (lowest) energy. One can define the best determinantal model according to other criteria [57], such as (a) that giving the best electron density or (b) that maximizing the overlap with the exact wave function. It seems that neither of these criteria can be formulated without assuming some knowledge of the exact wave function, making it difficult to give an a priori calculational scheme that would lead to them.

While we will leave discussion of the determination of the Hartree–Fock function for diatomic systems for a later chapter, we must mention here that the integrodifferential equations (68) and (69) are most conveniently dealt with in the form of matrix equations. These equations, presented and exploited by Roothaan [43a] and associated with his name, are obtained when the spin orbitals in the determinant are written as linear combinations of some fixed set of basis functions, as in Eq. (74):

$$\lambda_k = \sum_l c_{lk}\chi_l. \tag{74}$$

We must determine the coefficients c_{lk}. While this does not constitute a derivation, the equations which determine the coefficients may be obtained by substituting these expansions into (68) and (69), followed by multiplication by χ_m^* and integration. The result is written concisely in matrix form as

$$\mathbf{hc}_k + \mathbf{Gc}_k = \varepsilon_k \mathbf{Sc}_k \tag{75}$$

where $\mathbf{c}_k$ is the column vector of coefficients for λ_k, so that

$$(\mathbf{hc}_k)_l = \sum_m h_{lm}c_{mk}.$$

In (75),

$$h_{mn} = \langle \chi_m \mid h \mid \chi_n \rangle \tag{76a}$$

$$S_{mn} = \langle \chi_m \mid \chi_n \rangle \tag{76b}$$

and

$$G_{mn} = \sum_l \sum_i \sum_j c_{il}^* c_{jl} [\langle \chi_m(1)\chi_l(2) \mid g(1,2) \mid \chi_n(1)\chi_l(2) \rangle$$
$$- \langle \chi_m(1)\chi_l(2) \mid g(1,2) \mid \chi_l(1)\chi_n(2) \rangle]. \tag{76c}$$

6. *Density Matrix*

We have noted that a linear transformation can be performed among the occupied orbitals in any determinant without changing the determinantal function itself. The individual orbitals have no basic significance. All expectation values are sums over orbital contributions (one-electron operators) or contributions of pairs of orbitals (two-electron operators). As emphasized by Dirac [58], the determinantal wave function may be characterized by the first-order density matrix. The elements of the density matrix are defined by

$$\varrho(i, j) = \sum_{k}^{(occ)} \lambda_k(i)\lambda_k^*(j) \tag{77}$$

where i and j each represents a set of space and spin coordinates. The sum is over occupied spin orbitals. It is easy to see that $\varrho(i, j)$ is unchanged by a transformation

$$\lambda_k(i) = \sum_{l}^{(occ)} T_{kl}\lambda_l(j) \tag{78}$$

as long as the square matrix $\mathbf{T}$ is unitary, i.e.,

$$\sum_{l} (T_{kl})^*T_{ml} = \delta_{km}. \tag{79}$$

The indices of the matrix elements, i and j, are continuous variables rather than members of a discrete set. Thus, sums over indices are replaced by integrations, as in the matrix multiplication

$$\varrho^2(i, j) = \int d\tau_k \varrho(i, k)\varrho(k, j).$$

The diagonal element of $\boldsymbol{\rho}$ is the electron density at i, while the trace of $\boldsymbol{\rho}$ would be

$$\int d\tau_i \sum_k \lambda_k(i)\lambda_k(i)^* = \sum_k 1 = N.$$

All expectation values may be calculated from the density matrix. For example, for the one-electron operator h we construct the matrix $h(i, j)$ such that

$$\int h(i, j)f(j)\, d\tau_j = fh(i). \tag{80}$$

The integration is over space and spin coordinates j. If h is purely a multiplicative function, F, of spatial coordinates, $h(i, j)$ is $F(i)\,\delta(i, j)$. Now

$$\text{trace}(\boldsymbol{\rho}\mathbf{h}) = \int d\tau_k\, d\tau_i \varrho(i, k)h(k, i) = \sum_l \int d\tau_k \lambda_l^*(k)\, h\lambda_l(i).$$

Thus, the density matrix characterizes the determinant completely, and ϱ is invariant under transformations which leave the determinant invariant. The fundamental role of the density matrix was emphasized by Dirac [58] and subsequently in greater detail by Löwdin [59].

It is often convenient to go to a discrete representation of the density matrix. Let the functions $\{\phi_m\}$ be a complete orthonormal set in one-electron space (including spin) so that we can write

$$\lambda_k(i) = \sum_m c_m^{(k)} \phi_m(i) \tag{81}$$

Now knowledge of the λ_k is replaced by knowledge of the $c_m^{(k)}$. Substituting in (77), we obtain

$$\varrho(i, j) = \sum_k^{(occ)} \sum_{m,n} \phi_m(i)c_m^{(k)}c_n^{(k)*}\phi_n^*(j). \tag{82}$$

We see that the density matrix in the space of the functions ϕ_m is

$$\varrho_{mn} = \sum_k^{(occ)} c_m^{(k)}c_n^{(k)*} = \int d\tau_i\, d\tau_j\, \phi_m^*(i)\varrho(i, j)\phi_n(j). \tag{83}$$

If the $\{\phi_m{}^k\}$ are taken as the occupied and unoccupied orbitals of the determinant, i.e., the λ_k augmented by functions orthogonal to the λ_k to make the set complete, we have

$$\varrho_{mn} = \delta_{mn}\eta_m. \tag{84}$$

Here, η_m is 1 for an occupied orbital, 0 otherwise. Evidently, $\boldsymbol{\rho}^2 = \boldsymbol{\rho}$ in Eq. (84). This property is called idempotency. It must hold for all representations, since the new orbitals may be obtained from those used here by a unitary transformation, which leaves relations such as $\boldsymbol{\rho}^2 = \boldsymbol{\rho}$ unchanged.

Since the orbitals λ_k are unnecessary for calculations, one can object to any procedure which uses a particular set of orbitals which, once determined, are effectively recombined into the density matrix (77) or (83): It seems preferable to work directly with the density matrices from the beginning. Indeed, the Hartree–Fock procedure may be formulated in terms of the density matrix alone. The Hartree–Fock *equations* cannot be given in terms of the density matrix, since they are equations for orbitals. However, some of the manipulations in their derivation can be simplified by using the density matrix [60]. McWeeny [61] has long espoused the density matrix formulation in the Hartree–Fock procedure and other applications.

The matrix transcription of the eigenvalue equation of (65) in the representation of Eq. (81) is obtained by multiplying Eq. (65) by $\phi_m{}^*(2)$ and integrating over electron 2. We introduce the matrix elements of h^{HF} in the new basis

$$h_{ij}^{\mathrm{HF}} = \langle \phi_i \mid h^{\mathrm{HF}} \mid \phi_j \rangle,$$

expand the spin orbitals λ_k as in (81), and put c_{jk} for $c_j^{(k)}$ above. Then (65) becomes

$$\sum h_{ij}^{\mathrm{HF}} c_{jk} = \sum_i^{(\mathrm{occ})} \sum_j \varepsilon_{ij} c_{jk}.$$

The density matrix elements are given in Eq. (83). Now consider the energy of the determinant, Eq. (62). For the one-electron operator we have

$$\sum_{i=1}^{N} \sum_{k,l} c_{ki}^* h_{kl} c_{li} = \mathrm{trace}\ \boldsymbol{\rho}\mathbf{h} \tag{85}$$

where

$$h_{kl} = \langle \phi_k \mid h \mid \phi_l \rangle. \tag{86}$$

For the two-electron part we have, by expanding the orbitals

$$\sum_{i,j}^{(\mathrm{occ})} \sum_{k,l,m,n} c_{ki}^* c_{lj}^* c_{mi} c_{nj} (g_{klmn} - g_{klnm})$$

$$= \sum_{k,l,m,n} \varrho_{km} \varrho_{ln} (g_{klmn} - g_{klnm}). \tag{87}$$

Here,

$$g_{klmn} = \langle \phi_k(1)\phi_l(2) \mid g \mid \phi_m(1)\phi_n(2) \rangle. \tag{88}$$

Alternatively, we may introduce the matrix $\mathbf{G}$ by

$$G_{mk} = \sum_{ln} \varrho_{ln} (g_{klmn} - g_{klnm}) \tag{89}$$

and write our energy as

$$\langle \Phi \mid H \mid \Phi \rangle = \mathrm{trace}\ \boldsymbol{\rho}\mathbf{h} + \mathrm{trace}\ \boldsymbol{\rho}\mathbf{G}. \tag{90}$$

Obviously, $\mathbf{h} + \mathbf{G}$ is the matrix of the Hartree–Fock operator.

One could now consider varying ρ to minimize $\langle H \rangle_\Phi$, subject to the auxiliary conditions that $\rho^2 = \rho$ and that trace $\rho = N$. McWeeny [62] has developed a scheme for doing this. By working with the density matrix instead of the individual molecular orbitals, he is able to get more rapid convergence to the solution. In his method, both ρ and G vary in each step; in the conventional method, we fix G, solve for eigenfunctions, and then reconstruct G. Löwdin [59] has also indicated such a scheme.

The density matrix (77) is actually the first-order density matrix for a single-determinant wave function, one of a hierarchy of density matrices of various orders which may be defined for an arbitrary antisymmetric wave function. The density matrix of pth order, rather than the complete N-electron wave function itself, is necessary for computation of the expectation value of a symmetric p-electron operator. Thus the expectation value of the Hamiltonian operator requires only the first- and second-order density matrices. Since these are simpler objects than the wave function itself (which corresponds to the density matrix of Nth order), we might be able to simplify matters by dispensing with wave functions entirely. The difficulty is that the variational principle, for instance, refers to wave functions; if we want to vary density matrices to make the energy stationary, we must be able to state conditions which ensure that the density matrices that we work with are derivable from a wave function. This "N-representability problem" has resisted solution in a useful form [63]. Clinton and co-workers [64] have investigated the determination of the first order density matrix for small molecules by using theoretical and empirical information on molecular properties (see Subsection 3 and Sections C and D).

In general, we define the density matrix of order p corresponding to a normalized, antisymmetric N-particle wave function Ψ by

$$\Gamma^{(p)}(1', 2', \ldots, p' \mid 12, \ldots, p)$$

$$= [N/p!(N-p)!] \int \Psi^*(1'2', \ldots, p', p+1, \ldots, N)$$

$$\times \Psi(123, \ldots, p, p+1, \ldots, N)\, dx_{p+1}, \ldots, dx_N. \tag{91}$$

Here, i and x_i represent the space and spin coordinates for particle i. Equation (91) states that we integrate $\Psi^*\Psi$ over all but the first p coordinates to obtain a function of $2p$ sets of coordinates. Löwdin's article [65] discusses many of the properties of density matrices; we do not need a complete treatment here. Consider the symmetric m-particle operator

$$Q = \sum Q_{i_1 \cdots i_m}$$

where the sum is over all choices of m of the N particles. Each term in the sum has the same functional form. The expectation value of Q is

$$\binom{N}{m} = \frac{N!}{m!(N-m)!}$$

times that of $Q_{1\ldots m}$, by symmetry, and we then have

$$\langle Q \rangle = [N!/m!(N-m)!] \int [\Psi^*(1'2' \cdots m', m+1, \cdots N)Q_1 \cdots$$

$$\times \Psi(12 \ldots N)]_{x_1'=x_1 \cdots x_m'=x_m} dx_1 \cdots dx_N$$

$$= \int [Q_1 \ldots \Gamma(1' \cdots m' \mid 1 \cdots m)]_{x_1'=x_1, \cdots, x_m'=x_m}$$

after carrying out the integrations over $x_{q+1} \cdots x_N$ first. Only the mth order density matrix appears.

Now suppose Ψ is a single determinant composed of orthonormal spin orbitals. For $p = 1$, (91) becomes

$$\Gamma^{(1)}(1' \mid 1) = N \int [\mathscr{A}\{\prod \lambda_i(i')\}^* \mathscr{A}\{\prod \lambda_i(i)\}]_{2'=2,\ldots,N'=N} dx_2 \cdots dx_N \quad (92)$$

Let the determinants be expanded in minors of the first row, i.e.,

$$\mathscr{A}\{\prod \lambda_i(i)\} = N!^{-1/2} \sum_{k=1}^{N} (-1)^{k+1}\lambda_k(1)M_k(2 \cdots N).$$

The M_k are $(N-1)$-dimensional determinants. The integrations in (92) are over the orbitals appearing in the M_k, and the orthonormality property of the $\{\lambda_i\}$ makes the M_k orthogonal. Then Eq. (92) becomes

$$\Gamma^{(1)}(1' \mid 1) = N(N!)^{-1} \sum_{k=1}^{N} \lambda_k(1')^*\lambda_k(1) \int |M_k|^2 dx_2 \cdots dx_N$$

$$= (N-1)!^{-1} \sum_{k=1}^{N} \lambda_k(1')^*\lambda_k(1)(N-1)! \quad (93)$$

and we recover our previous result. We note that

$$\int \Gamma^{(1)}(1 \mid 1) dx_1 = \sum_{k=1}^{N} \int |\lambda_k(1)|^2 dx_1 = N$$

as follows from the definition (91). The case of determinants constructed from nonorthonormal spin orbitals was considered by Löwdin [65]. As might be anticipated, all higher-density matrices for a single determinant can be computed in terms of the one-particle density matrix, which means that in this case the one-particle matrix suffices to determine all properties. For instance, $\Gamma^{(2)}(1'2' \mid 12)$ can be computed by expansion in two-row minors, which we denote by N_{kl}

$$\Gamma^{(2)}(1' \mid 1) = [N(N-1)/2]N!^{-1} \sum_{k,l=1}^{N\ (k<l)} [\lambda_k(1')\lambda_l(2') - \lambda_k(2')\lambda_l(1')]^*$$

$$\times \sum_{m,n=1}^{N\ (m<n)} [\lambda_m(1)\lambda_n(2) - \lambda_m(2)\lambda_n(1)] \int N_{kl}^* N_{mn}\, dx_3 \cdots dx_N.$$

In the case of orthonormal orbitals, the $(N-2)$-row minors are orthogonal and $\int |N_{kl}|^2\, dx_s \cdots dx_N = (N-2)!$ Then

$$\Gamma^{(2)}(1' \mid 1) = \tfrac{1}{2} \sum_{k,l=1}^{N\ (k<l)} [\lambda_k(1')\lambda_l(2') - \lambda_k(2')\lambda_l(1')]^* [\lambda_k(1)\lambda_l(2) - \lambda_k(2)\lambda_l(1)]$$

$$= \tfrac{1}{2} [\Gamma^{(1)}(1' \mid 1)\Gamma^{(1)}(2' \mid 2) - \Gamma^{(1)}(1' \mid 2)\Gamma^{(1)}(2' \mid 1)]. \qquad (94)$$

7. Configuration Interaction

The properties of a single-determinant wave function are discussed further in Vol. 2, Ch. II, Sect. A. It will be pointed out that the orbitals which satisfy the unrestricted Hartree–Fock equations need not have the molecular symmetry, so that the unrestricted Hartree–Fock function may not belong to any particular representation of the molecular symmetry group. This inconvenient situation may be remedied by imposing conditions on the one-electron functions. For example, we can demand that certain pairs of spin orbitals be formed from the same spatial orbital, associated with α or β spin. Then a determinant which is an eigenfunction of total spin, and of molecular symmetry operators, may be formed in many cases. The minimum energy from such a function, referred to as a restricted Hartree–Fock function, is necessarily higher than that of the unrestricted Hartree–Fock function, because of the restrictions imposed. Even when one considers functions of this kind, however, there are situations where it is impossible to write an eigenfunction of spin and symmetry operators as a single determinant. Then one must write a linear combination of determinants, which differ from each other only in degenerate spin orbitals. The degenerate spin orbitals differ in spin orientation or direction of orbital angular momentum. Symmetry considerations alone suffice to define the correct linear combinations of the determinants in this case. Formally, we can define a projection operator (see Chapter II, Vol. 2) which converts any wave function into one having a desired symmetry. Operating on a single determinant, the projection operator produces a sum of determinants.

We may define a configuration by a set of occupation numbers for orbitals, all orbitals of the same energy being taken together. Thus the determinants needed to construct symmetry functions all belong to the same configuration. By a configuration interaction wave function (CI) we mean one which includes determinants coming from more than one configuration. This is an improvement on the single-configuration wave function, and, in principle, a CI function can approach the exact wave function.

It is easy to show that any antisymmetric function can be written as a sum of determinants, if these are constructed from a complete set of spin orbitals. Consider any antisymmetric function $\Psi(1 \cdots N)$. Fixing the values of the space spin coordinates $2, 3, \ldots, N$, expand Ψ in the complete set $\{\lambda_i(1)\}$. The expansion coefficients depend on $2, 3, \ldots, N$.

$$\Psi(1, 2, \ldots, N) = \sum_i \lambda_i(1) c_i(2 \cdots N) \tag{95}$$

Now fix $3 \cdots N$ in c_i and expand in the $\{\lambda_i(2)\}$

$$\Psi(1, 2, \ldots, N) = \sum_i \lambda_i(1) \sum_j \lambda_j(2) d_{ij}(3 \cdots N) \tag{96}$$

Repeating this, we have Ψ as a sum of products. Since Ψ is antisymmetrical, the antisymmetrizer may be applied to Eq. (96), leaving the left side invariant and converting all products into determinants (some of which are identical). Therefore,

$$\Psi(1, 2, \ldots, N) = \sum_k A_k \Phi_k \tag{97}$$

where the sum is over all sets of N indices, i.e., all ways of choosing N spin orbitals from the complete set λ_i. Similarly, if Ψ is an eigenfunction of spin with a certain eigenvalue, we can operate on Eq. (98) with a projection operator. This will transform each Φ_k into a sum of determinants having the desired spin property. We note that the passage from products to determinants could have been made by grouping together products in the expansion of Ψ differing only by permutation of the electrons among the same spin orbitals. The antisymmetry of Ψ guarantees that they will occur with the proper signs to form determinants. In the same way, all Φ_k in the same configuration may be grouped together in (97). The spin and angular momentum properties of Ψ require that this process yield functions of the proper spin and angular momentum symmetry.

In (95), Ψ was any antisymmetric function. If we take it as the exact wave function, we have shown that one can obtain the exact (nonrelativistic) energy by using configuration interaction wave functions like (97) in conjunction with the variational method. The necessity for combining determinants belonging to different configurations is due to the two-electron operators in H: a single-particle model, in which an electron is assigned to each spin orbital, cannot be exact. In a sense it could be said that the configuration interaction allows for a description of the correlation between electrons. The improvement in the energy in going beyond an antisymmetrized product function is often referred to as the correlation energy.

To this it may be objected that correlation between electrons of the same spin is already present in a single determinant. Indeed, in any antisymmetric function, the probability of finding two electrons of the same spin at the same spatial location vanishes. The same objection may be made to the definition of correlation energy as the difference between the energy of the simplest (restricted) Hartree–Fock function for the state (generally a single determinant including double occupation of orbitals) and the exact nonrelativistic energy. Clementi [66] has used the term "precorrelation" to refer to the correlation already present in the Hartree–Fock determinant. It might conceptually be more reasonable to start with the Hartree product, which is truly uncorrelated. It may also be objected that one must go beyond the single determinant in certain cases to ensure proper symmetry behavior of the function and the Fock operator. It is also useful to distinguish degeneracy and near-degeneracy effects from "true" correlations, which the simple definition of correlation energy does not do. Thus, a Hartree–Fock wave function for the atomic configuration $1s^2 2s^2 2p^n$ is nearly degenerate with one for $1s^2 2p^{n+2}$, and these wave functions would interact strongly in a CI calculation. The simple Hartree-Fock wave function would be a poor approximation and the "correlation energy" large.

On the other hand, problems arise with any attempt to define the energy due to correlation. In fact it seems best to follow current usage and use the definition promulgated by Löwdin [67]: The correlation energy is the difference between the expectation value of H for the appropriate symmetry-restricted Hartree–Fock function and the exact eigenvalue. Here, H is the fixed-nucleus, nonrelativistic Hamiltonian. (The separation of relativistic effects, implying additivity of relativistic and correlation energy, is not completely justified [66].) This energy is always positive, the variational principle guaranteeing that E^{HF} is an upper bound to the exact energy. Because of the precorrelation inherent in the Hartree–Fock function, the correlation energy depends strongly [66] on the number of pairs of electrons (the correlation between paired electrons is not properly treated in the

Hartree–Fock theory). Lowdin's discussion [67] of the correlation energy is of great value in clarifying the concepts involved, and Clementi [66] has given valuable results and discussion for the correlation energies of atoms.

If, in (97), Ψ is the exact function and the Φ_k are formed from the real and virtual Hartree–Fock spin orbitals in such a way that Φ_0 is the restricted Hartree–Fock determinantal wave function, the energy may be conveniently divided into the Hartree–Fock energy $\langle \Phi_0 \mid H \mid \Phi_0 \rangle$ and corrections to it. On the other hand, Ψ may be any antisymmetric wave function. Nakatsuji et al. [68] have expanded the unrestricted Hartree–Fock function in restricted Hartree–Fock determinants (see Volume 2, Chapter II, Section A), which could be classified according to degree of excitation (single-, double-, etc.) from the UHF function as well as by their total spin. It was then possible to elucidate the relation of the UHF function to determinantal functions produced in other ways.

In the configuration interaction method, one uses an expansion like (97) but with a finite number of terms. The individual Φ_K may be either single determinants or combinations of determinants set up to have the proper spin symmetry. If single determinants are used, and if one includes all determinants for each configuration, the proper combinations are obtained automatically. By using the correct combinations of determinants to start with, one has fewer terms and a secular equation (see page 101) of lower order. Minimizing $\langle \Psi \mid H \mid \Psi \rangle / \langle \Psi \mid \Psi \rangle$ with Ψ in the form of (97) is a linear variation problem and leads to the secular equation:

$$\mathbf{HC} = \mathbf{ESC} \tag{98}$$

The energy E is obtained from the secular determinant. Here, $H_{KL} = \langle \Phi_K \mid H \mid \Phi_L \rangle$ and $S_{KL} = \langle \Phi_K \mid \Phi_L \rangle$. A consequence of choosing the coefficients $\{c_i\}$ by the variational principle is that certain matrix elements vanish between Ψ and determinants of the spin orbitals employed. This may be referred to [69] as a generalized Brillouin theorem.

We now give formulas for the overlap and Hamiltonian matrix elements between determinantal wave functions, which enter (98). It is assumed that all the determinants are constructed from the same set of orthonormal spin-orbitals. Lowdin [65] has derived formulas for more general cases. First, we may use the arguments of Eqs. (50)–(51) to show that, for two determinantal functions $\Phi^{(1)}$ and $\Phi^{(2)}$ of the form (52) and for any operator Q which is symmetric to interchanges of electrons,

$$\langle \Phi^{(1)} \mid Q \mid \Phi^{(2)} \rangle = \left\langle \prod_{k=1}^{N} \lambda^{(1)}(k) \mid Q \mid \sum_{P} \delta_P P \left\{ \prod_{l=1}^{N} \lambda^{(2)}(l) \right\} \right\rangle. \tag{99}$$

Here, we denote by $\lambda_i^{(1)}$ one of the spin orbitals occupied in $\Phi^{(1)}$ and by $\lambda_i^{(2)}$ one of those occupied in $\Phi^{(2)}$. It is evident from (99) that when the two sets of spin orbitals are not identical the overlap $\langle \Phi^{(1)} \mid \Phi^{(2)} \rangle$ vanishes: For every permutation P there will be at least one electron in orthogonal spin orbitals on the left and on the right, so integration over that electron's coordinates will contribute a zero factor to the overlap. Similarly, if Q is a one-electron operator, (99) will vanish if the sets $\{\lambda_i^{(1)}\}$ and $\{\lambda_i^{(2)}\}$ differ in more than one spin orbital; if Q is two-electron, (99) will vanish when the sets differ by more than more than two spin orbitals; and so on.

Suppose $\lambda_k^{(1)}$ is identical to $\lambda_k^{(2)}$ except for $k = N$ ($\Phi^{(1)}$ and $\Phi^{(2)}$ differ by one spin orbital). Let Q first be a one-electron operator, $Q = \sum_i q(i)$, and consider the contribution of $q(1)$. For any permutation P, the contribution of electron N will be zero, since all the $\{\lambda_i^{(2)}\}$ are orthogonal to $\lambda_N^{(1)}$. The only $q(i)$ which does not give zero is for $i = N$. Here, to avoid getting zero we must use the identity permutation, so that the spin orbitals match for electrons other than N. Thus there is only one nonvanishing term in $\langle \Phi^{(1)} \mid Q \mid \Phi^{(2)} \rangle$, the matrix element of $q(N)$ between $\lambda_N^{(1)}$ and $\lambda_N^{(2)}$. To write the result we introduce a more convenient notation. Write Φ for $\Phi^{(2)}$ and $\Phi^{(a \rightarrow b)}$ for $\Phi^{(1)}$, to signify that the latter differs by the former in substitution of λ_b for λ_a. Then we have shown

$$\left\langle \Phi^{(a \rightarrow b)} \mid \sum_i q(i) \mid \Phi \right\rangle = \langle \lambda_b \mid q \mid \lambda_a \rangle. \tag{100}$$

Now suppose Q is a two-electron operator,

$$Q = \sum_{i,j}^{(i<i)} g(i, j).$$

In this case, it may be seen from (99) that, where neither i nor j is N, $g(i, j)$ makes no contribution to the matrix element. For any $i < N$, $g(i, N)$ gives a nonvanishing term if P affects no electrons other than i and N. Thus P is either the identity or the interchange of i and N, so the matrix element is

$$\left\langle \Phi^{(a \rightarrow b)} \mid \sum_{i,j}^{i<j} g(i, j) \mid \Phi \right\rangle$$
$$= \sum_{i=1}^{N-1} \langle \lambda_i(1)\lambda_b(2) \mid g(1, 2) \mid \lambda_i(1)\lambda_a(2) - \lambda_a(1)\lambda_i(2) \rangle \tag{101}$$

The sum is over the spin orbitals common to the two determinants.

If $\Phi^{(1)}$ differs from $\Phi^{(2)}$ by changes in the $(N - 1)$st and Nth spin orbitals, the same reasoning shows that only one term survives in the matrix element.

If we again put Φ for $\Phi^{(2)}$ and let $\Phi^{(1)}$ be obtained from Φ by the substitution of λ_a and λ_c for λ_b and λ_d, we may write the result as

$$\left\langle \Phi^{(a \to b, c \to d)} \mid \sum_{i,j}^{(i<j)} g(i,j) \mid \Phi \right\rangle$$
$$= \langle \lambda_b(1)\lambda_d(2) \mid g(1,2) \mid \lambda_a(1)\lambda_c(2) - \lambda_c(1)\lambda_a(2) \rangle \qquad (102)$$

The matrix element for a one-electron operator vanishes in this case. Note that it has always been assumed that the two determinants involved in the matrix elements have been written so there is the maximum possible coincidence of rows (spin orbitals). A reordering of the spin orbitals in a determinant could give an additional factor of -1.

Having discussed the matrix elements in (98), we turn to methods for its solution. In many cases the exact wave function may be approximated fairly well by a single determinant (or a set of determinants formed from a single configuration). If this determinant (or configuration) is used in an expansion like (97) its coefficient, assuming all Φ_k are normalized, will be large compared to the coefficients of the other configurations. For orthogonal spin orbitals the normalization condition is

$$\sum_K \mid C_K \mid^2 = 1 \qquad (103)$$

so that the other coefficients will be small compared to unity. This permits their estimation by methods based on perturbation theory. Not only can this estimation be used to avoid exact solution of very large secular equations, but even a rough estimate of the importance of different configurations allows us to eliminate many from consideration. Since the number of configurations that can be formed from n basis functions for an N-electron problem increases [70] roughly as n^N when $n \gg N$, this is of great importance. The estimation of the contributions of missing configurations corresponds to an estimate of the error.

The application of perturbation theory to the CI problem in a systematic way has been treated by Gershgorn and Shavitt [70]. They considered Eq. (98) for a finite number of configurations (such as all those that can be formed from a given set of one-electron functions). Taking $\mathbf{S} = \mathbf{1}$ for simplicity, this became

$$\mathbf{HC} = E\mathbf{C}. \qquad (104)$$

The exact solution of (104) would require evaluating all the matrix elements between all the configurations and then solving a secular equation of enormous size, neither of which is pleasant. We would like to be able to neglect some of the elements of H. This may be done by the Brillouin–Wigner expansion (Section B.1).

Suppose we have reason to believe that term "0" dominates the wave function so that a zeroth approximation to the energy is

$$E \sim \langle \Phi_0 | H | \Phi_0 \rangle \equiv E_0.$$

Then, as is shown in Section B.1 [Eq. (36)], the next approximation to the energy is given by

$$E = E_0 + \sum_{k \neq 0} H_{0k} H_{k0} (E - H_{kk})^{-1} \tag{105}$$

We may replace E by E_0 on the right side, since higher approximations bring in higher powers of the off-diagonal elements.

$$E = E_0 + \sum_{k}^{(k \neq 0)} | H_{0k} |^2 / (E_0 - H_{kk}). \tag{106}$$

The corresponding expression for C_k ($k \neq 0$) is, according to (35) of Section B.1,

$$C_k(k \neq 0) \cong C_0 H_{k0} / (E_0 - H_{kk}). \tag{107}$$

To generalize this, Gershgorn and Shavitt [70] partitioned the set of configurations into two parts, which we refer to as a and b. Part a includes $K + 1$ configurations, which are expected to contribute more to the final result than the rest (this might be the SCF function plus doubly excited configurations, for instance). The discussion of the preceding paragraph corresponds to $K = 0$. Part b includes all the other configurations. Suppose we first diagonalize the a submatrix of the Hamiltonian matrix to get an approximate solution. Let the $K + 1$ functions which diagonalize the a submatrix be represented by $\tilde{\Phi}_i$, with $\tilde{\Phi}_0$ that corresponding to the lowest energy, and use the $\tilde{\Phi}_i$ instead of the first $K + 1$ basis functions. In the new basis the approximate solution is $\tilde{C}_i = \delta_{i0}$, the tilde on the coefficients reminding us that we are writing the function in terms of the $\tilde{\Phi}_i$. Then the Brillouin–Wigner theory may be used.

Let $\tilde{H}_{kj} = \langle \tilde{\Phi}_k | H | \tilde{\Phi}_j \rangle$. Since $\tilde{C}_j$ for $j \neq 0$ is zero to a first approximation, and $\tilde{C}_j$ for $j > K + 1$ is expected to be small, the approximate solution [Eq. (107)] is

$$\tilde{C}_k(k > K + 1) \sim \tilde{H}_{k0} / (E - H_{kk}) \tag{108}$$

Here, E may be approximated as $\tilde{H}_{00} = \langle \tilde{\Phi}_0 | H | \tilde{\Phi}_0 \rangle$. The approximate energy is [see Eqs. (105) and (106)]

$$E = E_0 + \sum_{k>K+1} |\tilde{H}_{k0}|^2/(E_0 - H_{kk}). \tag{109}$$

Instead of $\mathbf{C}$, we can use in (108) any approximation to the projection of the exact C_i on the subspace $i \le K + 1$.

This projection is exactly $\mathbf{C}_a$, the solution to Eq. (47). Approximating it by $\mathbf{C}$ corresponds to using only $\mathbf{M}_{aa}$ in (47). For a better approximation to $\mathbf{C}_a$, we can include the diagonal part of $\mathbf{M}_{bb}$. Then we must put, in (47),

$$(\mathbf{M}_{ab}\mathbf{M}_{bb}^{-1}\mathbf{M}_{ba})_{ij} \rightarrow \sum_{k>K+1} H_{ik}H_{kj}/(H_{kk} - E)$$

and one finds $\mathbf{C}_a$ by diagonalizing the matrix $\mathbf{G}$ (which is of dimension $K + 1$), i.e., $(\mathbf{G} - E)\mathbf{C}_a = 0$, where

$$G_{ij} = H_{ij} - \sum_{k>K+1} H_{ik}H_{kj}/(H_{kk} - E). \tag{110}$$

With this $\mathbf{C}_a$, we can estimate C_i for $i > K + 1$, to judge whether the ith configuration should be included. The matrix elements needed for this calculation are the same as those needed for the preceding estimate (108), and using (110) is expected to be an improvement over (108) because it takes into account some of the effect of subspace b on the zero-order function in subspace a. Tests of these formulas were carried out by Gershgorn and Shavitt [70]. The superiority of the use of G [Eq. (110)] was demonstrated. The use of the simple expression (108) gave only relatively crude results.

If the dominance of a single determinant or configuration Φ_0 is to be made as great as possible, the best single determinant to be used as Φ_0 is some sort of Hartree–Fock function. Then the natural choice of spin orbitals from which to form the Φ_k is the set of occupied and virtual Hartree–Fock spin orbitals. Since all these spin orbitals are eigenfunctions of a Hermitian operator, they are orthogonal and should form a complete set. However, though it is convenient to use them in formal manipulations, there is a disadvantage to their use in actual calculations (see page 129).

Nesbet [71] has discussed the advantages of the unrestricted Hartree–Fock function as a starting point for a configuration interaction treatment. He assumed $\Phi_0 = \Phi^{\text{UHF}}$ and imagined all the Φ_i to be constructed from the occupied and virtual Hartree–Fock functions. Now Brillouin's theorem guarantees that H_{k0} vanishes unless Φ_k differs from Φ^{UHF} by two spin orbitals. Thus C_k should be small for singly excited configurations, since it vanishes in the first approximation [Eq. (107)]. If Φ_k is doubly excited C_k does not vanish in Eq. (107), but it can be argued that it is small because the energy of Φ_k is much higher than E_0. The energy expression (106) is now

$$E = E^{\text{UHF}} + \sum_i^{(i \neq 0)} |\langle \Phi_i | H | \Phi_0 \rangle|^2 / [\langle \Phi_i | H | \Phi_i \rangle - E_0]. \qquad (111)$$

It is possible that singly excited configurations, included in higher orders of perturbation theory, may be no more important than doubly excited configurations, which enter in first order.

The disadvantage in using the virtual orbitals to form the excited configurations stems from the fact that these orbitals have their greatest density far from the core or occupied orbitals, since they represent electrons moving in the field of the nuclei and all N electrons (not $N - 1$, as for occupied orbitals). The error in the Hartree–Fock function is mostly in the regions of configuration space where electrons of opposite spin come together, i.e., where the occupied orbitals are appreciable. The virtual orbitals are not very helpful in changing the wave function here. Alternatively we could say that there is very little differential overlap between the excited configurations and the Hartree–Fock function, so individual configurations have small interaction. A large number of configurations will have to be used. We can imagine forming wave packets from the virtual orbitals which have better localization properties. Then Φ^{UHF} could still be used as zero-order configuration, the advantages of Brillouin's theorem not being lost. Without changing the Hartree–Fock function, we are free to transform the occupied orbitals among themselves. Nesbet suggested [71] that a similar transformation be performed on the virtual orbitals. This use of a localized orbital basis would increase the importance of individual terms and decrease the number of terms that would have to be used.

Determination of the Hartree–Fock function is by no means a necessary intermediate step in carrying out a configuration interaction (CI). Indeed, if we could use all the configurations constructed from a given basis set, the energy obtained would be independent of how the determinants were formed. It also seems that if we could include all singly, doubly, and triply substituted configurations, the Hartree–Fock and other reasonable molecular orbital starting points would give essentially identical results. There would be no need to use a Hartree–Fock function as one of the Φ_i. The Hartree–Fock calculation would be a waste of time. However, using the Hartree–Fock makes it possible to work with fewer configurations.

It may be expected in general that (a) higher energy configurations (and in particular those differing from the ground state by many substitutions) contribute relatively little; (b) the singly excited configurations are less important than the doubly excited ones; (c) triply and quadruply excited

configurations, which cannot be mixed in in first order, are relatively unimportant. If the approximation to the ground state did not resemble the
self-consistent field (SCF) function, the singly excited configurations would
be of great importance. A study of convergence properties for a BH_3 CI
calculation by Pipano and Shavitt [72] noted that inclusion of singly excited configurations essentially allowed other molecular orbital (MO) functions to "catch up" with the SCF. While the energies of the MO and SCF
differed by 0.02 a.u., a CI with singly excited configurations gave energies
differing by 0.003 a.u. (the MO was, in fact, lower than the SCF here).
The triple and quadruple excitations gave closely additive contributions to
the energy, as expected from perturbation theory. Starting with a CI calculation including single and double configurations, perturbation theory could
be used [72] to calculate the contributions of triple and quadruple excitations. In particular, it is a good approximation to compute the coefficient
C_i for a quadruply excited determinant as a sum of products of coefficients
C_i for the doubly excited determinants. Thus the C_i for doubly excited
configurations can be used to select which quadruply excited configurations
to include. The four-electron excitation may be considered to occur in
steps of two, as appears in perturbation theory. Nesbet [71] gave a justification for this in terms of a paired-electron wave function describing
correlation in terms of localized orbitals. This becomes exact for weakly
interacting subsystems with strong internal correlations.

It is possible to define spin orbitals which make the CI expansion converge
optimally, in a sense to be specified later. This means fewer determinants
need be used. These orbitals, the "natural spin orbitals," were introduced
in connection with density matrices by Löwdin [65]. We have already
discussed the density matrices for a single determinant, and now we consider the density matrices for a function of more complicated form. Löwdin
[65] has given formulas for density and transition matrices between Slater
determinants. [In the transition matrix, Ψ^* and Ψ in (91) are replaced by
Φ_K^* and Φ_L, where Φ_K and Φ_L are two determinants.] From these formulas, one can compute the density matrix for a CI function. Consider the
first-order transition matrix between Φ_K and Φ_L, assuming orthonormal
spin orbitals. Expanding in minors as in Eqs. (92)–(94), we see that we
get zero unless Φ_K and Φ_L differ by only one spin orbital. If Φ_K and Φ_l
differ by the substitution of $\lambda_k \rightarrow \lambda_l$,

$$\Gamma_{KL}^{(1)}(1' \mid 1) = N(N!)^{-1} \int \lambda_k^*(1') M_k^K(2 \cdots N)^*$$

$$\times \lambda_l(1) M_l^L(2 \cdots N) \, dx_1 \cdots dx_N = \lambda_k^*(1')\lambda_l(1) \quad (112)$$

For the configuration interaction function

$$\Psi = \sum_k C_k \Phi_k \bigg/ \sum_k |C_k|^2 \tag{113}$$

we may write

$$\Gamma^{(1)}(1'|1) = \frac{\sum_{K,L} C_K^* C_L \Gamma_{KL}^{(1)}(1'|1)}{\sum_K |C_K|^2}.$$

(112) and (93) show that $\Gamma_{KL}^{(1)}(1'|1)$ is a bilinear form in the spin orbitals:

$$\Gamma_{KL}^{(1)}(1'|1) = \sum_{kl} \lambda_k^*(1')\lambda_l(1)\gamma_{lk} \tag{114}$$

From its definition, $\Gamma^{(1)}(1'|1)$ is Hermitian: $\Gamma^{(1)}(1'|1)^* = \Gamma^{(1)}(1|1')$. This implies that $\gamma_{lk} = \gamma_{kl}^*$, i.e., these coefficients form a Hermitian matrix. Hence there exists a unitary matrix $\mathbf{U}$ which diagonalizes γ. Let $\mathbf{n}$ be the resulting diagonal matrix, so that

$$\mathbf{U}^\dagger \gamma \mathbf{U} = \mathbf{n} \tag{115a}$$

$$\gamma_{kl} = \sum_m U_{km} n_m U_{ml}^\dagger. \tag{115b}$$

Defining a new set of spin orbitals by

$$\chi_k = \sum_l \lambda_l U_{lk}, \tag{116}$$

we see that the first-order density matrix may be written as

$$\Gamma^{(1)}(1'|1) = \sum_m \chi_m^*(1')\chi_m(1)n_m \tag{117}$$

The χ_m are called the natural spin orbitals associated with the wave function Ψ and the n_m the corresponding occupation numbers.

Because of the form of the λ_k, i.e., $\phi\alpha$ or $\phi\beta$, the density matrix can be written:

$$\Gamma^{(1)}(1'|1) = D_{\alpha\alpha}(\mathbf{r}_{1'}, \mathbf{r}_1)\alpha(1')\alpha(1) + D_{\alpha\beta}(\mathbf{r}_{1'}, \mathbf{r}_1)\alpha(1')\beta(1)$$

$$+ D_{\beta\alpha}(\mathbf{r}_{1'}, \mathbf{r}_1)\beta(1')\alpha(1) + D_{\beta\beta}(\mathbf{r}_{1'}, \mathbf{r}_1)\beta(1')\beta(1). \tag{118}$$

If we deal with spin-free operators, it is sufficient to consider the spinless density matrix, obtained from $\Gamma^{(1)}(1'|1)$ by putting $1 = 1'$ in the spin functions and integrating over spin coordinates. Then (118) yields $D_{\alpha\alpha}(\mathbf{r}_{1'}, \mathbf{r}_1) + D_{\beta\beta}(\mathbf{r}_{1'}, \mathbf{r}_1)$. Comparing (118) with (114), we see that $D_{\alpha\alpha}$

may be written

$$D_{\alpha\alpha}(\mathbf{r}_{1'}, \mathbf{r}_1) = \sum_{k,l} \phi_k(\mathbf{r}_{1'})\phi_l(\mathbf{r}_1)\, d_{lk}$$

where the sum is over the $\{\phi_i\}$ associated with α-spin in forming the $\{\lambda_i\}$. We may diagonalize $\mathbf{d}$ as in Eq. (115) and derive natural orbitals (spinless). We now return to the case where spin has not been integrated out.

Since γ_{kl} is Hermitian, the n_m must be real. Let us compute

$$\gamma_{kk} = \int dx_1 dx_{1'} \lambda_k(1')\Gamma^{(1)}(1' \mid 1)\lambda_k^*(1). \tag{119}$$

Only diagonal terms ($K = L$) contribute to γ_{kk} [see (113)–(114)]

$$\gamma_{kk} = \sum_{K}^{(k)} \mid C_K \mid^2 \Big/ \sum_{K} \mid C_K \mid^2 \tag{120}$$

where the first sum is over configurations including λ_k. Equation (120) shows $0 \leq \gamma_{kk} \leq 1$ whenever the basis is orthonormal. The χ_k are orthonormal, since they are obtained by a unitary transformation of the orthonormal λ_K. Here, n_m is the diagonal element (off-diagonal elements are zero), so $0 \leq n_m \leq 1$. The value of n_m can be 1 only when all configurations include χ_m. If all n_m are unity, we have a single determinant. We can say that the effect of correlation is to depress the occupation numbers of the natural orbitals below unity.

The value of the use of natural orbitals can be seen in a single example. If in a single-determinant wave function each orbital were expanded in some basis set, we would have something looking like a CI function, and its first-order matrix would look like (114). By finding the natural orbitals and occupation numbers, we would know that, in reality, the function was equivalent to a single determinant and we should be able to write this "CI function" in an extremely concise form. Similarly, if only a finite number, say M, if the n_m are nonzero, the natural orbital expansion is finite, and only a finite number of determinants of natural orbitals, $M!/(N![M - N]!)$, are needed.

It would seem that something similar obtains even when such a radical simplification is not possible, i.e., the natural orbitals should allow a contraction in the number of determinants needed. For the configuration expansion (115), where the determinants Φ_k are formed from the orthonormal spin orbitals $\{\lambda_k\}$, we have $0 \leq \gamma_{kk} \leq 1$, with more rapid convergence possible when more of the γ_{kk} are 0 or 1. Löwdin took [59, 65] the closeness of the nonnegative quantity

$$\Theta = N^{-1} \sum_k \gamma_{kk}(1 - \gamma_{kk}) \tag{121}$$

to zero as a measure of the quality of the set $\{\lambda_k\}$ in expressing Ψ concisely. Since $\sum_k \gamma_{kk} = N$, $\Theta - 1 = -N^{-1} \sum_k \gamma_{kk}^2$, i.e., $\Theta < 1$. The best basis functions are those which maximize $\sum_k \gamma_{kk}^2$. From (115b)

$$\sum_{k,l} \gamma_{kl}\gamma_{lk} = \sum_{k,m} U_{km}n_m{}^2 U_{mk}^\dagger = \sum_m n_m{}^2 \tag{122}$$

and the first member is $\sum_k \gamma_{kk}^2$ plus other nonnegative terms. Therefore $\sum_k \gamma_{kk}^2 \leq \sum_m n_m{}^2$, and the natural spin orbitals minimize Θ for a given Ψ.

The use of natural orbitals to speed up configuration interaction calculations was exploited by Bender and Davidson [73]. Given a CI function, we can find natural orbitals and express the CI function in terms of their determinants. By not including the natural orbitals with very small occupation numbers, we can cut down markedly on the number of determinants without changing the wave function much. We have essentially the same function, but in a more concise form. The procedure of Bender and Davidson [73] was first to guess a set of molecular orbitals, from which they constructed a CI wave function (not using all possible configurations). For this function, they calculated the natural orbitals, and used those with highest occupancy to construct a new configuration interaction function. Then new natural orbitals were obtained and the procedure was repeated until successive iterations changed nothing. The final natural orbitals should resemble those for the full CI, which need never be carried out. If all the molecular orbitals are expanded in some fixed basis set, as is generally the case for molecular calculations, all the natural orbitals will also be within this subspace. Note that once the integrals of the one- and two-electron parts of H have been calculated over the basis set, no more integrals need be calculated. At each stage in the iteration, it is necessary only to transform to a new basis of orbitals, which are linear combinations of the initial basis set. Bender and Davidson used 50 configurations at each stage of the iteration. The natural orbitals resembled the occupied SCF orbitals, but not the virtual ones, in accordance with our preceding discussion. For the starting molecular orbitals, it was best to solve the Roothaan equations. "Correlation orbitals," chosen to diagonalize the SCF exchange integral operator were added to the occupied Roothaan orbitals. The full set of orbitals was orthonormal.

Calculations were performed by Bender and Davidson [73] for HeH at $3a_0$ and for LiH at $3.015a_0$, and later [74] for HF at $1.7328a_0$. The energies for HeH were significantly better than those from any other method. For HeH, the energy was within 0.002 a.u. of experiment, while the slope dE/dR agreed with $R^{-1}(T + E)$ (see Section C) to several figures. For LiH, the role of the natural orbitals with respect to in–out, left–right, and angular correlation was discussed. In this case, a 16-configuration natural orbital function gave as good on energy as a previous 28-configuration function. The LiH energy differed from experiment by 0.01 a.u. The HF calculation gave 80% of the difference between exact and SCF energies.

The possibility of doing something more radical suggests itself. We may write a configuration interaction wave function in terms of the natural orbitals, and derive equations to determine simultaneously the coefficients of the determinants and the expansion coefficients of the natural orbitals in the basis functions. Such equations are discussed by several authors [65, 75]. Their complexity suggests that it may be easier to carry out CI by the usual methods and extract natural orbitals, if desired, subsequently. We must note that natural orbitals may be derived for any wave function, not necessarily of configuration interaction form. Indeed, one advantage of the natural orbital expansion is that it is independent of the form of the initial wave function. Any wave function can be analyzed in terms of its natural orbitals and their occupation numbers, and this facilitates comparison of functions of different form. Much work of this nature has been done [76, 76a, b].

Without invoking natural orbitals, we can write a configuration interaction wave function and demand that its energy be stationary with respect to variation of both the orbitals and the coefficients. This would be a generalization of the Hartree–Fock idea (one determinant). Methods of this type are referred to as multiconfiguration SCF or as extended Hartree–Fock theories. A simpler way of proceeding would be to derive equations for the orbitals, the coefficients being fixed. With the best orbitals, the CI may be carried out, giving a new set of coefficients. Repeating the process until the coefficients and orbitals do not change is a sort of super-SCF method, and may be practical in some cases.

The general formalism in which both coefficients and spin orbitals are to be determined is quite complicated. If the variational principle is invoked, one may expect that certain matrix elements vanish between the CI function and other linear combinations of determinants. These relations may be referred to [69] as generalized Brillouin theorems. They can be used in setting up the effective potential for determination of the spin orbitals. As in the case of the Hartree–Fock function itself, the Brillouin theorem implies that the Hellmann–Feynman theorem holds [77]. Then (see Section D), if u is a parameter entering the Hamiltonian, the derivative of the expectation values of the Hamiltonian with respect to u is equal to the expectation value of the derivative. The consequence of this is that the change in the energy error due to a change in such a parameter (like the internuclear distance) is likely to be small. Potential energy curves should thus be parallel to the true ones. This is not always true for arbitrary CI wave functions [78].

An important special case of multiconfiguration SCF, investigated by Veillard [79], is that of two configurations, when a near-degeneracy makes one-configuration results inaccurate (as in the ns^2np^N and np^{N+2} atomic situation). The results were extended [79] to the case of many configurations, where one determinant corresponded to a doubly occupied closed-shell wave function and the others were derived from it by excitation of two electrons at a time. Since the excited and the ground-state orbitals were optimized simultaneously, the energy of the first configuration would be higher than the Hartree–Fock energy, but the multiconfiguration function would have a lower energy than the function formed from the Hartree–Fock and double excitations. The equations to be solved were coupled pseudoeigenvalue equations, most easily handled by expanding the orbitals in some basis set, and solving for the expansion coefficients and the mixing coefficients for the configurations.

We have noted above that, if one starts from a Hartree–Fock function, the coefficients of singly excited configurations will be small. This suggests that a good approximate wave function is

$$\Psi = \Psi^{\mathrm{HF}} + \sum C_k \Phi_k \tag{123}$$

where the Φ_k are all doubly excited configurations. Nesbet [71, 80] proposed a trial function of this form, using Hartree–Fock orbitals, and derived the equations for the coefficients. Four-particle excitations, which become important for large systems, are not included in (123), which describes correlations between electrons of parallel spin as well as those paired. There is a close relation between (123) and correlated-pair functions (see next section) when these are expanded in an orbital basis. Going a step further, Nesbet [71] has proposed the use of equations which determine pair functions for two electrons at a time moving in the average field of the other electrons.

Multiconfiguration SCF calculations [81] are becoming more common. For example, Wahl and Das [82] have developed such a formalism and implemented it with a number of calculations (see Section B of Chapter II, Volume 2).

At the same time, progress has been made in the development of algorithms for the solution of very large secular equations, taking advantage of two simplifications which appear in the calculations of interest to us here: (a) the matrix is real and symmetrical and (b) the lowest root is being sought. For discussion of such schemes, we refer to the articles of Nesbet and Shavitt [83].

8. *Pair Functions and Correlated Wave Functions*

The importance—to both the molecule and its calculator—of formation of electron pairs was mentioned in the preceding discussion and is also substantiated by much empirical and theoretical evidence. The concept of electron pairing has, of course, long played an important role in theoretical chemistry, and arguments from perturbation theory (see Section B.3) show why pair structure should be important. It has already been pointed out that electrons of parallel spin are automatically kept apart by the Pauli principle which is built into an antisymmetric wave function (Fermi hole) so that their electrostatic interaction is decreased. The problem is with electrons of opposite spin, which also should avoid each other to minimize interelectronic repulsion (Coulomb hole). It may be conjectured that *pairing of electrons of opposite spins* uses the Fermi hole to produce the Coulomb hole. Consider the interaction of a pair of electrons of opposite spins, localized in some region of space, with a third electron. The exclusion principle will keep the third electron away from the electron in the pair having the same spin, and hence automatically from the other electron. This may help to explain the successes of the restricted Hartree–Fock theory. In a more accurate theory, the two electrons in the pair, while occupying on the average the same region of space, must avoid each other. Their motions must be correlated so that the position of one depends strongly on the instantaneous position of the other, and this is beyond the power of a Hartree–Fock function.

It must be noted that one cannot speak unambiguously of intrapair and interpair contributions to the correlation energy. The former will be more important relative to the latter when the orbitals are spatially localized. But such a localization can be performed without changing the wave function, since the choice of the orbitals is arbitrary to the extent of a similarity transformation of orbitals of each spin. The correlation energy is invariant under this transformation, but intrapair and interpair contributions change.

It may be expected that a wave function which considers the electrons by pairs and describes well the correlation within each pair of electrons would be an accurate description of the system. Instead of building the wave function from one-electron functions (spin orbitals) one uses two-electron or pair functions, often referred to as geminals. One would like to deal with a set of two-electron problems, and solve for the geminals, but the coupling between these two-electron problems can never be rigorously removed. Here we will be concerned with variational calculations of such wave functions and their application to diatomic molecules.

If the energy really could be written as a sum of pair contributions one could deal with a set of two-electron problems. In 1951, Lennard-Jones and Pople [84] formulated a treatment for paired electrons within molecules. Correlation was introduced between, for instance, the motions of the two electrons in a bond. It was assumed that the orbitals had been chosen to represent bonds, atomic orbitals, etc. If these were localized in space, one could, in a first approximation, neglect the details of the Coulombic interactions between an electron in one orbital and the electrons in other orbitals which are mainly in other regions of space. Then one could consider two electrons in the same orbital. For a bond in a homonuclear diatomic molecule, this is essentially the hydrogen molecule problem with effective nuclear charges. The effective nuclear charges express the screening of the nuclei by the other electrons. Lennard-Jones and Pople [84] wrote down the Schrödinger equation, separated off a spin singlet factor, and considered successively better approximations to the exact solution, starting with one of molecular orbital (i.e., product function) form.

For the homonuclear case, this meant

$$\Psi^{(1)} = f_{s,0}(1)f_{s,0}(2) \tag{124}$$

where $f_{s,0}$ is symmetrical to inversion. The exact two-electron function must have also a part antisymmetric to inversion of one electron, as was seen from the expansion of $1/r_{12}$ in bicentric coordinates. Then $\Psi^{(1)}$ became

$$\Psi^{(2)} = f_{s,0}(1)f_{s,0}(2) - f_{a,0}(1)f_{a,0}(2), \tag{125}$$

which is a configuration interaction and gives correlation. This can be seen by rewriting $\Psi^{(2)}$ in terms of nonsymmetric orbitals

$$\chi_A = 2^{-1/2}(f_{s,0} + f_{a,0}); \qquad \chi_B = 2^{-1/2}(f_{s,0} - f_{a,0}).$$

We recover

$$\Psi^{(2)} = \chi_A(1)\chi_B(2) + \chi_B(1)\chi_A(2)$$

which, if the χ_A are essentially atomic orbitals, is the valence bond function. Thus $\Psi^{(2)}$ tends to keep the electrons on different atoms, which may be referred to as left–right condition. Angular correlation, that is with respect to the azimuthal angle Φ around the figure axis, was next included. The form of the $1/r_{12}$ operator requires that the exact solution include terms depending on the angles ϕ_1 and ϕ_2 for the two electrons by way of $\cos[\nu(\phi_2 - \phi_1)]$. Thus the two-electron wave function could now be written

$$\Psi^{(3)} = \sum_{\nu=0}^{\infty} \{f_{s,\nu}(1)f_{s,\nu}(2) - f_{a,\nu}(1)f_{a,\nu}(2)\} \cos[\nu(\phi_2 - \phi_1)] \qquad (126)$$

The terms for $\nu \neq 0$ are largest when $\phi_2 - \phi_1 = \pi$, the electrons then being on opposite sides of the figure axis.

Calculations corresponding to $\Psi^{(1)}$, $\Psi^{(2)}$, and $\Psi^{(3)}$ were carried out by Lennard-Jones and Pople [84] for H_2. The improvement in going from $\Psi^{(1)}$ to $\Psi^{(2)}$ was substantial, amounting to $\frac{1}{6}$ the total energy. The angular correlation gave an additional improvement of almost $\frac{1}{3}$ that gained by using $\Psi^{(2)}$. At this point, it was suggested that an explicit dependence on r_{12} be introduced instead of on $(\phi_2 - \phi_1)$. Explicit use of r_{12} leads to a wave function which can be written [85, 85a] as an expansion like (126) since in ellipsoidal coordinates

$$r_{12} = \tfrac{1}{2}R[Q - S\cos(\phi_2 - \phi_1)]^{1/2}$$

with

$$Q = \lambda_1^2 + \lambda_2^2 + \mu_1^2 + \mu_2^2 - 2 - 2\lambda_1\lambda_2\mu_1\mu_2$$
$$S = 2[(\lambda_1^2 - 1)(\lambda_2^2 - 1)(1 - \mu_1^2)(1 - \mu_2^2)]^{1/2}.$$

One could substitute, for $\Psi^{(3)}$, $\Psi^{(1)}$ multiplied by a power series in r_{12}. All kinds of correlation would be taken care of by this.

We will use the term "correlated wave function" to mean one in which one or more interelectronic distances enter explicitly. The power of correlated wave functions was shown by Hylleraas' work [86] on helium and that of James and Coolidge [87] on the hydrogen molecule. A comparison between correlated wave functions and configuration interaction wave functions in perturbation calculations for He, together with expansions sharing some characteristics of both, has recently been given by Byron and Joachain [88].

The use of r_{12} is convenient to describe the way in which the position of electron 1 depends on the instantaneous position of electron 2. This dependence is sometimes referred to as short-range correlation. Schwartz

[88a] has also indicated reasons for the superiority of correlated functions to configuration interaction. Explicit dependence on the r_{ij} is necessary to eliminate poles in the Hamiltonian [89–91], and gives the correct cusp behavior of the wave function in regions of configuration space corresponding to small r_{ij} (see our discussion in Section E.1). The cusp leads to an infinite kinetic energy contribution for $r_{ij} \to 0$, which cancels the singularity in the interelectronic repulsion. Use of a sum of configurations to approximate the cusp could lead to slow convergence, since a discontinuity in slope must be built up from functions continuous in r_{ij}. However, the idea that the superior convergence of energy with correlated wavefunctions over CI for He and H_2 is due to the cusp and the consequent correct description of short-range correlation has been challenged by Gilbert [92]. He argued that the volume of configuration space involved is small, so that the regions of small r_{ij} can contribute but little to the correlation energy, and emphasized the shape of the Coulomb hole, i.e., the difference between the pair distribution functions for Hartree–Fock and "exact" wave functions. The size of the Coulomb hole is several a_0, but it can be described simply as a function of r_{12} (rapidly converging power series), and not in terms of space-fixed coordinates. The power of correlated wave functions was ascribed to this [92].

It is to be expected that correlated wave functions would give good approximations to the exact wave function for atoms and molecules with more than two electrons. There were some early attempts to use correlated functions; for example, for Li by James and Coolidge [93] and for atoms like Be by Fock et al. [94]. The difficulty in using a correlated function for a multielectronic system is that the evaluation of the expectation value of the energy leads to expressions which cannot be reduced to one- and two-electron integrals. For instance, if Ψ depends on r_{12} the expectation value of the Hamiltonian will involve $\int |\Psi|^2 r_{13}^{-1}\, d\tau$ which involves at least three electrons simultaneously. Integrals involving more than two electrons are difficult to calculate accurately and efficiently for atoms, and almost intractable for molecules, at present. It may be noted that the variational method is notorious for the way in which "small" errors in integrals may grow into catastrophes in the final result. The problem is thus to arrange the calculations so that the multielectron integrals do not arise, or arise in a harmless way, or in small number.

Several stratagems may be employed. First, the interelectronic distances or their inverses can be expanded, in the Neumann or a related series [85a], as sums of products of functions of one electron. The troublesome integral becomes a sum of products of two-electron integrals. The amount of work

that must be done now depends on how easily we can evaluate these and/or
how many of them we need for the desired accuracy (convergence of the
series). In fact, the wave function is really of the configuration interaction
type with the use of correlated pair functions being used as a device to
select important configurations and to group configurations together. Second, one can use some technique which does not require the ability to
accurately evaluate integrals, like a local energy method (Section E). The
perturbation theory approach is a third possible way out. As we discuss in
Section B.3, one may start with a single determinant and calculate correlation corrections to it in terms of pair functions, which may themselves be
correlated, without dealing with multielectron integrals. Three-electron
integrals are needed for the second-order energy, but small errors here do
not lead to serious difficulties. Finally, one could attempt to arrange a
correlated trial function so that three- and four-electron integrals do not
enter, or enter in a relatively harmless way. It is with methods which do
this that we shall be mainly concerned for the rest of this section.

An important simplification in this regard is obtained if the pair functions
are assumed to be strongly orthogonal. Strong orthogonality means that
the product of two functions with one common argument, integrated over
this single argument, gives zero. The meaning of the strong orthogonality
condition has been discussed by Löwdin [95]. He considered functions
$F_1(1, 2)$, and $F_2(1, 2, \ldots, p)$ which were strongly orthogonal:

$$\int d\tau_1 F_1^*(1, 2)F_2(1, 2', 3', \ldots, p) = 0.$$

The pair function F_1 could be either symmetric or antisymmetric. If it is
imagined to be expanded in some set of one-electron functions Ψ_k,

$$F_1(1, 2) = \sum_{k,l} \Psi_k(1)\mathbf{A}_{kl}\Psi_k^*(2),$$

the symmetry or antisymmetry implies $\mathbf{A}^\dagger = \mathbf{A}$ or $\mathbf{A}^\dagger = -\mathbf{A}$. In either case,
$\mathbf{A}$ may be diagonalized by a unitary transformation. Applying this trans-
formation to the basis set we obtain the natural orbitals χ_k, and the expansion is

$$F(1, 2) = \sum_k \chi_k(1)a_k\chi_k^*(2).$$

Let F_2 be expanded in the χ_k (the expansion coefficients are functions of
$p - 1$ electronic coordinates). Then

$$\int d\tau_1 F_1^*(1, 2)F_2(1, 2', 3', \ldots, p') = \sum_k a_k^* C_k(2', 3', \ldots, p')$$

where

$$C_k(2', 3', \ldots, p') = \int \chi_k{}^*(1) F_2(1, 2', \ldots, p') \, d\tau_1.$$

Thus, strong orthogonality implies $C_k = 0$ whenever a_k is nonzero. The Hilbert space is divided into two subspaces, corresponding to the occupied and unoccupied natural orbitals of F_1, with F_2 belonging wholly to the latter. If there are two real pair functions which are strongly orthogonal, their natural orbitals span two mutually orthogonal subspaces. If one has several pair functions, all mutually strongly orthogonal, the restrictions on the natural orbitals involved may be of great importance in keeping the wave function from approaching the exact.

However, strong orthogonality leads to a great simplification of the energy formulas, so that it is often employed. For example, the early formalisms of Hurley *et al.* [96] and of Fock *et al.* [94] used such a constraint. McWeeny and Steiner [97] have discussed many of the properties of wave functions built from strongly orthogonal parts, and reviewed work involving them. When one turns to actual integral evaluation, the strong orthogonality requirement may be troublesome (see Section B.3): if we obtain strongly orthogonal pairs by an orthogonalization process, formally two-electron integrals, involving the pair function, may become in reality three-electron [98].

Many of the methods using pair functions relate to Sinanoğlu's many-electron theory. Sinanoğlu* has generalized the results of perturbation theory to higher orders to give a "cluster expansion" for the exact wave function. From this, he has formulated a theory of correlation energies for atoms and molecules, the concepts of which have enjoyed much success. A review article [99] presents the developments of this theory up to 1964, and also discusses other many-particle theories and other attempts to include correlation in the wave function. Stanton [100] has given a lucid and careful discussion and review of this formalism.

If the exact wave function is written in the configuration expansion, one may choose a single determinant (hopefully, an approximation to the exact wave function) and classify all the other determinants according to the spin orbitals in which they differ from this reference determinant. Then one may consider together, for instance, all configurations in which λ_i and λ_j are replaced by other spin orbitals. This sum of determinants may be

---

* For references see Sinanoğlu [99], Stanton [100], Sinanoğlu [101], and Silverstone and Sinanoğlu [102]. Related discussion will be found on pp. 174–177.

written as

$$\mathscr{A}\left\{\frac{u_{ij}(i,j)}{\lambda_i(i)\lambda_j(j)}\prod_k\lambda_k(k)\right\} = U_{ij} \tag{127}$$

where $u_{ij}(i,j)$ is now an "exact pair" function or two-electron cluster. The exact function is then written as a sum of contributions of clusters of various number of electrons. When certain strong orthogonality constraints (see below) are introduced, the cluster functions are uniquely defined. Further, the clusters have "linked" and "unlinked" parts: the four-electron cluster u_{ijkl}, which enters the exact wave function in the term

$$U_{ijkl} = \mathscr{A}\left\{\frac{u_{ijkl}(i,j,k,l)\prod_m\lambda_m(m)}{\lambda_i(i)\lambda_j(j)\lambda_k(k)\lambda_l(l)}\right\}$$

includes terms like $u_{ij}u_{kl}$, where u_{ij} is the two-electron cluster of (127). It may be argued, partly from perturbation theory and partly on physical grounds [101–103], that, when the reference determinant is the Hartree–Fock, the two-electron clusters are of predominant importance. For instance, u_{ijkl} is almost exclusively due to the unlinked contribution of two such two-electron clusters, as can be shown from the results of configuration interaction calculations. Physically, this corresponds to simultaneous "collisions" of two pairs of electrons. Also, one-electron and three-electron clusters play a relatively small role.

If indeed such clusters are unimportant, one could consider a variational wave function which included the u_{ij} only. [See Eq. (123) above.] The energy correction to the Hartree–Fock energy breaks down into pair contributions plus many-electron terms, which one can expect to be very small. Neglecting the many-electron terms, each of the pair contributions may be separately minimized, yielding two-electron equations for the u_{ij}. We may treat these equations like any two-electron problems, using, for example, correlated wave functions. Actually, the nonlocal nature of the Hartree–Fock Hamiltonian, plus orthogonality constraints, mean the problems are formally but not computationally two-electron: Three-electron integrals come in [103].

Once the wave function is determined, the contribution of the other clusters, and the terms neglected in the theory, can be computed. Few *ab initio* calculations have been performed by this method. The paper of Geller *et al.* [103] is important in this regard. They show that, while good results can be obtained for the correlation energy of a four-electron system (Be atom), the many-electron terms are not negligible. Their importance grows with the number of electrons (to perhaps 20% of the correlation energy for Ne atom), so their neglect in deriving the pair equations variationally is unjustified. This means that the energies obtained are not bounded by the exact energy from below. (Other workers [104] have likewise inserted explicit correlation into wave functions without protection of the variational principle; physical intuition serves here.) Finally, Geller *et al.* [103] also note that, where one demands a wavefunction which is an eigenfunction of S^2 and other angular momentum operators, the pair equations may be coupled.

As was done by Sinanoğlu, Szász [105] grouped the terms of a configuration interaction wave function so as to write the exact wave function as a reference determinant plus corrections which were one-electron, two-electron, ... N-electron in nature. He proposed that a good approximate wave function could be the Hartree–Fock function plus terms corresponding to doubly excited configurations as in Eq. (123). In terms of pair functions, (123) is written

$$\Psi = \Psi^{\text{HF}} + \sum_{i,j} U_{ij}$$

and the pair functions are to be calculated from the variational principle. If U_{ij} is written as the two-electron function

$$\lambda_i(i)\lambda_j(j) - \lambda_i(j)\lambda_j(i)$$

multiplied by a function of r_{ij}, the wave function may be written

$$\Psi = \mathscr{A}\left\{\prod_i \lambda_i(i)\left[1 + \sum_{j,k}^{i<k} W_{jk}(j,k)\right]\right\} \tag{128}$$

where W_{jk}, the correlation factor, involves r_{jk}. Szász showed [106] that all integrals in the energy expression for this wave function can be reduced to two- and three-electron integrals, and that these were tractable for atomic problems. On physical grounds one would not expect to need $\frac{1}{2}N(N-1)$ correlation factors for the N-electron problem. Thus, for Be, only the two intrashell correlations were included [106], the wave function being taken as

$$\mathscr{A}\{\lambda_1(1)\lambda_2(2)\lambda_3(3)\lambda_4(4)[1 + c_1 r_{12} + c_2 r_{34}]\}$$

with λ_i the previously determined Hartree–Fock orbitals for Be. Szász and Byrne [107] give a review of the analysis and summarize results of calculations by this method.

The separated pair ideas have been applied to diatomic molecules, using experience gained from extensive work on atoms, references to which will be found in the articles to be cited later. Schaefer [108] has argued, however, that the treatment of correlation in terms of pair contributions should work less well for molecules than for atoms. Much work on molecules has concentrated on LiH, which is similar to Be (its united atom) in pair structure.

In LiH, there are two pairs whose significance is evident: One is the inner-shell core of Li and the other the pair of electrons forming the bond. The four-electron wave function may be written

$$\Phi = \mathscr{A}\{\Phi_A(1, 2)\Phi_B(3, 4)\} \tag{129}$$

where Φ_A and Φ_B are usually taken to be products of symmetric spatial functions u_A, u_B with singlet spin functions. The strong orthogonality between the pair functions demands strong orthogonality between the spatial functions themselves. The normalization condition

$$\langle \Phi_A(1, 2) \mid \Phi_A(1, 2)\rangle = \langle \Phi_B(1, 2) \mid \Phi_B(1, 2)\rangle = 1$$

implies

$$\langle u_A \mid u_A \rangle = \langle u_B \mid u_B \rangle = 1.$$

The energy for the wave function of Eq. (129) consists of three parts, corresponding to Φ_A, Φ_B, and their interaction:

$$E = E_A + E_B + E_{AB} \tag{130}$$

Here,

$$E_A = \int u_A{}^*(1, 2)(h(1) + h(2) + r_{12}^{-1})u_A(1, 2) \, d\tau_1 \, d\tau_2 \tag{131a}$$

$$E_B = \int u_B{}^*(3, 4)[h(3) + h(4) + r_{34}^{-1}]u_B(3, 4) \, d\tau_1 \, d\tau_2 \tag{131b}$$

and

$$E_{AB} = \int u_A{}^*(1, 2)u_B{}^*(3, 4)(4 - 2\mathscr{P}_{13})r_{13}^{-1}u_A(1, 2)u_B(3, 4) \, d\tau_1 \, d\tau_2 \, d\tau_3 \, d\tau_4. \tag{131c}$$

It may be shown in general [97, 109] that the total electronic energy for a system described by strongly orthogonal geminals is a sum of energies of the individual geminals plus interaction terms which can be written as Coulomb and exchange integrals over the natural orbitals of the geminals involved.

An important question to be dealt with is the effect on the energy of imposing the strong orthogonality constraint. This was investigated by Csizmadia *et al.* [110], who built both pair functions from numerically determined Hartree–Fock atomic orbitals for H–$1s$, Li–$1s$, Li–$2s$, and Li–$2p$. The core was described by the unchanged Li–$1s$ orbitals:

$$\Phi_{\mathrm{A}}(1, 2) = \Phi^{\mathrm{Li}}_{1s}(1)\Phi^{\mathrm{Li}}_{1s}(2)\mathscr{S}(1, 2)$$

and the bonding pair function was built from the other orbitals. For strong orthogonality, it was necessary only to orthogonalize ϕ^{H}_{1s} to ϕ^{Li}_{2s}. In the simplest approximation, the spatial part of Φ_{B} was taken as $u_{\mathrm{B}} = \phi_{\mathrm{B}}(1)\phi_{\mathrm{B}}(2)$, a product of identical one-electron functions: then, Eq. (129) actually is a single determinant. The one-electron function was a linear combination of the three orbitals orthogonal to ϕ^{Li}_{1s}:

$$\phi_{\mathrm{B}} = c_1\phi^{\mathrm{Li}}_{2s} + c_2\phi^{\mathrm{Li}}_{2p} + c_3\phi^{\mathrm{Li}}_{1s}.$$

The energy was minimized with respect to two independent coefficients (since ϕ_{B} was normalized), and the results were very close to those obtained from an SCF (restricted Hartree–Fock) calculation using these basis functions. Thus the strong orthogonality constraint does not constitute a harmful restriction in this case. The calculation, which is simpler than the SCF calculation, also served to test the formalism, which is in terms of density matrices and can be generalized to more complicated pair functions.

The first improvement to the simple wave function was the addition of pair functions in which Φ_{A} is unchanged but Φ_{B} is constructed from other linear combinations of the orbitals in ϕ_{B}. The resulting wave function is still in the form (129), but the spatial part of Φ_{B} may be written as a sum of products of orbitals of the form of ϕ_{B}. Φ_{B} now corresponds to $\Psi^{(2)}$ of (125).

Ebbing and Henderson [76a] gave a more general trial function of the from (129). The spatial parts of Φ_{A} and Φ_{B} were written in their natural expansions:

$$u_{\mathrm{A}}(1, 2) = \sum_k a_k{}^{\mathrm{A}}\chi_k{}^{\mathrm{A}}(1)\chi_k{}^{\mathrm{A}}(2) \tag{132a}$$

$$u_{\mathrm{B}}(3, 4) = \sum_k a_k{}^{\mathrm{B}}\chi_k{}^{\mathrm{B}}(B)\chi_k{}^{\mathrm{B}}(4). \tag{132b}$$

(The expansion of an antisymmetric two-electron function in its natural spin orbitals may be written, if the function is a singlet, as an expansion of the form (132) times the singlet spin function. For a triplet, related choices of the natural spin orbitals can be used [109] to write the function as a triplet spin function times a sum of antisymmetric space functions.) Strong orthogonality means $\{\chi_k{}^{\mathrm{A}}\}$ and $\{\chi_k{}^{\mathrm{B}}\}$ form disjoint sets. When the natural orbitals are given, minimization of the energy with respect to the coefficients in (132) gives two coupled secular equations. After determining the occupation coefficients in this way, one must vary the orbitals themselves, while keeping the coefficients fixed, to minimize the energy. Following this, new coefficients, corresponding to the new natural orbitals, must be computed.

Ebbing and Henderson [76a] constructed the $\{\chi_k^A\}$ and the $\{\chi_k^B\}$ as linear combinations of seven orthogonal molecular orbitals of σ symmetry, including the two SCF orbitals. It turned out that the lowest natural orbital of Φ_A was mostly the lower SCF orbital and that for Φ_B the other SCF orbital. The results of the natural orbital calculation were compared to those of a configuration interaction calculation which allowed some single and double excitations: those maintaining double occupation of either the inner or the outer molecular orbital. The strong orthogonality constraint for LiH decreased the calculated correlation energy by only 2%. It was admitted that LiH is a favorable case.

Silver et al. [109] have recently reviewed the formalism of separated pairs with natural orbital expansion of the geminals, and derived the equations determining the natural orbitals and occupation coefficients. The natural spin orbitals of all the geminals satisfy a single integrodifferential equation of the pseudoeigenvalue type, with the operator involving all the natural orbitals. An iterative method, for both orbitals and coefficients, is needed. Some results of these calculations are discussed in Sect. C of Vol. 2, Ch. II.

In some cases, only one electron pair is of interest, allowing for considerable simplification. Thus Fock et al. [94] were able to carry through a calculation correlating the pair of valence electrons of Be at an early date. This could apply to a diatomic, where we want to discuss the pair of electrons in a bond. We will show some of the analysis for this. Let us write the wave function as

$$\Psi = \mathscr{A}\{\lambda_1(1) \cdots \lambda_N(N)\Phi(N + 1, N + 2)\} \tag{133}$$

where the λ_i $(i = 1, \ldots, N)$ are Hartree–Fock spin orbitals describing the core electrons and Φ, assumed strongly orthogonal to the λ_i, describes the valence electrons. We may choose Φ antisymmetric: $\Phi(1, 2) = -\Phi(2, 1)$. If the core consists of two electrons this is a special case of (129). The energy of this wave function is

$$\langle \Psi \mid H \mid \Psi \rangle = \left\langle \prod_{i=1}^{N} [\lambda_i(i)]\Phi(N + 1, N + 2) \,\middle|\, \sum_{k<l} g_{kl} + \sum_k h(k) \,\middle|\, \right.$$
$$\left. \times \sum_P \delta_P P \prod_{i=1}^{N} [\lambda_i(i)]\Phi(N + 1, N + 2) \right\rangle \tag{134}$$

where $g_{kl} = 1/r_{kl}$ and $h(k)$ is the one-electron Hamiltonian (kinetic and nuclear attraction energies) for electron k. It is straightforward to reduce this expression, taking advantage of the strong orthogonality, to

$$
\begin{aligned}
\langle \Psi \mid H \mid \Psi \rangle = CE_c + 2\langle \Phi(N+1, N+2) \mid & h(N+1) + h(N+2) \\
& + g_{N+1,N+2} \mid \Phi(N+1, N+2) \rangle \\
& + 4 \sum_{k=1}^{N} \langle \lambda_k(k)\Phi(N+1, N+2) \mid g_{k,N+1} \mid \lambda_k(k)\Phi(N+1, N+2) \\
& \quad - \lambda_k(N+1)\Phi(k, N+2) \rangle
\end{aligned}
\tag{135}
$$

Here, $C = 2\langle \Phi \mid \Phi \rangle$ and E_c is the core energy. The last term in (135), which is the effect of the core on the valence electron pair, may be written

$$
\begin{aligned}
4\langle \Phi(N+1, N+2) \mid \mathscr{F}(N+1) \mid \Phi(N+1, N+2) \rangle \\
= 2\langle \Phi(N+1, N+2) \mid \mathscr{F}(N+1) + \mathscr{F}(N+2) \mid \Phi(N+1, N+2) \rangle
\end{aligned}
$$

where

$$
\mathscr{F}(j) = \sum_{k=1}^{N} \lambda_k(k) r_{kj}^{-1} (1 - \mathscr{P}_{kj}) \lambda_k(k).
\tag{136}
$$

Combining (136) with the one-electron Hamiltonian $h(j)$ to get the Hartree–Fock Hamiltonian $h^{HF}(j)$ [Eq. (64) and (65)], we have

$$
\begin{aligned}
\langle \Psi \mid H \mid \Psi \rangle = CE_c + 2\langle \Phi(N+1, N+2) \mid & h^{HF}(N+1) + h^{HF}(N+2) \\
& + g_{N+1,N+2} \mid \Phi(N+1, N+2) \rangle.
\end{aligned}
\tag{137}
$$

To minimize E_c, the λ_i should be the Hartree–Fock functions. To minimize the rest of (137) with λ_i fixed, we must also require the second term to be stationary with respect to changes in Φ keeping C constant. But this means simply that $\Phi(1, 2)$ should be an eigenfunction of $h^{HF}(1) + h^{HF}(2) + 1/r_{12}$.

Now if one considers a function $\bar{\Phi}$ which is not strongly orthogonal to the core spin orbitals, one could make it so using a projection operator $\mathscr{P}$, as in Eqs. (174)–(175). We require then that $\mathscr{P}\bar{\Phi}(1, 2)$ be an eigenfunction of $h^{HF}(1) + h^{HF}(2) + 1/r_{12}$. But using $\bar{\Phi}$ for Φ in (133) changes nothing: Any contribution of λ_i will drop out. Thus, the best pair function, irrespective of strong orthogonality conditions, is that which minimizes the expectation value of the projected Hamiltonian

$$
H_M = \mathscr{P}[h^{HF}(1) + h^{HF}(2) + 1/r_{12}]\mathscr{P}
$$

keeping $\langle \bar{\Phi} \mid \mathscr{P}\mathscr{P} \mid \bar{\Phi} \rangle$ constant. Since the projected Hamiltonian appearing in the expectation value is formally multi-electron, no simplification has been accomplished by this formal manipulation. However, Szász et al. [98] have made the following suggestion. If $\bar{\Phi}$ is a simple product of orbitals, the Hamiltonian H_M may be written down explicitly. They argue that this Hamiltonian, since it includes the effects of orthogonalization, should also be roughly correct when $\bar{\Phi}$ is of a more general form. This Hamiltonian is, while long to write down, truly two-electron. One then should seek eigenfunctions of it, in the form of explicitly correlated wavefunctions. This conjecture, like others mentioned in this section, forms a current topic of investigation.

Bender et al. [111] have discussed the problems in the application of geminal or pair methods to molecular calculations. They suggested that approximations which would simplify the calculations could be introduced, provided the geminals were carefully chosen and sufficient testing of the consequences was done for a variety of systems. The implementation of such a program seems to be possible with present computing methods. Some results of pair calculations are discussed in Chapter II, Vol. 2.

REFERENCES

1. J. Goodisman and W. Klemperer, *J. Chem. Phys.* **38**, 721 (1963); **46**, 4552 (1967).
2. C. Eckart, *Phys. Rev.* **36**, 878 (1930).
3. H. Conroy, *J. Chem. Phys.* **47**, 930 (1967).
4. F. Weinhold, *J. Chem. Phys.* **46**, 2448 (1969).
5. H. F. Weinberger, *J. Res. Nat. Bur. Stand. Sect. B* **64**, 217 (1960).
6. D. H. Weinstein, *Proc. Nat. Acad. Sci. U. S.* **20**, 529 (1934).
7. A. F. Stevenson, *Phys. Rev.* **53**, 199 (1938).
8. G. Temple, *Proc. Roy. Soc. Ser. A* **119**, 276 (1928).
9. J. Goodisman and D. Secrest, *J. Chem. Phys.* **45**, 1515 (1966).
10. M. Walmsley, *Proc. Phys. Soc. London* **91**, 785 (1967).
11. N. Bazley and D. W. Fox, *Phys. Rev.* **120**, 144 (1960); **124**, 483 (1961); *J. Res. Nat. Bur. Stand. Sect. B* **65**, 105 (1961); *J. Math. Phys.* **4**, 1147 (1963).
12. N. Bazley and D. W. Fox, *Phys. Rev.* **148**, 90 (1966).
13. J. Goodisman and D. Secrest, *J. Chem. Phys.* **41**, 3610 (1964).
14. I. T. Keaveny and R. E. Christofferson, *J. Chem. Phys.* **50**, 80 (1969).
15. R. G. Gordon, *J. Chem. Phys.* **48**, 4984 (1968).
16. J. H. Choi and D. W. Smith, *J. Chem. Phys.* **45**, 4425 (1966).
17. B. P. Johnson and C. A. Coulson, *Proc. Phys. Soc. London* **84**, 263 (1964).
18. P. Jennings, *J. Chem. Phys.* **46**, 2442 (1967).
19. W. H. Miller, *J. Chem. Phys.* **42**, 4305 (1965); **50**, 2758 (1969).
20. E. B. Wilson, Jr., *J. Chem. Phys.* **43**, S172 (1965).
21. S. T. Epstein, Theor. Chem. Inst., Rep. WIS-TCI-361. Univ. of Wisconsin, Madison, Wisconsin, 1969.

22. N. W. Bazley and D. W. Fox, *Rev. Mod. Phys.* **35**, 712 (1963).
23. P.-O. Löwdin, *Annu. Rev. Phys. Chem.* **11**, 107 (1960).
24. N. W. Bazley and D. W. Fox, *J. Math. Phys.* **7**, 413 (1966).
25. P. Jennings and E. B. Wilson, Jr., *J. Chem. Phys.* **45**, 1847 (1966); **47**, 2130 (1967).
26. M. H. Alexander, *J. Chem. Phys.* **51**, 5650 (1969).
27. P. S. C. Wang, *J. Chem. Phys.* **51**, 4767 (1969); F. Weinhold, *J. Phys. A* **1**, 305, 535 (1968).
28. J. C. Y. Chen and A. Dalgarno, *Proc. Phys. Soc. London* **85**, 399 (1965).
29. J. O. Hirschfelder, W. Byers-Brown, and S. T. Epstein, *Advan. Quantum Chem.* **1**, 255 (1964).
30. J. C. Y. Chen and S. Ehrenson, *J. Chem. Phys.* **44**, 1020 (1966).
31. A. Mukherji and M. Karplus, *J. Chem. Phys.* **38**, 44 (1963).
32. R. F. W. Bader and G. A. Jones, *J. Chem. Phys.* **38**, 2791 (1963).
33. Y. Rasiel and D. R. Whitman, *J. Chem. Phys.* **42**, 2124 (1965).
34. T. A. Weber and N. C. Handy, *J. Chem. Phys.* **50**, 2214 (1969).
35. W. L. Clinton, A. J. Galli, G. A. Henderson, G. B. Lamers, L. J. Massa, and J. Zarur, *Phys. Rev.* **177**, 27 (1969).
36. W. Byers Brown, *J. Chem. Phys.* **44**, 567 (1966); D. P. Chong and W. Byers Brown, *Ibid.* **45**, 392 (1966).
37. D. P. Chong and Y. Rasiel, *J. Chem. Phys.* **44**, 1819 (1966); M. L. Benston and D. P. Chong, *Mol. Phys.* **12**, 487 (1967); **13**, 199 (1967); D. P. Chong, *J. Chem. Phys.* **47**, 4907 (1967).
38. G. A. Henderson and J. Zarur, *Bull. Amer. Phys. Soc.* **16**, 647 (1971); W. L. Clinton, A. J. Galli, G. A. Henderson, G. B. Lamers, L. J. Massa, and J. Zarur, *Phys. Rev.* **177**, 7, 27, (1969).
39. J. C. Y. Chen, *J. Chem. Phys.* **43**, 3673 (1965).
40. J. O. Hirschfelder, *J. Chem. Phys.* **33**, 1462 (1960); J. O. Hirschfelder and C. A. Coulson, *Ibid.* **36**, 941 (1962).
41. A. Wallis, D. L. S. McElwain, and H. O. Pritchard, *Int. J. Quantum Chem.* **3**, 711 (1969).
41a. P.-O. Löwdin, *J. Math. Phys.* **3**, 969 (1962).
42. J. C. Slater, *Phys. Rev.* **35**, 210 (1930).
43. V. Fock, *Z. Phys.* **61**, 126 (1930).
43a.C. C. J. Roothaan, *Rev. Mod. Phys.* **23**, 69 (1951).
44. J. Koutecký and V. Bonačič, *J. Chem. Phys.* **55**, 2408 (1971).
45. W. R. Wessel, *J. Chem. Phys.* **47**, 3253 (1967).
46. C. C. J. Roothaan and P. S. Bagus, *Methods Comput. Phys.* **2**, 62 (1963).
47. H. F. King and R. E. Stanton, *J. Chem. Phys.* **50**, 3789 (1969).
48. T. Koopmans, *Physica (Utrecht)* **1**, 104 (1934).
49. V. Fock, *Z. Phys.* **81**, 195 (1933).
50. W. G. Laidlaw and F. W. Birss, *Theor. Chim. Acta* **2**, 181, 186 (1964).
51. S. T. Epstein, *J. Chem. Phys.* **41**, 1045 (1964).
52. C. Møller and M. S. Plesset, *Phys. Rev.* **46**, 618 (1934).
53. L. Brillouin, *Actual. Sci. Ind.* **71** (1933); **159**, **160** (1934); L. D. Carlson and D. R. Whitman, *Int. J. Quantum Chem.* S1, 81 (1967).
54. D. J. Thouless, "The Quantum Mechanics of Many-Body Systems," Chapter III.1. Academic Press, New York, 1961.
55. R. LeFebvre and C. M. Moser, *J. Chim. Phys. Physicochim. Biol.* **53**, 393 (1956).

56. R. LeFebvre, *J. Chim. Phys. Physicochim. Biol.* **54**, 168 (1957).
57. W. Kutzelnigg and V. H. Smith, Jr., *J. Chem. Phys.* **41**, 896 (1964).
58. P. A. M. Dirac, *Proc. Cambridge Phil. Soc.* **27**, 240 (1937).
59. P.-O. Löwdin, *Phys. Rev.* **97**, 1490 (1965).
60. J. Lennard-Jones, *Proc. Roy. Soc. Ser. A* **198**, 1, 14 (1949).
61. R. McWeeny and B. T. Sutcliffe, "Methods of Molecular Quantum Mechanics." Academic Press, New York, 1969. See particularly Sects. 4.2, 4.3, 5.1, 5.2, and 5.3.
62. R. McWeeny, *Proc. Roy. Soc. Ser. A* **235**, 496 (1956).
63. C. A. Coulson, *Rev. Mod. Phys.* **32**, 170 (1960); A. J. Coleman, *Int. J. Quantum Chem.* S4, 355 (1970).
64. W. L. Clinton, J. Nakhleh, F. Wunderlich, A. J. Galli, L. J. Massa, G. A. Henderson, J. V. Prestia, G. B. Lamers, and J. Zarur, *Phys. Rev.* **177**, 1, 7, 13, 19, 27 (1969).
65. P.-O. Löwdin, *Phys. Rev.* **97**, 1474 (1955).
66. E. Clementi, *J. Chem. Phys.* **38**, 2248 (1963); **39**, 175 (1963); **42**, 2783 (1965); E. Clementi and A. Veillard, *Ibid.* **44**, 3050 (1966).
67. P.-O. Löwdin, *Advan. Chem. Phys.* **2**, 207 (1959); *In* "Quantum Theory of Atoms, Molecules, and the Solid State (P. O. Löwdin, ed.), Academic Press, New York, 1966; p. 601. L. Carroll, "Complete Works," Random House, New York; p. 214.
68. H. Nakatsuji, H. Kato, and T. Yonezawa, *J. Chem. Phys.* **51**, 3175 (1969).
69. B. Levy and G. Berthier, *Int. J. Quantum Chem.* **2**, 307 (1968).
70. Z. Gershgorn and I. Shavitt, *Int. J. Quantum Chem.* **2**, 751 (1968).
71. R. K. Nesbet, *Advan. Chem. Phys.* **9**, 321 (1965).
72. A. Pipano and I. Shavitt, *Int. J. Quantum Chem.* **2**, 741 (1968).
73. C. F. Bender and E. R. Davidson, *J. Phys. Chem.* **70**, 2675 (1966).
74. C. F. Bender and E. R. Davidson, *J. Chem. Phys.* **47**, 360 (1967).
75. W. Kutzelnigg, *Theor. Chim. Acta* **1**, 327, 343 (1963); C. E. Reid and Y. Ohrn, *Rev. Mod. Phys.* **35**, 445 (1963).
76. E. R. Davidson and L. L. Jones, *J. Chem. Phys.* **37**, 2966 (1962); S. Hagstrom and H. Shull, *Rev. Mod. Phys.* **35**, 624 (1963).
76a. D. D. Ebbing and R. D. Henderson, *J. Chem. Phys.* **42**, 2225 (1965).
76b. G. P. Barnett, J. Linderberg, and H. Shull, *J. Chem. Phys.* **43**, S80 (1965).
77. D. F.-T. Tuan, *J. Chem. Phys.* **51**, 607 (1969).
78. R. E. Stanton, *J. Chem. Phys.* **36**, 1298 (1962).
79. A. Veillard, *Theor. Chim. Acta* **4**, 22 (1966); A. Veillard and E. Clementi, *Ibid.* **7**, 133 (1967).
80. R. K. Nesbet, *Phys. Rev.* **109**, 1632 (1958).
81. P.-O. Löwdin, *Phys. Rev.* **97**, 1509 (1955).
82. G. Das and A. C. Wahl, *J. Chem. Phys.* **44**, 87 (1966); A. C. Wahl and G. Das, *Advan. Quantum Chem.* **5**, 261 (1970).
83. R. K. Nesbet, *J. Chem. Phys.* **43**, 311 (1965); I. Shavitt, *J. Comput. Phys.* **6**, 124 (1970)
84. J. Lennard-Jones and J. A. Pople, *Proc. Roy. Soc. Ser. A* **210**, 190 (1951).
85. J. Goodisman, *J. Chem. Phys.* **43**, 3037 (1965).
85a. N. Cressy and K. Ruedenberg, *Int. J. Quantum Chem.* **3**, 493 (1969).
86. E. A. Hylleraas, *Z. Phys.* **54**, 347 (1929); **65**, 209 (1930).
87. H. M. James and A. S. Coolidge, *J. Chem. Phys.* **1**, 825 (1933).
88. F. W. Byron, Jr., and C. J. Joachain, *Phys. Rev.* **157**, 1 (1967).
88a. C. S. Schwartz, *Phys. Rev.* **126**, 1015 (1962).
89. K. Jandowski and W. Woźnicki, *Bull. Acad. Pol. Sci.* **13**, 249 (1965).

90. C. C. J. Roothaan and A. W. Weiss, *Rev. Mod. Phys.* **32**, 194 (1960).
91. W. Kołos and C. C. J. Roothaan, *Rev. Mod. Phys.* **32**, 205 (1960).
92. T. L. Gilbert, *Rev. Mod. Phys.* **35**, 491 (1963).
93. H. M. James and A. S. Coolidge, *Phys. Rev.* **49**, 676 (1936).
94. V. Fock, M. Wesselow, and M. Petrashen, *Zh. Eksp. Teor. Fiz.* **10**, 723 (1940).
95. P.-O. Löwdin, *J. Chem. Phys.* **35**, 78 (1961).
96. A. C. Hurley, J. Lennard-Jones, and J. A. Pople, *Proc. Roy. Soc. Ser. A* **220**, 46 (1953).
97. R. McWeeny and E. Steiner, *Advan. Quantum Chem.* **2**, 93 (1965).
98. L. Szász, G. McGinn, and J. Schroeder, *Z. Naturforsch.* **a 22**, 2109 (1967).
99. O. Sinanoğlu, *Advan. Chem. Phys.* **6**, 315 (1964).
100. R. E. Stanton, *J. Chem. Phys.* **42**, 2353 (1965).
101. O. Sinanoğlu, *Proc. Nat. Acad. Sci. U.S.* **47**, 1217 (1961).
102. H. J. Silverstone and O. Sinanoğlu, *J. Chem. Phys.* **46**, 854 (1967).
103. M. Geller, H. S. Taylor, and H. B. Levine, *J. Chem. Phys.* **43**, 1727 (1965).
104. C. A. Coulson and I. Fischer, *Phil. Mag.* **40**, 386 (1949); W. Kołos, *Acta Phys. Pol.* **16**, 257, 267 (1957).
105. L. Szász, *Phys. Rev.* **126**, 169 (1962).
106. L. Szász, *Z. Naturforsch.* **a 15**, 909 (1960); *J. Chem. Phys.* **35**, 1072 (1961).
107. L. Szász and J. Byrne, *Phys. Rev.* **158**, 34 (1967).
108. H. F. Schaefer, III, "The Electronic Structure of Atoms and Molecules," p. 182 et seq. Addison-Wesley, Reading, Massachusetts, 1972.
109. D. M. Silver, E. L. Mehler, and K. Ruedenberg, *J. Chem. Phys.* **52**, 1174 (1970); E. L. Mehler, K. Ruedenberg, and D. M. Silver, *Ibid.* **52**, 1181, 1206 (1970).
110. I. G. Csizmadia, B. T. Sutcliffe, and M. P. Barnett, *Can. J. Chem.* **42**, 1645 (1964).
111. C. F. Bender, E. R. Davidson, and F. D. Peat, *Phys. Rev.* **174**, 75 (1968).

B. Perturbation Theory

Through the use of perturbation theory, one can go from the solution of one Schrödinger equation to that of another, generating a sequence of corrections to eigenfunctions and eigenvalues. The difference between the Hamiltonians of the two problems is known as the perturbation, and need not represent a physical operator in the sense that there is a physical process by which it may be turned on or off. Perturbation theory is particularly useful when one problem is more easily soluble than another, when the solutions to the two problems are not very different, or when the perturbation represents a physical effect in which one is interested, superposed on a perhaps less interesting problem. For example, we are interested in the *change* in energy with change in R, so that perturbation theory seems like a logical way to treat potential energy curves (cf. Subsection 4 and Vol. 2, Ch. II). However, Hirschfelder [1] has cautioned that (*a*) R is not a natural parameter of perturbation theory for small internuclear distances because of the $R^3 \ln R$ term (see Vol. 2, Ch. I, Sect. C) and (b) R^{-1} is not a natural perturbation parameter for large internuclear distances because exponentials in R occur. Claverie [2] has discussed the latter problem, to which we will return in Volume 2, Chapter I, Section B in detail.

Two reviews by Hirschfelder and co-workers [3, 4] contain discussions of perturbation theory, emphasizing recent developments, particularly as they apply to atomic and molecular problems. In what follows, we shall by no means mention all the topics touched on in these articles.

Much of this chapter deals with perturbations on the Hartree–Fock function, because of its central importance in calculations of molecular electronic structure. In Subsection 2, we consider one-electron perturbations, such as arise in calculation of the induction and dispersion forces discussed in Vol. 2, Ch. I, Sect. A. In Subsection 3, we are concerned with perturbative approaches for the correction of an approximate wave function of determinantal form to the exact wave function. The use of perturbation theory to make the R-dependence of the energy explicit is presented in Subsection 4, and the double perturbation theory for the treatment of two simultaneous perturbations is taken up in Subsection 5. We begin with a summary of basic formulas.

1. General Formulas

We write H, the Hamiltonian for which eigenfunctions and eigenvalues are desired, as

$$H = H_0 + V, \tag{138}$$

where H_0 refers to the "unperturbed" problem, which is considered to be completely solved, giving a set of eigenfunctions $\Psi_k^{(0)}$ and corresponding eigenvalues $E_k^{(0)}$

$$H_0 \Psi_k^{(0)} = E_k^{(0)} \Psi_k^{(0)}. \tag{139}$$

To develop the Rayleigh–Schrödinger perturbation theory, which we shall use most frequently, we consider

$$H(\gamma) = H_0 + \gamma V$$

seeking its eigenfunctions

$$H(\gamma)\Psi_k(\gamma) = E_k(\gamma)\Psi_k(\gamma). \tag{140}$$

We are interested particularly in the solutions for $\gamma = 1$, but require that as γ goes from 0 to 1, so that the Hamiltonian goes from the unperturbed problem to the full Hamiltonian, the energies and eigenfunctions of $H(\gamma)$ go smoothly from those of H_0 to those of $H = H_0 + V$. The energies E_k and eigenfunctions depend parametrically on γ. If the dependence is analytic, we may write power series

$$\Psi_k = \Psi_k^{(0)} + \gamma\Psi_k^{(1)} + \cdots \tag{141a}$$

$$E_k = E_k^{(0)} + \gamma E_k^{(1)} + \cdots \tag{141b}$$

It is general practice to assume analyticity until problems arise (see Vol. 2, Ch. I, Section C). Substituting in (140) and equating powers of γ, we find the equations of the Rayleigh–Schrödinger perturbation theory,

$$(H_0 - E_k^{(0)})\Psi_k^{(0)} = 0 \tag{142}$$

$$(H_0 - E_k^{(0)})\Psi_k^{(1)} + (V - E_k^{(1)})\Psi_k^{(0)} = 0 \tag{143}$$

$$(H_0 - E_k^{(0)})\Psi_k^{(2)} + (V - E_k^{(1)})\Psi_k^{(1)} - E_k^{(2)}\Psi_k^{(0)} = 0 \tag{144}$$

$$(H_0 - E_k^{(0)})\Psi_k^{(3)} + (V - E_k^{(1)})\Psi_k^{(2)} - E_k^{(2)}\Psi_k^{(1)} - E_k^{(3)}\Psi_k^{(0)} = 0 \tag{145}$$

of which (142) is satisfied by hypothesis. We assume Ψ normalized to all orders in γ, which implies

$$\langle \Psi_k^{(0)} \mid \Psi_k^{(0)}\rangle = 1 \tag{146}$$

$$\langle \Psi_k^{(0)} \mid \Psi_k^{(1)}\rangle + \langle \Psi_k^{(1)} \mid \Psi_k^{(0)}\rangle = 0 \tag{147}$$

$$\langle \Psi_k^{(0)} \mid \Psi_k^{(2)}\rangle + \langle \Psi_k^{(1)} \mid \Psi_k^{(1)}\rangle + \langle \Psi_k^{(2)} \mid \Psi_k^{(0)}\rangle = 0 \tag{148}$$

and so on. From Eqs. (142)–(145) one can show (PILAR, Sect. 10.4, LEVINE, Sect. 9.2)

$$E_k^{(1)} = \langle \Psi_k^{(0)} \mid V \mid \Psi_k^{(0)}\rangle, \tag{149}$$

$$E_k^{(2)} = \langle \Psi_k^{(0)} \mid V - E_k^{(1)} \mid \Psi_k^{(1)}\rangle, \tag{150}$$

$$E_k^{(3)} = \mathrm{Re}\langle \Psi_k^{(0)} \mid V - E_k^{(1)} \mid \Psi_k^{(2)}\rangle.$$

Taking advantage of the hermiticity of $V - E_k^{(1)}$ to let it operate on $\Psi_k^{(0)}$, and then using (143) and (144), we get

$$E_k^{(3)} = -\mathrm{Re}\langle \Psi_k^{(1)} \mid H_0 - E_k^{(0)} \mid \Psi_k^{(2)}\rangle$$
$$= \langle \Psi_k^{(1)} \mid V - E_k^{(1)} \mid \Psi_k^{(1)}\rangle - E_k^{(2)}\mathrm{Re}\langle \Psi_k^{(1)} \mid \Psi_k^{(0)}\rangle \tag{151}$$

so that $E_k^{(3)}$ may be determined from $\Psi_k^{(1)}$. In general, the nth-order wave function yields the energy to $(2n + 1)$st order. This is obvious [3, Pt. X] since the wave function to nth order is in error in $(n + 1)$st order and hence produces an error of $2(n + 1)$ order in the energy. For the expectation value of a property other than the energy,

$$\langle \Psi_k \mid Q \mid \Psi_k \rangle = \langle \Psi_k^{(0)} \mid Q \mid \Psi_k^{(0)} \rangle + 2 \operatorname{Re} \langle \Psi_k^{(0)} \mid Q \mid \Psi_k^{(1)} \rangle \qquad (152)$$

to terms of second order, so we need the first-order terms in the wave function to get the corrections.

In the case of degeneracy, there is an additional complication. Suppose $E_k^{(0)}$ were part of a degenerate level to which $E_l^{(0)}$ also belongs. Multiplying Eq. (143) by $\Psi_l^{(0)*}$ and integrating over configuration space yields

$$\langle \Psi_l^{(0)} \mid V \mid \Psi_k^{(0)} \rangle - 0 \qquad (153)$$

which will not, in general, hold. But any linear combination of the degenerate zero-order functions of the form $\tilde{\Psi}_k^{(0)} = \sum a_i^{(k)} \Psi_i^{(0)}$ is also a possible zero-order function. We must specify the correct linear combinations to make $\Psi_k(\gamma)$ go over smoothly to $\Psi_k^{(0)}$ as $\gamma \to 0$ (analytical dependence on γ). The determination of the coefficients $a_i^{(k)}$, (EYRING, Sect. 7b, LEVINE, Sect. 9.6, PILAR, Sect. 10.5), is by substitution into (143), multiplication by $\Psi_j^{(0)*}$, and integration, which yields

$$\sum_i \langle \Psi_j^{(0)} \mid V \mid \Psi_i^{(0)} \rangle a_i^{(k)} = E_k^{(1)} a_i^{(k)} .$$

This is the secular equation, an eigenvalue equation in the d-dimensional subspace ($d =$ degree of degeneracy). The first-order energies (there will be d of them, but not all need be different) come from solution of the secular determinant

$$\left| \langle \Psi_j^{(0)} \mid V \mid \Psi_i^{(0)} \rangle - E_k^{(1)} \delta_{ij} \right| = 0 \qquad (154)$$

as do the coefficients $a_i^{(k)}$ corresponding to each. The linear combinations will yield mutually orthogonal functions (eigenfunctions of the secular equation with different eigenvalues). These new linear combinations will obey (153) and our equations will be well behaved. In general, we assume that this has been done, i.e., the zero-order functions are chosen to diagonalize V. In many cases, this may be done on the basis of symmetry alone: If V commutes with some symmetry operator, the basis functions are chosen as eigenfunctions of the operator with different eigenvalues.

It is sometimes advantageous to treat eigenfunctions for which the zero-order energies are close to equal as degenerate. This is because the coefficients of such functions in the first-order wave function [Eq. (157), and their contributions to the second-order energy [Eq. (158)] will become large. Thus we try to include much of the contributions of nearby states at the outset, by diagonalizing V over the almost-degenerate levels. The secular equation becomes

$$| V_{ij} - (E - E_i^{(0)}) \delta_{ij} | = 0$$

where $V_{ij} = \langle \Psi_i^{(0)} | V | \Psi_j^{(0)} \rangle$ [compare Eq. (154)].

If there are only two functions in the almost-degenerate set, the secular equation yields

$$E_{\pm} = \tfrac{1}{2}(E_1^{(0)} + E_2^{(0)} + V_{11} + V_{22} \pm [(V_{22} - V_{11} + \Delta)^2 + 4 | V_{12} |^2]^{1/2}) \quad (155)$$

where Δ is $E_2^{(0)} - E_1^{(0)}$. Where Δ is large, the radical becomes, taking $\Delta > 0$,

$$\Delta[1 + 2\Delta^{-1}(V_{22} - V_{11}) + \Delta^{-2}(V_{22} - V_{11})^2 + 4\Delta^{-2} | V_{12} |^2]^{1/2}$$
$$\sim \Delta[1 + \Delta^{-1}(V_{22} - V_{11}) + \tfrac{1}{2}\Delta^{-2}(V_{22} - V_{11})^2 + 2\Delta^{-2} | V_{12} |^2$$
$$- \tfrac{1}{2}\Delta^{-2}(V_{22} - V_{11})^2]$$

so $E_+ = E_2^{(0)} + V_{22} + | V_{12} |^2/\Delta$ and $E_- = E_1^{(0)} + V_{11} - | V_{12} |^2/\Delta$. We recover the perturbation expansions [Eqs. (139) and (158)]. For $\Delta = 0$ we of course obtain from (155) the result of degenerate perturbation theory. Corresponding to $E_{\pm}$ are $\Psi_{\pm}^{(0)}$, linear combinations of $\Psi_1^{(0)}$ and $\Psi_2^{(0)}$. Note that they are mutually orthogonal and orthogonal to all $\Psi_k^{(0)}$ except for $k = 1$ or 2. $\langle \Psi_+^{(0)} | H_0 | \Psi_-^{(0)} \rangle$ does not vanish, but $\langle \Psi_+^{(0)} | H_0 + V | \Psi_-^{(0)} \rangle$ does.

Now let us put $V \to \gamma V$ in (155) and allow γ to grow from 0 to 1. Remembering $\Delta > 0$, we see that, for $\gamma = 0$, E_+ is $E_2^{(0)}$ and E_- is $E_1^{(0)}$. If γ now increases, both E_+ and E_- will be analytic functions of γ so long as the radical in (155) never vanishes. This is guaranteed if $V_{12} \neq 0$. Then $E_2^{(0)}$ goes into E_+, $E_1^{(0)}$ into E_-. For all values of γ, evidently $E_+ > E_-$. This is the noncrossing rule, discussed in Chapter I, Section A, Subsection 4.

It is often convenient to imagine $\Psi_k^{(1)}$ (and higher corrections) to be expanded in the zero-order eigenfunctions (LEVINE, Sect. 9.2, EYRING, Sect. 7a)

$$\Psi_k^{(1)} = \sum_i c_i{}^k \Psi_i^{(0)} \quad (156)$$

To satisfy (147), we must take $c_k{}^k = 0$. Substitution of (156) into (143) yields

$$c_i{}^k = \langle \Psi_i^{(0)} | V | \Psi_k^{(0)} \rangle / (E_k^{(0)} - E_i^{(0)}) \quad (157)$$

so that (150) becomes

$$E_k^{(2)} = \sum_{i \neq k} \frac{\langle \Psi_k^{(0)} | V | \Psi_i^{(0)} \rangle \langle \Psi_i^{(0)} | V | \Psi_k^{(0)} \rangle}{E_k^{(0)} - E_i^{(0)}} \tag{158}$$

and (151)

$$E_k^{(3)} = \sum_{i \neq k} \sum_{j \neq k}$$

$$\times \frac{\langle \Psi_k^{(0)} | V - E_k^{(1)} | \Psi_i^{(0)} \rangle \langle \Psi_i^{(0)} | V - E_k^{(1)} | \Psi_j^{(0)} \rangle \langle \Psi_j^{(0)} | V - E_k^{(1)} | \Psi_k^{(0)} \rangle}{(E_k^{(0)} - E_i^{(0)})(E_k^{(0)} - E_j^{(0)})} \tag{159}$$

From a computational viewpoint, the energy expressions are not too useful unless the sums converge rapidly. We can define an "average energy denominator" $\bar{E}$ such that

$$E_k^{(2)} = \bar{E}^{-1} \sum_{i \neq k} \langle \Psi_k^{(0)} | V | \Psi_i^{(0)} \rangle \langle \Psi_i^{(0)} | V | \Psi_k^{(0)} \rangle$$

$$= \bar{E}^{-1}[\langle \Psi_k^{(0)} | V^2 | \Psi_k^{(0)} \rangle - |\langle \Psi_k^{(0)} | V | \Psi_k^{(0)} \rangle|^2] \tag{160}$$

where the closure or completeness relation for the $\Psi_i^{(0)}$ (including $i = k$) has been invoked. At this point, we must attempt to estimate $\bar{E}$ (dangerous) or use the formula for nonquantitative thinking.

Except for some few cases [3, 4], we cannot produce an exact solution to the first-order equation. Closely related to finding such a solution is summing the series (158). There are a number of cases where sums over excited states can be evaluated explicitly [3, Pt. X], and these have been discussed by Dalgarno and others [5]. In some cases, we can get by with a few terms in (158). Since the convergence of an expansion depends on the basis functions, we can consider using another basis set to expand $\Psi^{(1)}$, i.e., put $\Psi^{(1)}$ equal to $\sum_k d_k \Phi_k$ for insertion in Eq. (143), which then is to be transformed into an equation for the expansion coefficients $\{d_k\}$. In order to satisfy the orthogonality condition $\langle \Psi^{(0)} | \Psi^{(1)} \rangle = 0$, the basis functions must be orthogonalized to $\Psi^{(0)}$: We expand in the functions

$$\tilde{\Phi}_k = \Phi_k - \Psi^{(0)} \langle \Psi^{(0)} | \Phi_k \rangle$$

instead. The zero-order function $\Psi^{(0)}$ is assumed nondegenerate. Putting

$$\Psi^{(1)} = \sum_k c_k \tilde{\Phi}_k \tag{161}$$

into (143), multiplying by $\tilde{\Phi}_j{}^*$ and integrating, we obtain

$$\sum_k M_{jk} c_k = N_j. \tag{162}$$

Here,

$$M_{jk} = \langle \tilde{\Phi}_j | H_0 - E^{(0)} | \tilde{\Phi}_k \rangle \qquad (163a)$$

$$N_j = \langle \tilde{\Phi}_j | V - E^{(1)} | \Psi^{(0)} \rangle. \qquad (163b)$$

If expansion (161) runs over the complete (infinite) set of $\tilde{\Phi}_k$, $\Psi^{(1)}$ will be the exact first-order function. The hope is that the set can be chosen so that a finite number of functions suffices to give a good approximation to $\Psi^{(1)}$. When $\mathbf{M}$ is a finite matrix,

$$c_k = \sum_l (\mathbf{M}^{-1})_{kl} N_l \qquad (164)$$

where $\mathbf{M}^{-1}$ is the inverse of $\mathbf{M}$. We have arranged things so $\mathbf{M}$ has no zero eigenvalues, and the inverse exists. Then

$$E^{(2)} = \sum_{k,l} N_k^* (\mathbf{M}^{-1})_{kl} N_l \qquad (165)$$

and

$$E^{(3)} = \sum_{k,l,m,n} N_k^* (\mathbf{M}^{-1})_{kl} \langle \tilde{\Phi}_l | V - E^{(1)} | \tilde{\Phi}_m \rangle (\mathbf{M}^{-1})_{mn} N_n. \qquad (166)$$

One may expect that, as the set $\{\Phi_k\}$ becomes larger, we approach the exact expressions for the second-order and third-order energies. If the functions Φ_k are chosen to be orthonormal and to diagonalize the matrix of H_0, M_j becomes $(\varepsilon_k - E^{(0)}) \delta_{jk}$ and (165) becomes

$$E^{(2)} = \sum_{k,l} \langle \Psi^{(0)} | V | \tilde{\Phi}_k \rangle (\varepsilon_k - E^{(0)})^{-1} \delta_{jk} \langle \tilde{\Phi}_l | V | \Psi^{(0)} \rangle$$

$$= \sum_k \frac{|\langle \Psi^{(0)} | V | \tilde{\Phi}_k \rangle|^2}{\varepsilon_k - E^{(0)}}$$

This becomes (158) when the basis set becomes complete.

The result (162) for a finite basis set is also derivable from a variational principle, associated with Hylleraas [6]. Consider the functional

$$\mathscr{F}(\Phi) = \langle \Phi | H_0 - E^{(0)} | \Phi \rangle + 2\langle \Phi | V - E^{(1)} | \Psi^{(0)} \rangle. \qquad (167)$$

If $\mathscr{F}(\Phi)$ is to be stationary to arbitrary variations in the trial function Φ, we find that Φ must satisfy Eq. (143). Then Φ is the exact $\Psi^{(1)}$, and $\mathscr{F}(\Phi)$ becomes $E^{(2)}$. By writing $\Phi = \Psi^{(1)} + \varepsilon X$ with $\langle \Psi^{(1)} | X \rangle = 0$ as in our discussion of the variational principle itself, it is easily seen that

$$\mathscr{F}(\Phi) = \langle \Psi^{(1)} | V - E^{(1)} | \Psi^{(0)} \rangle + \langle \varepsilon X | H_0 - E^{(0)} | \varepsilon X \rangle.$$

The first-order terms vanish, as is characteristic of a variation problem. The term in ε^2 is evidently positive if $E^{(0)}$ is the lowest eigenvalue of H_0. Thus the functional $\mathscr{F}(\Phi)$ for the second-order energy in the case of a ground state obeys a minimum principle. Lower bounds to $E^{(2)}$ are also possible [2, 3, 7]. For the particular limited variation function represented by (161), we form $\mathscr{F}(\Phi)$ as a quadratic form in the c_k and differentiate with respect to each separately. Then we recover (162) and $\mathscr{F}(\Phi)$ becomes (165). Variational principles also exist for higher-order energies and wave functions [8, 9]. Musher [10] has shown how they may be derived directly from the differential equations, rather than from energy functionals. $\Psi^{(0)}$ and the higher-order wave functions could be expanded in some basis set, as well as $\Psi^{(1)}$. This converts Eqs. (142)–(145) into matrix equations. Here $E_k^{(0)}$ is the eigenvalue of the matrix of H_0 and $\Psi_k^{(0)}$ is expressed in terms of expansion coefficients which constitute the corresponding eigenvector. Representing this column vector as $\mathbf{c}_k^{(0)}$, we have

$$E_k^{(1)} = \mathbf{c}_k^{(0)\dagger}\mathbf{V}\mathbf{c}_k^{(0)}$$

$$(\mathbf{H}_0 - E_k^{(0)}\mathbf{S})\mathbf{c}_k^{(1)} + (\mathbf{V} - E_k^{(1)}\mathbf{S})\mathbf{c}_k^{(0)} = 0$$

and so on, each operator being replaced by the corresponding matrix. $\mathbf{S}$ is the overlap matrix for the basis set. The matrix eigenvalue equation

$$\mathbf{Hc} = E\mathbf{Sc} \qquad (168)$$

which is the transcription of the Schrödinger eigenvalue equation is being solved by the Rayleigh–Schrödinger perturbation theory.

Another perturbation approach to the problem of finding eigenvalues of $\mathbf{H}$ is possible. It deals directly with the matrix elements, and does not necessitate a division of the Hamiltonian into two parts. Another order of smallness, defined in terms of matrix elements of the full Hamiltonian, becomes relevant. The formalism derived for such a situation, which we now discuss, is referred to as Brillouin–Wigner perturbation theory [11]. We have used some of the formulas previously in our discussion of configuration interaction. Suppose one function in the basis set Φ_0 is an approximation to Ψ so that the coefficients c_i for $i \neq 0$ may be expected to be small compared to c_0. For convenience, take c_0 equal to unity. Rewrite (168) as follows:

$$(H_{00} - ES_{00}) + \sum_{i \neq 0} c_i(H_{0i} - ES_{0i}) = 0 \qquad (169a)$$

$$(H_{k0} - ES_{k0}) + \sum_{i \neq 0} c_i(H_{ki} - ES_{ki}) = 0 \qquad (k \neq 0) \qquad (169b)$$

We have used the abbreviations

$$S_{ij} = \langle \Phi_i \mid \Phi_j \rangle, \qquad H_{ij} = \langle \Phi_i \mid H \mid \Phi_j \rangle.$$

The convergence of the Brillouin–Wigner series for the energy will depend on the smallness of the off-diagonal quantities $H_{ij} - ES_{ij}$ compared to the diagonal $H_{ii} - ES_{ii}$. For the case where the basis functions are eigenfunctions of some Hamiltonian H_0 with $H - H_0$ small, we can see why the off-diagonal quantities should be small.

Neglecting the summation in (169a), we obtain a first approximation to the energy

$$E' = H_{00}/S_{00}. \tag{170}$$

The neglected terms are of the first order of smallness. The next approximation to E is obtained by using in (169a) a first approximation to the c_i. This comes from (169b):

$$c_k(H_{kk} - ES_{kk}) = -(H_{k0} - ES_{k0}) - \sum_{l \neq k, 0} c_l(H_{kl} - ES_{kl}). \tag{171}$$

The summation is neglected since each factor in each term is small, and the result is used in (169a) to give

$$E'' = \frac{H_{00}}{S_{00}} - \sum_{i \neq 0} \frac{(H_{0i} - ES_{0i})(H_{i0} - ES_{i0})}{S_{00}(H_{ii} - ES_{ii})}. \tag{172}$$

By a continuation of this procedure, higher corrections to E may be generated. It will be noted that the expressions for the energy from E'' on are actually implicit equations for E and must be solved iteratively. If the first-order approximation (170) for E is used in (173), the result is

$$E'' \approx \frac{H_{00}}{S_{00}} - \sum_{i \neq 0} \frac{H_{0i}S_{00} - H_{00}S_{0i}}{(H_{ii}S_{00} - H_{00}S_{ii})S_{00}^2}.$$

Now if H may be written as $H_0 + V$ with V small, and the Φ_i are eigenfunctions of H_0 with eigenvalues ε_i, the diagonal matrix element H_{ii} in the second-order term may be approximated as ε_i, and

$$E'' \approx \varepsilon_0 + \langle \Psi_0 \mid V \mid \Psi_0 \rangle - \sum_{i \neq 0} \frac{|\langle \Psi_0 \mid V \mid \Psi_i \rangle|^2}{\varepsilon_i - \varepsilon_0} \tag{173}$$

We recover the result of Rayleigh–Schrödinger perturbation theory, which may be regarded as an approximation to the Brillouin–Wigner theory [12].

Another way of looking at the Brillouin–Wigner energies is that they are solutions to the secular determinant

$$| H_{ij} - ES_{ij} |$$

obtained by successive approximation. Here E' results from neglecting all matrix elements except those on the diagonal, and E'' from including in addition those in the zeroth row and zeroth column. It is also possible to derive the Brillouin–Wigner theory in terms of differential equations, which are formally simpler than those of the Rayleigh–Schrödinger theory, and thence to derive variation–perturbation methods for approximation of the solutions [12]. The comparison of the two methods is carried out very clearly by Hirschfelder *et al.* [3]. The convergence problems for the Rayleigh–Schrödinger and Brillouin–Wigner series have been investigated [3]. Most applications of interest to us use the former, except where secular equations are involved.

Another perturbative approach was developed by Silverman and van Leuven [13], and named "perturbation–variation theory." ("Variation–perturbation" generally refers to perturbation theory calculations employing Hylleraas variational principles.) Here, an *approximate* solution to $H_0 + \gamma V$ was assumed to depend on γ by way of embedded parameters $\{a_i\}$, whose values are determined variationally. The optimal values of the $\{a_i\}$ were written as power series in γ,

$$a_i(\gamma) = a_i + \gamma a_i^{(1)} + \gamma^2 a_i^{(2)} + \cdots$$

giving the wave functions and expectation values as series in γ, and equations were derived for computing higher $a_i^{(n)}$ in terms of preceding ones. The zero-order equation is the variational problem corresponding to H_0, and the higher-order equations are linear in the $\{a_i^{(n)}\}$. If γ were a function of R (see Subsection 4), the R-dependence of the energy could be made explicit. While little application has been made of this theory, it seems useful for the analysis of approximate functions.

Many perturbation theories appear as special cases of the partitioning theory [14, 15] developed by Löwdin. For an orthonormal basis, Eq. (47) may be written, since $\mathbf{M}_{ab} = \mathbf{H}_{ab}$,

$$[\mathbf{H}_{aa} + \mathbf{H}_{ab}(E\mathbf{1}_{bb} - \mathbf{H}_{bb})^{-1}\mathbf{H}_{bb}]\mathbf{c}_a = E\mathbf{c}_a. \tag{174}$$

To make manipulations more transparent, this result may be written in terms of projection operators (McWEENY, Sects. 3.1, 3.6). Suppose the operator O operating on an arbitrary function selects the part in the sub-space a and P the part in the subspace b, so Of is orthogonal to any basis function in b for f arbitrary, while Pf is orthogonal to any basis function in a. If the full set of functions, a plus b, is complete, $O + P = 1$, the unit operator, which leaves all functions unchanged. Furthermore, $OP = PO = O$, and $O^2 = O$, $P^2 = P$. Corresponding to the matrices H_{aa}, H_{ab}, H_{ba}, and H_{bb} are the operators OHO, OHP, PHO, and PHP.

Where the a subspace consists of one basis function φ, E is simply given by the quantity in the square bracket in (174). The corresponding operator equation is

$$E = \langle \varphi \mid OHO + OHTHO \mid \varphi \rangle \qquad (175a)$$

where

$$T = P[\alpha O + P(E - H)P]^{-1}P \qquad (175b)$$

corresponds to the inverse of $E\mathbf{1}_{bb} - \mathbf{H}_{bb}$. Here α is an arbitrary nonzero constant, needed to make the inverse operator well defined. For a proof of Eq. (175a, b), see Löwdin [14]. Extension to degenerate levels is possible. The Brillouin–Wigner perturbation theory may be derived from (175a, b) by an expansion of the inverse operator T, and the Rayleigh–Schrödinger theory by an expansion when H is written as $H_0 + V$ with V small.

A lower-bound formula is obtained if we try to find a solution to (175a, b) by an iterative method. Denoting the right side of (175a) by $f(E)$, it may be shown that

$$f' = df/dE = -\langle \varphi \mid HP[\alpha O + P(E - H)P]^{-2}PH \mid \varphi \rangle$$

which is clearly always negative. Suppose we start with a guess $E^{(0)}$ for E and use (175a) to obtain a second estimate $E^{(1)}$ according to

$$E^{(1)} = f(E^{(0)}).$$

If $E^{(0)} = E + \varepsilon^{(0)}$, we have, by the mean-value theorem,

$$E^{(1)} = f(E) + \varepsilon^{(0)}f'(E + \theta\varepsilon^{(0)})$$

with $0 \leq \theta \leq 1$. But $f(E) = E$ and f' is always negative. Thus, if $E^{(0)}$ is greater than E so that $\varepsilon^{(0)}$ is positive, $E^{(1)}$ will be less than E, and vice versa. This is referred to as the bracketing theorem. From an upper bound on the energy we can obtain a lower bound [14]. There are other applications of the partitioning approach which we have not touched on here, and it seems likely that it will find application in energy calculations in the future.

2. *Hartree–Fock Perturbation Theory*

Returning to Rayleigh–Schrödinger theory, we are particularly interested in the case where the zero-order function is a Hartree–Fock function. Taking the perturbation as the difference between the exact and Hartree–Fock Hamiltonians, we have a scheme for correcting the Hartree–Fock function, i.e., introducing correlation (see Subsection 3). Another important situation is where the perturbation is one-electron in nature,

$$V = \sum_{i=1}^{N} h'(i)$$

From (149) we immediately have the first correction to the energy

$$E^{(1)} = \sum_{i=1}^{N} \langle \lambda_i \mid h' \mid \lambda_i \rangle. \tag{176}$$

We suppress the index k (which numbers the many-electron states) for clarity. If we want corrections to the Hartree–Fock orbitals due to V, there are at least two ways to proceed, as shown by Dalgarno [16].*

In Method I, we attempt to minimize the expectation value of $H = H_0 + V$ keeping a trial function of determinantal form. To order the corrections in the strength of h', we put $V \to \gamma V$ in evaluating $\langle H \rangle$, and write each spin orbital as $\lambda_i^{(0)} + \gamma \lambda_i^{(1)} + \gamma^2 \lambda_i^{(2)} + \cdots$, with $\lambda_i^{(1)}$, $\lambda_i^{(2)}$, etc. independently variable, and take terms in each order separately. This should be equivalent to solving the Hartree–Fock equations for $H = H_0 + \gamma V$ as a function of γ by expanding all the orbitals and orbital energies in γ, substituting into the Hartree–Fock equation, and solving each order in γ successively. In Method II, we consider V as a perturbation to the Hartree–Fock Hamiltonian, and solve $H^{\mathrm{HF}} + V$ by perturbation theory as usual.

Method II is simpler. In this case, V being one-electron and the zero-order eigenfunction a single determinant, the first-order wavefunction will necessarily be in the form of a sum of singly excited determinants. In each, one of the occupied spin orbitals for H^{HF} is replaced by a first-order correction, thus

$$\Psi^{(1)} = \sum_{j=1}^{N} \mathscr{A}\left\{ \prod_{i}^{(i \neq j)} \lambda_i(i) \lambda_j^{(1)}(j) \right\} \tag{177}$$

---

* For additional references, see Langhoff *et al.* [17].

where

$$\Psi^{(0)} = \Psi^{\mathrm{HF}} = \mathscr{A}\left\{\prod_i \lambda_i(i)\right\}.$$

Because of the antisymmetrizer in (177), we may add to $\lambda_j^{(1)}$ any admixture of λ_i without changing $\Psi^{(1)}$, so we may demand that the correction $\lambda_j^{(1)}$ be orthogonal to λ_i for $j \neq i$. Then, by taking $\lambda_j^{(1)}$ orthogonal to λ_j, we assure $\langle \Psi^{(1)} \mid \Psi^{(0)} \rangle = 0$. In what follows, we introduce the abbreviations

$$b(i) \equiv h'(i) - \langle \lambda_i \mid h' \mid \lambda_i \rangle, \qquad a(i) \equiv h_0(i) - \varepsilon_i$$

where the original Hartree–Fock Hamiltonian is

$$H^{\mathrm{HF}} = \sum_i h_0(i)$$

and the ε_i are the original orbital energies. Since $H_0 - E^{(0)}$ and $V - E^{(1)}$ commute with the antisymmetrizers, we may rewrite (143) as

$$\mathscr{A}\left\{\sum_i a(i) \sum_j \left[\prod_j^{(k \neq j)} \lambda_k(k)\right]\lambda_j^{(1)}(j)\right\}$$

$$+ \mathscr{A}\left\{\sum_i b(i) \prod_k \lambda_k(k)\right\} = 0.$$

This will be solved if the corrections $\lambda_j^{(1)}$ satisfy

$$[a(i)\lambda_i^{(1)}(i) + b(i)\lambda_i(i) + f_i(i)] \prod_k^{(k \neq i)} \lambda_k(k) = 0 \tag{178}$$

[note that $a(i)\lambda_i(i) = 0$], where $f_i(i)$ is any function such that the last term in (178) is annihilated by the antisymmetrizer. We may then write

$$f_i(i) = \sum_k^{(k \neq i)} c_k \lambda_k(i). \tag{179}$$

The last term in (178) is necessary because (178) is otherwise inconsistent: Multiplication by $\lambda_k(i)$ and integration over the coordinates i makes the first term vanish but not the second. To make (178) consistent, we must put $c_k = -\langle \lambda_k(i) \mid b(i) \mid \lambda_i(i) \rangle$. Then the $\lambda_i^{(1)}$ satisfy

$$a(i)\lambda_i^{(1)} + \left(h'(i) - \langle \lambda_i \mid h' \mid \lambda_i \rangle\right)\lambda_i = \sum_k^{(k \neq i)} \langle \lambda_k \mid h' \mid \lambda_i \rangle \lambda_k \tag{180}$$

where $\langle \lambda_k \mid \lambda_i^{(1)} \rangle = 0$ for all k. We can write more generally $\bar{\lambda}_i^{(1)} = \lambda_i^{(1)} +$

$\sum_j d_j \lambda_j$, where $\lambda_i^{(1)}$ is the part of $\bar{\lambda}_i^{(1)}$ orthogonal to all λ_i, so $d_j = \langle \lambda_j \mid \bar{\lambda}_i^{(1)} \rangle$. Then solving for $\lambda_i^{(1)}$ in terms of $\bar{\lambda}_i^{(1)}$ and substituting yields an equation for $\bar{\lambda}_i^{(1)}$, identical to Dalgarno's [16] Eq. 58. That equation is

$$a(i)\bar{\lambda}_i^{(1)}(i) + \sum_k (\varepsilon_k - \varepsilon_i)\lambda_k \langle \lambda_k \mid \bar{\lambda}_i^{(1)} \rangle + b(i)\lambda_i$$
$$\overset{(k \neq i)}{= \sum_k} \langle \lambda_k \mid h' \mid \lambda_i \rangle \lambda_k \tag{181}$$

The second-order energy [Eq. (150)] is a sum of terms, each the matrix element of a one-electron Hamiltonian between Ψ^{HF} and a singly excited determinant, so, according to Eq. (100),

$$\langle \Psi^{(0)} \mid V \mid \Psi'^{(1)} \rangle - \sum_{k=1}^N \langle \lambda_k \mid h' \mid \lambda_k^{(1)} \rangle. \tag{182}$$

The third order energy [Eq. (151)] is slightly more complicated.

Returning to Method I, we have to minimize the expectation value of $H = H_0 + V$ for the trial function

$$\Psi = \mathscr{A} \left\{ \prod_{i=1}^N \lambda_i^\gamma \right\} \tag{183a}$$

where

$$\lambda_i^\gamma = \lambda_i^{(0)} + \gamma \lambda_i^{(1)} + \gamma^2 \lambda_i^{(2)} + \cdots \tag{183b}$$

This will give the Hartree–Fock equations

$$\left(\bar{h}(i) + \gamma h'(i) + \sum_j [J_j^\gamma(i) - K_j^\gamma(i)] \right)(\lambda_i^{(0)} + \gamma \lambda_i^{(1)} + \cdots)$$
$$= (\varepsilon_i^{(0)} + \gamma \varepsilon_i^{(1)} + \gamma^2 \varepsilon_i^{(2)} + \cdots)(\lambda_i^{(0)} + \gamma \lambda_i^{(1)} + \cdots)$$

where $\bar{h}$ is the truly one-electron part of H_0. The orbital energies are functions of γ, as are the Coulomb and exchange operators, defined by

$$J_j^\gamma(i)f(i) = \langle \lambda_j^\gamma(j) \mid 1/r_{ij} \mid \lambda_j^\gamma(j) \rangle_j f(i)$$
$$K_j^\gamma(i)f(i) = \langle \lambda_j^\gamma(j) \mid 1/r_{ij} \mid f(j) \rangle_j \lambda_j^\gamma(i)$$

the integration being over j. Clearly the terms zero order in γ yield

$$h_0(i)\lambda_i^{(0)}(i) = \varepsilon_i^{(0)}\lambda_i^{(0)}(i)$$

so we identify $\lambda_i^{(0)}$ with λ_i of the previous discussion, solutions to the

Hartree–Fock problem in the absence of V. The first-order equation determines $\lambda_i^{(1)}$ and $\varepsilon_i^{(1)}$:

$$\bar{h}(i)\lambda_i^{(1)}(i) + h'(i)\lambda_i^{(0)}(i) + \sum_j [\langle \lambda_j^{(0)}(j) \mid 1/r_{ij} \mid \lambda_j^{(0)}(j)\rangle_j \lambda_i^{(1)}(i)$$

$$- \langle \lambda_j^{(0)}(j) \mid 1/r_{ij} \mid \lambda_i^{(1)}(j)\rangle_j \lambda_j^{(0)}(i)$$

$$+ \langle \lambda_j^{(1)}(j) \mid 1/r_{ij} \mid \lambda_j^{(0)}(j)\rangle_j \lambda_i^{(0)}(i) - \langle \lambda_j^{(1)}(j) \mid 1/r_{ij} \mid \lambda_i^{(0)}(j)\rangle_j \lambda_j^{(0)}(i)$$

$$+ \langle \lambda_j^{(0)}(j) \mid 1/r_{ij} \mid \lambda_j^{(1)}(j)\rangle_j \lambda_i^{(0)}(i)$$

$$- \langle \lambda_j^{(0)}(j) \mid 1/r_{ij} \mid \lambda_i^{(0)}(j)\rangle \lambda_j^{(1)}(j)] = \varepsilon_i^{(0)}\lambda_i^{(1)}(i) + \varepsilon_i^{(1)}\lambda_i^{(0)}(i) \qquad (184)$$

Dalgarno [16] gives the second-order equation as well as formulas for $\varepsilon_i^{(1)}$ and $\varepsilon_i^{(2)}$. Here again, if we write $\Psi(\gamma) = \Psi^{(0)} + \gamma\Psi^{(1)} + \cdots$, we find the second-order energy is determined from (182), provided that we assume, as we may, $\lambda_i^{(1)}$ orthogonal to all λ_i.

Unquestionably, (184) is messier than (180), and not at all equivalent, so the $\lambda_i^{(1)}$ of Methods I and II are in general different. There is, first of all, a self-consistency problem in (184); all the $\lambda_i^{(1)}$ and $\varepsilon_i^{(1)}$ must be determined at once, iteratively perhaps. In particular, eqs. (184) for each $\lambda^{(1)}$ are coupled, unlike eqs. (180). Dalgarno [16], however, compares $E^{(0)} + E^{(1)} + E^{(2)}$ from the two methods with the corresponding quantity obtained by perturbing the exact Hamiltonian H with V. This may be done by double perturbation theory (Subsection 5) with the second perturbation being $W = H - H^{\mathrm{HF}}$. Using (184), $E^{(0)} + E^{(1)} + E^{(2)}$ is correct through first order in W, but this is not the case for Method II. Method I should then give more accurate results, as is shown by examples [17].

Since Hartree–Fock functions for molecules are usually of the Roothaan type, i.e., expansions in a limited basis, one requires the limited-basis version of the Hartree–Fock perturbation theory. Such a formalism was worked out by Stevens et al. [18].

If the uncoupled theory is used (Method II) the formulas may be written as the matrix transcriptions of Dalgarno's equations. It is convenient to write them in the basis formed by the occupied and virtual Hartree–Fock orbitals (in the absence of the perturbation), rather than in the basis orbitals from which they were constructed. The former are obtained from the latter by a nonsingular linear transformation. Since the first-order orbitals $\lambda_i^{(1)}$ can be taken orthogonal to all the occupied orbitals, we may write, with p the size of the basis,

$$\lambda_i^{(1)} = \sum_{k=N+1}^{p} C_{ki}\lambda_k \qquad (185)$$

and determine the C_{ki}. Inserting (185) into (180), and noting that λ_k is an eigenvector with eigenvalue ε_k of the matrix corresponding to h_0, we have after multiplication by λ_l^* and integration

$$(\varepsilon_l - \varepsilon_i)C_{li} + \langle \lambda_l \mid h' \mid \lambda_i \rangle = 0$$

Note that $l > N$ here and that the sum on the right side of (180) is over the occupied spin orbitals. Substituting in (182), we have

$$E^{(2)} = \langle \Psi^{(0)} \mid V \mid \Psi^{(1)} \rangle = \sum_{k=1}^{N} \sum_{l=N+1}^{p} \langle \lambda_k \mid h' \mid \lambda_l \rangle C_{lk} \qquad (186)$$

assuming real functions.

A similar matrix transcription could be made of Method I, or one could proceed, as do Stevens *et al.* [18], by expansions in the orders of the perturbation parameter of all expansion coefficients and orbital energies, followed by grouping together all terms of each order. This is equivalent to minimizing $H_0 + V$ over a determinantal function. The results will not be given here, as they appear as a special case of some we will obtain later in Subsection 4.

Langhoff *et al.* [17] discuss the relation between Methods I and II and introduce another, intermediate between the two: It is hopefully simpler than I and more accurate than II. In Eq. (184), if all terms involving first-order functions except that being calculated ($\lambda_i^{(1)}$) are dropped, we obtain, after rearrangement,

$$[\bar{h}(i) - \varepsilon_i^{(0)}]\lambda_i^{(1)}(i) + [h'(i) - \varepsilon_i^{(1)}]\lambda_i^{(0)}(i)$$

$$+ \sum_j [\langle \lambda_j^{(0)}(j) \mid 1/r_{ij} \mid \lambda_j^{(0)}(j) \rangle_j \lambda_i^{(1)}(i) - \langle \lambda_j^{(0)}(j) \mid 1/r_{ij} \mid \lambda_i^{(1)}(j) \rangle_j \lambda_i^{(0)}(i)]$$

$$+ \langle \lambda_i^{(0)}(j) \mid 1/r_{ij} \mid \lambda_i^{(1)}(j) \rangle_j \lambda_i^{(0)}(i) - \langle \lambda_i^{(0)}(j) \mid 1/r_{ij} \mid \lambda_i^{(0)}(i) \rangle_j \lambda_i^{(1)}(i) = 0.$$

$$(187)$$

Here the sum may be limited to $j \neq i$ and the last two terms dropped, since the four terms thus removed add to zero. Now the equations for the first-order corrections $\lambda_i^{(1)}$ are uncoupled. If we eliminate all first-order corrections to the Fock operator Eq. (187) becomes

$$[h_0(i) - \varepsilon_i^{(0)}]\lambda_i^{(1)}(i) + [h'(i) - \varepsilon_i^{(1)}]\lambda_i^{(0)}(i) = 0$$

where h_0 incorporates $\bar{h}$ and the Fock operator. This is Dalgarno's Method II, Eq. (180). Another simplification, in which an approximation to the Fock operator is introduced in (180) and (187), is also suggested [17].

Here, exchange terms are reduced to local interactions. This greatly reduces computation times. The four methods are applied to a number of atomic problems and compared for accuracy and speed.

Results of calculations by methods of this section will be given in Chapter I, Section A, of Volume 2.

3. *Improving a Determinantal Function*

We now want to discuss perturbations on the Hartree–Fock which have two-electron parts, such as the difference between the exact and Hartree–Fock Hamiltonians. (In this case, the corrections to the energy after the first are the correlation energy.) The fact that the perturbation involves only two electrons at a time seems to mean that physically we can consider electrons two by two, and deal only with two-electron problems. This would be a considerable simplification. In fact, this can be accomplished for the first-order wave function. The formalism was worked out by Sinanoğlu. He first considered [19] the bare-nucleus starting point, i.e., where the entire interelectronic repulsion was taken as perturbation, so the zero-order Hamiltonian corresponds to noninteracting electrons moving in the field of the nuclei (the Pauli principle of course leads to a correlation between them). A lot of work has been done for atoms using this approach [8, 9, 20–23]. Since many of the features of the Hartree–Fock perturbation theory are already seen here, and since we can also consider unperturbed Hamiltonians intermediate between bare-nucleus and Hartree–Fock (where some of the effect of other electrons is taken into account in H_0), we consider this first.

Here we have

$$H_0 = \sum_i \bar{h}(i) \tag{188a}$$

$$\bar{h}(i) = -\tfrac{1}{2}\nabla_i^2 - \sum_\alpha Z_\alpha/r_{\alpha i} \tag{188b}$$

where $r_{\alpha i}$ is the distance of the ith electron from nucleus α of charge Z_α. The perturbation is the interelectronic repulsion:

$$V = \sum_{i<j=1}^N g(i, j) \tag{189a}$$

$$g(i, j) = 1/r_{ij}. \tag{189b}$$

Since H_0 is one-electron, its eigenfunctions may be constructed as products of the one-electron eigenfunctions of $\bar{h}$. The antisymmetry requirement means that we take the determinant

$$\Psi^{(0)} = \mathscr{A} \prod_k \lambda_k(k) \tag{190}$$

where λ_k is a spin orbital and

$$\bar{h}(k)\lambda_k(k) = \varepsilon_k{}^0 \lambda_k(k). \tag{191}$$

Clearly,

$$H_0\Psi^{(0)} = \left(\sum_{i=1}^{N} \varepsilon_i{}^0 \right)\Psi^{(0)}.$$

Unless $\Psi^{(0)}$ is already an eigenfunction of total spin and of other symmetry operators (see Sections A.1 and B.1 of Chapter II, Vol. 2), the correct zero-order eigenfunction would be a sum of determinants like (190). The formalism of this section does not seem to have been worked through for anything but the nondegenerate case, where a single determinant suffices. This may be connected with the fact that, when several determinants are needed, we cannot identify (before antisymmetrization) one electron with each orbital, and this identification is partly responsible for the simplicity of what follows.

We consider the nondegenerate case only. Since we are interested in the ground state, we use the N spin orbitals with lowest orbital energies. The zero- and first-order energies are

$$E^{(0)} = \sum_{k=1}^{N} \varepsilon_k{}^0 \tag{192}$$

$$E^{(1)} = \sum_{i<j=1}^{N} (\bar{J}_{ij} - \bar{K}_{ij}). \tag{193}$$

$\bar{J}_{ij}$ is the Coulomb integral between spin orbitals λ_i and λ_j,

$$\bar{J}_{ij} = \langle \lambda_i(1)\lambda_j(2) \mid g(1, 2) \mid \lambda_i(1)\lambda_j(2) \rangle \tag{194a}$$

and $\bar{K}_{ij}$ the exchange integral,

$$\bar{K}_{ij} = \langle \lambda_i(1)\lambda_j(2) \mid g(1, 2) \mid \lambda_j(1)\lambda_i(2) \rangle \tag{194b}$$

$\bar{K}_{ij}$ vanishes unless λ_i and λ_j have parallel spins. We now consider Eq. (143) for the first order wave function, and show that it may be reduced to a set of pair problems; i.e., that $\Psi^{(1)}$ may be written in terms of the λ_i and functions u_{ij}, which are obtained as solutions to two-electron problems.

Sinanoğlu [19] uses Green's functions for this; we shall express the argument somewhat differently.

Since V and $E^{(1)}$ are totally symmetric to exchanges of electrons, Eq. (143) can be written

$$(H_0 - E^{(0)})\Psi^{(1)} = \mathscr{A}\left\{(E^{(1)} - V)\prod_k \lambda_k(k)\right\}. \tag{195}$$

Thus the left side of (195) is antisymmetric, which implies, because $H_0 - E^{(0)}$ is totally symmetric and commutes with $\mathscr{A}$, that $\Psi^{(1)}$ must be antisymmetric. We put

$$\Psi^{(1)} = 2^{-1/2}\mathscr{A}\Psi \tag{196}$$

(the reason for the $2^{-1/2}$ will appear later). Thus a sufficient (but not necessary) condition for the solution of (195) is

$$2^{-1/2}(H_0 - E_0)\Phi = (E^{(1)} - V)\prod_k \lambda_k(k). \tag{197}$$

In fact, we can add to the right side of (197) any function F that will be annihilated by the antisymmetrizer; that is, we can replace the right side by anything which, after antisymmetrization, gives an identical result. The Φ which solves the resulting equation, inserted into (196), will satisfy (195). With Sinanoğlu, we define

$$e_i = \bar{h}(i) - \varepsilon_i \tag{198a}$$

and

$$m_{ij} = -g(i, j) + \bar{J}_{ij} - \bar{K}_{ij} \tag{198b}$$

which takes advantage of the fact that, in (197), each electron is associated with a spin orbital. We now have to solve

$$2^{-1/2}\left(\sum_i e_i\right)\Phi = \left(\sum_{i<j} m_{ij}\right)\prod_k \lambda_k(k) + F. \tag{199}$$

Except for the presence of F, Φ could be taken as a sum of terms, each obtained from $\prod_k \lambda_k(k)$ by replacing a pair of spin orbitals by a pair function.

$$2^{-1/2}\Phi = \sum_{i<j}\left[\frac{u_{ij}(i, j)}{\lambda_i(i)\lambda_j(j)}\prod_k \lambda_k(k)\right]. \tag{200}$$

The pair functions would satisfy

$$(e_k + e_l)u_{kl}(k, l) = m_{kl}\lambda_k(k)\lambda_l(l). \tag{201}$$

But (201) is not well posed because of the exchange degeneracy. To use the language of degenerate perturbation theory, (201) is the equation for a first-order wave function, with the zero-order function, $\lambda_k(k)\lambda_l(l)$, one of a degenerate set. The other partner is $\lambda_k(l)\lambda_l(k)$, and the perturbation g is not diagonalized by these functions, since on multiplying (201) by $\lambda_k(l)\lambda_l(k)$ and integrating:

$$\langle \lambda_l(k)\lambda_k(l) \mid e_k + e_l \mid u_{kl}(k, l)\rangle = -\langle \lambda_l(k)\lambda_k(l) \mid g(k, l) \mid \lambda_k(k)\lambda_l(l)\rangle$$

Since e_k and e_l are Hermitian, the left side vanishes but the right side does not. The symmetric and antisymmetric combinations

$$2^{-1/2}(\lambda_k(k)\lambda_l(l) \pm \lambda_l(k)\lambda_k(l))$$

should be used instead of the simple products. This is accomplished by replacing m_{ij} in (199) by

$$2^{-1/2}\mathscr{B}_{ij}m_{ij} = \tfrac{1}{2}(1 - \mathscr{T}_{ij})m_{ij}$$

where $\mathscr{T}_{ij}$ interchanges electrons i and j. Operation of the antisymmetrizer $\mathscr{A}$ on this gives the same result as before:

$$\mathscr{A}\left\{2^{-1\,2}\sum_{i<j}\mathscr{B}_{ij}m_{ij}\prod_k\lambda_k(k)\right\} = \mathscr{A}\left\{\sum_{i<j}m_{ij}\prod_k\lambda_k(k)\right\}$$

We are using some of the arbitrariness expressed by F, Eq. (199). Now, however, the pair functions satisfy, instead of (201),

$$(e_k + e_l)u_{kl}(k, l) = m_{kl}\,\mathscr{B}_{kl}\{\lambda_k(k)\lambda_l(l)\} \tag{202}$$

and (200) is replaced by

$$\Phi = \sum_{i<j}\left[\frac{u_{ij}(i, j)}{\lambda_i(i)\lambda_j(j)}\prod_k\lambda_k(k)\right] \tag{203}$$

Note that (assuming orthonormal spin orbitals) $\mathscr{B}_{kl}\{\lambda_k(k)\lambda_l(l)\}$ is normalized. Furthermore, the degenerate zero-order function orthogonal to this is the symmetric function

$$[\lambda_k(k)\lambda_l(l) + \lambda_l(k)\lambda_k(l)]$$

and the matrix element of g between this and $\mathscr{B}_{kl}\{\lambda_k(k)\lambda_l(l)\}$ vanishes by symmetry. Thus (202) is well posed as far as exchange degeneracy is concerned.

There is, however, a difficulty due to spin degeneracy, if λ_k and λ_l differ in both space and spin parts. (Similar problems arise due to orbital degeneracy, especially for atoms, but we do not consider them here.) Here again, one can use the arbitrariness contained in F, and eventually show that the first-order wave function is given exactly by

$$\Psi^{(1)} = 2^{-1}{}^2\mathscr{A}\left\{ \sum_{i<j} \frac{u_{ij}(i, j)}{\lambda_i(i)\lambda_j(j)} \prod_k \lambda_k(k) \right\} \tag{204}$$

where the u_{ij} are constructed from the usual singlet and triplet spin functions and spatial functions which are determined from equations involving two electrons only [19]. Furthermore, these two-electron equations, just as the original many-electron first-order equation, are derivable from a variational principle like (167), so that approximate solutions may be generated by choosing parameters in a trial function f. If f is chosen as a linear combination of two-electron functions, the coefficients are obtained from (161)–(164). If these functions are determinants of one-electron functions, u_{kl} and hence $\Psi^{(1)}$ will be a sum of determinants, each differing from $\Psi^{(0)}$ by one or two spin orbitals. However, the two-electron functions can involve the interelectron distance r_{kl} directly. Then $\Psi^{(1)}$ would be a correlated wavefunction, where interelectronic distances appear explicitly.

For the bare-nucleus case [24], it turns out that the first-order wavefunction for the many-electron problem can be constructed from the first-order solutions for the ground and excited states of the two-electron molecule with the same nuclear charges. (Furthermore, a scaling such as that used to generate the $1/Z$ expansion for atoms [22, 23] (see Section C.1) can be used to show [25] that from the solutions to the problem for nuclear charges Z_A and Z_B we can generate solutions for any molecule with nuclear charges in the same ratio.) The zero-order energy (192) can in fact be written [24, 25] as a sum of the zero-order energies for the states of the two-electron molecule used to build up the many-electron wave function. The first-order energy may also be written as a sum of contributions of these pairs, but the second-order energy [see Eqs. (150 and (204)] is no longer a sum of pair contributions. The second-order energies of the two-electron problems appear, but there are other terms [24, 25], involving three and four electrons. The latter break down into products of two-electron terms, while the three-electron terms are of the form

$$\langle u_{ij}(1, 2)\lambda_k(3) \mid g(2, 3) \mid \lambda_i(1)\lambda_j(2)\lambda_k(3) \rangle.$$

Some of these correspond to one-electron corrections to the zero-order functions, while others are "exclusion terms." These are corrections arising from the fact that in (204) the pair functions are inserted into the "sea" of occupied orbitals, which removes contributions of all λ_k ($k \neq i, j$) from u_{ij} [2, 26, 27].

It must be noted that the two-electron states treated by the bare-nucleus theory may include autoionizing states; that is, the bare-nucleus states are discrete states imbedded in a continuum. This occurs for the zero order pair $\lambda_k(1)\lambda_k(2)$ when $2(\varepsilon_k^0 - \varepsilon_1^0) > \bar{\varepsilon} - \varepsilon_1^0$, where ε_1^0 is the energy of the lowest-lying bare-nucleus spin orbital and $\bar{\varepsilon}$ the onset of the continuum. Then $\lambda_k(1)\lambda_k(2)$ is degenerate with a pair corresponding to occupation of λ_1 and a continuum orbital. The two-electron equation for the first order wavefunction is incorrect, since the perturbation has not been diagonalized over the degenerate zero-order functions. There are several ways out of this difficulty [28]. In the many-electron system, the autoionization is prevented by the fact that the continuum state is not allowed by the Pauli principle since λ_1 is already doubly filled. In other words, when the pair function is inserted into the many-electron wave function Φ [Eq. (203)] and Φ is antisymmetrized [Eq. (196)], the (infinite) contribution of the continuum state to the pair function will be canceled off. Thus, we can demand that the pair function u_{ij} be strongly orthogonal to all occupied spin orbitals except λ_i and λ_j. But if one assures strong orthogonality by explicitly removing contributions of λ_i and λ_j from a trial function f_{ij}, integrals like $\langle f_{ij}(1, 2) \,|\, 1/r_{12} \,|\, \lambda_i(1)\lambda_j(2) \rangle$, which appear in the calculation for the pair functions, are now in fact no longer two-electron [29]. Suppose $f_{ij}(1, 2)$ is replaced by $f_{ij}(1, 2) - \lambda_k(1)\langle \lambda_k(1') \,|\, f_{ij}(1, 2)\rangle \cdots$. The integral would include the term

$$\langle f_{ij}(1', 2)\lambda_k(1) \,|\, 1/r_{12} \,|\, \lambda_k(1')\lambda_j(2)\lambda_i(1)\rangle.$$

If $f_{ij}(1', 2)$ is correlated, i.e., contains terms in $r_{1'2}$, this cannot be reduced to two-electron integrals without, for instance, expanding $1/r_{12}$ in something like a Neumann expansion. The problem is only formally two-electron. We can thus content ourselves with a configuration expansion for the trial function, as was done by Byron and Joachain [30], or use other perhaps not completely rigorous methods [28]. In fact, it has been argued [31] that the CI form may not be very disadvantageous for certain pairs.

In principle, we could calculate the third-order energy from the first-order wave function according to (151). In practice, the bookkeeping quickly becomes quite complicated, and the number of integrals to evaluate large, as the number of electrons increases. For the three-electron atom,

Seung and Wilson [20] have carried this task to completion, but the molecular calculations [24, 25] stopped with $E^{(2)}$. If the perturbation method is to be useful, the energy series (141b) must converge rapidly. The bare-nucleus theory for H_2 is satisfactory [22] (error of 0.01 a.u. when the series is truncated at $E^{(2)}$)—here, we add, a large number of terms can be calculated in the energy series [22]—but results for He_2 [24] are already disappointing.

It is possible to improve on the bare-nucleus perturbation theory by improving the zero-order Hamiltonian, thus giving better convergence in the series for E and Ψ, without complicating the formalism or the calculations. Let the zero-order Hamiltonian be any local Hamiltonian (containing no integral operators). We will assume, although it is not necessary, that it consists of the usual kinetic energy operator plus an arbitrary potential energy operator. Instead of (188),

$$h_0(i) = - \tfrac{1}{2}\nabla_i^2 - f(\mathbf{r}_i). \tag{205}$$

The perturbation Hamiltonian is now

$$\bar{V} = \sum_{i<j=1}^{N} g(i, j) - \sum_i v(i) \tag{206a}$$

where

$$v(i) = \sum_\alpha Z_\alpha/r_{\alpha i} - f(\mathbf{r}_i). \tag{206b}$$

$\bar{V}$ is to be made effectively small, so that f differs from the bare nucleus potential by something that simulates the interelectronic repulsion. A screened-nucleus potential (Coulombic interactions with *effective* Z_α) would be a first possibility. The function f may also be taken as something simpler than the bare-nucleus molecular Hamiltonian, such as that for a one-center system without interelectronic repulsion. Then the zero-order function is built up from hydrogen-like orbitals and the additional nuclear–electronic interaction terms appear, together with the interelectronic repulsion, in the perturbation. Montgomery *et al.* [32] performed variation–perturbation calculations for the second-row hydrides with this starting point. If we rewrite the perturbation Hamiltonian as

$$\bar{V} = \sum_{i<j=1}^{N} g(i, j) - (N - 1)^{-1} \sum_{i<j=1}^{N} [v(i) + v(j)] \tag{207}$$

the entire bare-nucleus formalism may be taken over using the one-electron zero-order Hamiltonian (207) and putting for $g(i, j)$

$$h(i, j) = g(i, j) - (N - 1)^{-1}[v(i) + v(j)]. \tag{208}$$

The zero-order wave function will be a determinant formed from the spin orbitals which are eigenfunctions of h_0, Eq. (205). The best such determinant (in the sense of minimizing $E^{(0)} + E^{(1)}$) is the Hartree–Fock, but we should be able to approximate the nonlocal Hartree–Fock potential by a local function.

The power of such perturbation theories would be greatly enhanced if h_0 were different for different electrons. Then we could freely use physical intuition in choosing a zero-order wave function and subsequently constructing a Hamiltonian of which it is an eigenfunction. With the formalism of this section, the zero-order Hamiltonian must be symmetric in the electrons. However, the Distinguishable Electron Method, being developed by Kirtman [33], promises to give the necessary flexibility. It has already been applied to some atomic and molecular problems.

The idea of using perturbation theory starting from the Hartree–Fock function was used by Møller and Plesset [34] in proving Brillouin's theorem, but was worked through in detail by Sinanoğlu in an influential series of papers [26, 27, 35–37]. Even in the absence of calculations using the formalism, Sinanoğlu's work has had much influence on the thinking of quantum chemists. The physical ideas emerging from these papers have proved fruitful in other calculations. [See pp. 141–143.] Here we are concerned with the Hartree–Fock perturbation theory as a tool for calculations on diatomic molecules. At this writing, no such actual calculation is known to the author. The reasons may lie in the discussion to follow.

In our previous notation [Eqs. (205)–(208)], take

$$v(1) = \sum_{k=1}^{N} \langle \lambda_k^{\mathrm{HF}}(2) \mid 1/r_{12} \mid (1 - \mathscr{S}_{12}) \lambda_k^{\mathrm{HF}}(2) \rangle_2 \tag{209}$$

where the subscript on the bracket means the integration is over the space and spin coordinates of electron 2, and the interchange of 1 and 2 is to be performed before the integration. The first-order energy is given in this case by [see Eq. (194)]

$$\langle \Psi^{\mathrm{HF}} \mid V \mid \Psi^{\mathrm{HF}} \rangle = \sum_{i>j=1}^{N} (\bar{J}_{ij} - \bar{K}_{ij}) - \sum_{i=1}^{N} \langle \lambda^{\mathrm{HF}}(i) \mid v(i) \mid \lambda^{\mathrm{HF}}(i) \rangle$$

$$= \sum_{i>j=1}^{N} (\bar{J}_{ij} - \bar{K}_{ij}) - \sum_{i,k=1}^{N} (\bar{J}_{ik} - \bar{K}_{ik}) \tag{210}$$

The terms for $i = k$ vanish in the second sum, making

$$E^{(1)} = -\sum_{i<j=1}^{N} (\bar{J}_{ij} - \bar{K}_{ij}). \qquad (211)$$

The first-order wave function may be written as $U' - W$, where U' satisfies

$$(H_0 - E^{(0)})U' + \left[\sum_{i>j=1}^{N} (g_{ij} - \bar{J}_{ij} + \bar{K}_{ij})\right]\Psi^{\mathrm{HF}} = 0 \qquad (212)$$

and W

$$(H_0 - E^{(0)})W + \left[-2\sum_{i>j=1}^{N} (\bar{J}_{ij} - \bar{K}_{ij}) + \sum_{i=1}^{N} v(i)\right]\Psi^{\mathrm{HF}} = 0. \qquad (213)$$

The difference of (212) and (213) is just Eq. (143), the usual equation for the first-order wave function. The reason for making this separation is that W may be obtained from U', while U' is the first order wave function of the bare-nucleus perturbation theory, but *starting from the spin orbitals* $\{\lambda_k^{\mathrm{HF}}\}$ instead of the bare-nucleus spin orbitals, and with the Hartree–Fock one-electron Hamiltonian instead of h_0. Considering this problem to be solved, put

$$U' = 2^{-1/2}\mathscr{A}\left\{\sum_{i>j=1}^{N} \frac{u_{ij}(i, j)}{\lambda_i^{\mathrm{HF}}(i)\lambda_j^{\mathrm{HF}}(j)} \prod_k \lambda_k^{\mathrm{HF}}(k)\right\} \qquad (214)$$

[see Eqs. (196) and (203)]. We will assume here that the pair functions u'_{ij} are strongly orthogonal to all the occupied Hartree–Fock spin orbitals except for λ_i^{HF} and λ_j^{HF}, and also that

$$\langle u'_{ij}(1, 2) \mid \lambda_i^{\mathrm{HF}}(i)\lambda_j^{\mathrm{HF}}(j) - \lambda_j^{\mathrm{HF}}(i)\lambda_i^{\mathrm{HF}}(j)\rangle = 0. \qquad (215)$$

This is the significance of the primes here. These orthogonality conditions may be demanded with no loss of generality (see pages 108 and 153).

In Eq. (213), we regroup terms in $\sum_i v(i)$ [somewhat as we did for the local perturbation in Eq. (207)], noting that the terms for $i = k$ in the double sum vanish,

$$\sum_{i=1}^{N} v(i) = \sum_{i>k=1}^{N} [\langle \lambda_i^{\mathrm{HF}}(2) \mid 1/r_{2k} \mid (1 - \mathscr{P}_{2k})\lambda_i^{\mathrm{HF}}(2)\rangle_2$$
$$+ \langle \lambda_k^{\mathrm{HF}}(2) \mid 1/r_{2i} \mid (1 - \mathscr{P}_{2i})\lambda_k^{\mathrm{HF}}(2)\rangle_2] \qquad (216)$$

The square bracket has four terms: the Coulomb potential of spin orbital i acting on electron k, which we denote by $S_i(k)$, the exchange potential of spin orbital i acting on electron k, which we denote by $R_i(k)$, and $S_k(i)$

and $R_k(i)$. Then (213) is

$$(H_0 - E^{(0)})W + \left[\sum_{i>j=1}^{N} (S_i(j) - R_i(j) - \bar{J}_{ij} + \bar{K}_{ij} \right.$$
$$\left. + S_j(i) - R_j(i) - \bar{J}_{ji} + \bar{K}_{ji}) \right] \Psi_{\mathrm{HF}} = 0. \qquad (217)$$

Let the square bracket be written as $-\sum_{i>j=1}^{N} k_{ij}$. Comparing this to (199), we see we have essentially the same thing, with k_{ij} replacing m_{ij}. Thus we can put

$$W = 2^{-1/2} \mathscr{A} \left[\sum_{i<j=1}^{N} \frac{w_{ij}(i, j)}{\lambda_i^{\mathrm{HF}}(i) \lambda_j^{\mathrm{HF}}(j)} \prod_k \lambda_k^{\mathrm{HF}}(k) \right] \qquad (218)$$

where, analgously to (202),

$$[h_0(k) - \varepsilon_k{}^0 + h_0(l) - \varepsilon_l{}^0] w_{kl}(k, l) = k_{kl} \mathscr{B}_{kl} \{\lambda_k^{\mathrm{HF}}(k) \lambda_l^{\mathrm{HF}}(l)\} \qquad (219)$$

with $\varepsilon_k{}^0$ and $\varepsilon_l{}^0$ Hartree–Fock orbital energies. Treating (219) as we treated (202), it is straightforward to show

$$w_{kl}(k, l) = \lambda_k^{\mathrm{HF}}(k) \langle \lambda_k^{\mathrm{HF}}(n) \mid u'_{kl}(n, l) \rangle - \lambda_k^{\mathrm{HF}}(l) \langle \lambda_k^{\mathrm{HF}}(n) \mid u'_{kl}(n, k) \rangle$$
$$+ \lambda_l^{\mathrm{HF}}(l) \langle \lambda_l^{\mathrm{HF}}(n) \mid u'_{kl}(k, n) \rangle - \lambda_l^{\mathrm{HF}}(k) \langle \lambda_l^{\mathrm{HF}}(n) \mid u'_{kl}(l, n) \rangle \qquad (220)$$

The antisymmetry of the u'_{kl} has been used.

Finally, we may combine [27] U' (equation 223) and W to give

$$\Psi^{(1)} = 2^{-1/2} \mathscr{A} \left[\sum_{i<j=1}^{N} \frac{u_{ij}(i, j) - w_{ij}(i, j)}{\lambda_i^{\mathrm{HF}}(i) \lambda_j^{\mathrm{HF}}(j)} \prod_k \lambda_k^{\mathrm{HF}}(k) \right]$$
$$= 2^{-1/2} \mathscr{A} \left[\sum_{i<j=1}^{N} \frac{\hat{u}_{ij}(i, j)}{\lambda_i^{\mathrm{HF}}(i) \lambda_j^{\mathrm{HF}}(j)} \prod_k \lambda_k^{\mathrm{HF}}(k) \right]. \qquad (221)$$

This means the pair functions for the first-order wave function in the Hartree–Fock case may be obtained by solving for the pair functions using the full interelectronic repulsion as perturbation and then carrying out strong orthogonalization of each u'_{ij} to all the occupied spin orbitals λ_i^{HF}. Alternatively, it shows that, if one solves for $\Psi^{(1)}$ directly without partitioning into U' and W, the pair functions will automatically be strongly orthogonal to the occupied spin orbitals [30], which is essentially Brillouin's theorem.

If we calculate $E^{(2)}$ using $\Psi^{(1)}$ of (221), using the strong orthogonality, we obtain a sum over pair contributions:

$$E^{(2)} = \sum_{i<j=1}^{N} \langle \mathscr{B}_{ij} \{\lambda_i(i) \lambda_j(j)\} \mid g(i, j) \mid \hat{u}_{ij}(i, j) \rangle \qquad (222)$$

The third order energy is rather complicated and we shall not give formulas here. Byron and Joachain [30] have discussed it for the four-electron atom, where the six pair functions yield 36 terms. They decided to neglect many of these.

It is possible to formulate perturbation theory on the Hartree–Fock function exclusively in terms of the density matrix. This includes the case of one-electron perturbations as well as the case just discussed. McWeeny and co-workers have been very active in developing and applying such forma-lisms.* Musher [10] has derived equations for the pair functions which are somewhat simpler than Sinanoğlu's.

In recent years, there has been a great deal of success in calculating correlation energies by many-body theories. Here, we are using perturba-tion theory to obtain corrections to a single-determinant wave function (usually the Hartree–Fock), including higher-order energies than second. We must go beyond second order if we want energies to chemical accuracy (1 kcal mole^{-1}) even for a system as small as Be. Effectively, all higher-order wave functions are expressed as expansions in a convenient set of basis functions, but one deals directly with the expansion for the energy and other expectation values. These would be rather complicated collections of one- and two-electron integrals. They can be handled because of several features of these calculations. First, the terms are represented as diagrams, which helps in removing terms which cancel and also in identifying those which make important contributions. The cancellations between certain terms are necessary, as pointed out by Brueckner [38, 39], so that the energy be proportional to the number of particles as this number becomes large. In some cases, sets of diagrams of various orders of perturbation theory can be combined and their contributions summed. When one can estimate that certain contributions will be small, approximations can be introduced. The resulting formulas often lend themselves to simple interpretations in terms of, e.g., pair functions. It is even possible to derive equations deter-mining such pair contributions which are simpler than those obtained from cluster-type wave functions [40].

By choosing a basis set, such as the occupied and virtual Hartree–Fock orbitals, where most or all of the excited functions form a continuum, one replaces summations over intermediate states by integrations, which are more tractable. For atoms, the angular coordinates may be handled simply and sums over states become simply integrations over radial quantum numbers. When we turn to diatomic and other molecules, the lack of

* For references see McWeeny and Sutcliffe [15].

spherical symmetry complicates the calculations. Thus, in Kelly's work on H_2 and HF, a one-center basis set (single-center or united atom expansion) was used [41]. For HF, for instance, the unperturbed Hamiltonian corresponded to the neon atom, and the difference in nuclear potential between Ne and HF was included in the perturbation. 92% of the correlation energy was obtained. Energy calculations not using a single-center expansion have appeared only for H_2. Schulman and Kaufman [42] calculated the second-order energy using many-body diagram techniques and a basis set constructed from Gaussian functions on the two nuclei. The energy $E^{(0)} + E^{(1)} + E^{(2)}$ was -1.880 a.u., hardly closer to the correct energy (1.8888 a.u.) than $E^{(0)} + E^{(1)} + E^{(2)}$ from the bare-nucleus perturbation theory (-1.899 a.u.). Kelly's one-center calculation [41] included the important contributions to $E^{(3)}$ and obtained -1.890 a.u. As shown in Chapter I, Section A, Volume 2, the London dispersion forces can be computed in terms of *atomic* polarizabilities for imaginary frequencies, and these quantities have been successfully calculated by many-body theory [43].

A detailed explication of the formulas of the many-body theory would take us far afield. Books on the subject are plentiful [44], and a number of reviews of their application to atomic and molecular problems have appeared [40, 45, 46]. The relations between the many-body perturbation theory and pair theories such as those of Sinanoğlu (already mentioned and discussed in Section A.8), Szász, Nesbet, and others are being elucidates [40, 46]. However, there has been little success in using the many-body theory for $U(R)$.

4. *Perturbation Theory for Change of R*

As should be evident at this point, most calculations aimed at obtaining $U(R)$ actually produce a sequence of energies for different values of R, each corresponding to a separate calculation. The R-dependence is not explicit. Among the exceptions are the united atom expansions (Chapter I, Section C, Volume 2) and the calculation of long-range forces (Chapter I, Section A, Volume 2). Perturbation theory in principle allows us to develop the energy as a function of R since, starting from the wave function and energy for some particular value of R, the change in internuclear separation can be taken as a perturbation. This was done in the development of the Born–Oppenheimer theory in Chapter I, but the step from this formalism to numerical calculations was not taken for a long time.

Except for a limited amount of work on very small molecules, perturbation calculations of the change of energy with R have produced little in the way of accurate numerical results. Rather, their principal use has been to derive formal expressions for such quantities as force constants, which, combined with physical intuition, have aided in the interpretation of experimental results. Here, we include formulas like equations (243), (247), and (268). It seems possible, however, that calculations according to Hartree–Fock perturbation theory may be useful where accurate Hartree–Fock wavefunctions are available and force constants are of particular interest, so that a brief discussion of its application to the problem is given [Eq. (249)–(266)]. The calculation of force constants directly, rather than from a series of energies for different values of R (Volume 2, Chapter I, Section C.1) has always been an appealing idea.

We first discuss some general features which arise in calculating the change in energy of a diatomic system with change in internuclear distance, using perturbation theory.

As in the Born–Oppenheimer case, it must be noted that the Hamiltonian is not written simply as $H_0 + \gamma V$ with γ the parameter used to "order" the terms, but as a power series in γ, so the perturbation itself has first-order, second-order, etc. terms [see Eq. (17) of Chapter I]. To make this specific to the two-nucleus case, let us write for the electronic Hamiltonian, including internuclear repulsion,

$$
\begin{aligned}
H = H(0,0) &+ (\partial H/\partial r_A)_0 \, \delta r_A + (\partial H/\partial r_B)_0 \, \delta r_B \\
&+ \tfrac{1}{2}[(\partial^2 H/\partial r_A{}^2)_0 \, \delta r_A{}^2 + 2(\partial^2 H/\partial r_A \, \partial r_B)_0 \, \delta r_A \, \delta r_B \\
&+ (\partial^2 H/\partial r_B{}^2)_0 \, \delta r_B{}^2] + \cdots
\end{aligned}
\tag{223}
$$

where δr_A and δr_B, the displacements of the two nuclei from some reference position, are taken as positive in the directions tending to increase R. The eigenvalue of H is similarly expanded:

$$
\begin{aligned}
E = E(0,0) &+ (\partial E/\partial r_A)_0 \, \delta r_A + (\partial E/\partial r_B)_0 \, \delta r_B \\
&+ \tfrac{1}{2}[(\partial^2 E/\partial r_A{}^2)_0 \, \delta r_A{}^2 + 2(\partial^2 E/\partial r_A \, \partial r_B)_0 \, \delta r_A \, \delta r_B \\
&+ (\partial^2 E/\partial r_B{}^2)_0 \, \delta r_B{}^2] + \cdots
\end{aligned}
\tag{224}
$$

In (223) and (224), the subscripts 0 refer to evaluation at the reference position. Now an increase of R by δR can be carried out in many different ways, i.e., δr_A and δr_B could both be positive, one could be negative and the other positive but larger in magnitude, either could be zero, and so on. [This arbitrariness comes up in the discussion (Section D) of the Hellmann–Feynman theorem.] This gives relations between the coefficients in (224).

For instance, E must be unchanged if $\delta r_A = -\delta r_B$, so that $(\partial E/\partial r_A)_0$ must equal $(\partial E/\partial r_B)_0$, and $2(\partial^2 E/\partial r_A\,\partial r_B)_0$ must equal $(\partial^2 E/\partial r_A{}^2) + (\partial^2 E/\partial r_B{}^2)$. The derivative $(\partial E/\partial r_A)_0$ is the force on nucleus A in the direction of decreasing R. If the reference configuration were the equilibrium internuclear distance, this would vanish. Under these circumstances, the quadratic force constant is given by

$$k_e = \left(\frac{\partial^2 E}{\partial R^2}\right)_0 = \left(\frac{\partial^2 E}{\partial r_A{}^2}\right)_0 = \left(\frac{\partial^2 E}{\partial r_B{}^2}\right)_0. \tag{225}$$

To go over to the language of perturbation theory, we put $\delta r_A = c_A\,\delta R$ and $\delta r_B = c_B\,\delta R$ (c_A and c_B now specify the process by which R is changed by δR) and take δR as our ordering parameter.

$$H = H_0 + H_1\,\delta R + H_2(\delta R)^2 + \cdots \tag{226}$$

$$E = E^{(0)} + E^{(1)}\,\delta R + E^{(2)}(\delta R)^2 + \cdots \tag{227}$$

$$\Psi = \Psi^{(0)} + \Psi^{(1)}\,\delta R + \Psi^{(2)}(\delta R)^2 + \cdots \tag{228}$$

Here, H_0 is the molecular Hamiltonian H for some reference position of the nuclei,

$$H_1 = c_A(\partial H/\partial r_A)_0 + c_B(\partial H/\partial r_B)_0, \tag{229}$$

and

$$H_2 = \tfrac{1}{2}\left[c_A{}^2\left(\frac{\partial^2 H}{\partial r_A{}^2}\right)_0 + 2c_A c_B\left(\frac{\partial^2 H}{\partial r_A\partial r_B}\right) + c_B{}^2\left(\frac{\partial^2 H}{\partial r_B{}^2}\right)\right]. \tag{230}$$

Substituting into the eigenvalue equation [see equations (138)–(151) of this chapter, (17)–(25) of Chapter I], we have

$$H_0\Psi^{(0)} = E^{(0)}\Psi^{(0)} \tag{231}$$

which is assumed to be solved with $\Psi^{(0)}$ normalized, and again

$$E^{(1)} = (\partial E/\partial R)_0 = \langle \Psi^{(1)} \mid H_1 \mid \Psi^{(0)}\rangle. \tag{232}$$

This is just the Hellmann–Feynman theorem (Section D). The first-order wave function is again determined by (143), which is now written

$$(H_0 - E^{(0)})\Psi^{(0)} + (H_1 - E^{(1)})\Psi^{(1)} = 0 \tag{233}$$

but the second-order energy has two terms:

$$E^{(2)} = \langle \Psi^{(0)} \mid H_1 - E^{(1)} \mid \Psi^{(1)}\rangle + \langle \Psi^{(0)} \mid H_2 \mid \Psi^{(0)}\rangle. \tag{234}$$

The first term is Eq. (151), the perturbation H_1 coming in in the second
order of perturbation theory, the second is the second-order perturbation
H_2 coming in in the first-order of perturbation theory. This is Eq. (25)
of the Born–Oppenheimer treatment (Chapter I). The force constant k_e is
twice $E^{(2)}$.

Relation (234) is sometimes encountered in a slightly different form,
where the higher-order energies and wave functions are written as nuclear
displacements multiplied by derivatives of the energy and wave function
with respect to nuclear displacements, as we did in equations (223) and (224).
We shall see examples of this in Section D. Byers Brown [47] considered
formulas of this kind for the general polyatomic molecule. If χ and χ'
represent any nuclear coordinates, he wrote

$$-\frac{\partial^2 E_n}{\partial\chi\partial\chi'} = \left\langle \psi_n \left| \frac{\partial^2 H}{\partial\chi\partial\chi'} \right| \psi_n \right\rangle + 2\sum_{m\neq n} \frac{\langle \psi_n | \partial H/\partial\chi | \psi_m\rangle\langle\psi_m | \partial H/\partial\chi' | \psi_n\rangle}{E_n - E_m}.$$

(235)

Here, ψ_n is the eigenfunction of H with eigenvalue E_n. Byers Brown showed
how (235) could be derived from the equations of perturbation theory, or
by differentiating $(H - E_n)\psi_n = 0$. A number of special cases of (235)
will be met with below. In his derivation, Byers Brown wrote

$$\partial\psi_n/\partial\chi = \sum_m c_{nm}\psi_m$$

and noted that, if the eigenfunctions are to remain orthonormal under a
nuclear displacement, the c_{nm} form an anti-Hermitian matrix. According
to the Hartree–Fock perturbation theories, Section 2, the derivative of a
Hartree–Fock orbital with nuclear displacement may also be written as a
sum over other Hartree–Fock orbitals. However, one cannot conclude
from this that the derivative of the density matrix with nuclear displacement
vanishes [48]. Using the definition of equation (77),

$$\frac{\partial\varrho(i,j)}{\partial\chi} = \sum_k^{(k\leq N)} \left\{ \frac{\partial\lambda_k(i)}{\partial\chi}\lambda_k(j) + \lambda_k(i)\frac{\partial\lambda_k(j)}{\partial\chi} \right\}$$

(236)

where the spin orbitals are assumed to be real. Inserting

$$\partial\lambda_k/\partial\chi = \sum_m^{(m>N)} d_{km}\lambda_m$$

we may write the derivative of the density matrix as

$$\partial\varrho(i,j)/\partial\chi = \sum_k^{(k<N)} \sum_m^{(m>N)} \{d_{km}\lambda_m(i)\lambda_k(j) + d_{km}\lambda_k(i)\lambda_m(j)\}.$$

(237)

The indices k and m may not be interchanged in the second term because the limits on their respective summations differ.

We now return to Eq. (234). Taking $\Psi^{(0)}$ and $\Psi^{(1)}$ orthogonal, we expand $\Psi^{(1)}$ in the excited eigenfunctions of H_0, denoted by $\Psi_k^{(0)}$. These are the excited electronic functions corresponding to the unperturbed (reference) position. Then

$$E^{(2)} = \sum_{k>0} \frac{|\langle \Psi^{(0)} | H_1 | \Psi_k^{(0)} \rangle|^2}{E^{(0)} - E_k^{(0)}} + \langle \Psi^{(0)} | H_2 | \Psi^{(0)} \rangle. \qquad (238)$$

$E^{(2)}$ will not depend on c_A and c_B individually so long as

$$c_A = 1 - c_B \qquad (239)$$

Inserting (229), (230), and (239) in $E^{(2)}$ and setting terms in c_B and c_B^2 equal to zero, we find (using k for $\Psi_k^{(0)}$)

$$\mathrm{Re}\left[\sum_{k>0} \frac{\langle 0 | (\partial H/\partial r_A)_0 | k \rangle \langle k | (\partial H/\partial r_B)_0 - (\partial H/\partial r_A)_0 | 0 \rangle}{E^{(0)} - E_k^{(0)}} \right]$$
$$+ \frac{1}{2} \left\langle 0 \left| \left(\frac{\partial^2 H}{\partial r_A \partial r_B} \right)_0 - \left(\frac{\partial^2 H}{\partial r_A^2} \right) \right| 0 \right\rangle = 0 \qquad (240)$$

and

$$\sum_{k>0} \frac{|\langle 0 | (\partial H/\partial r_B)_0 - (\partial H/\partial r_A)_0 | k \rangle|^2}{E^{(0)} - E_k^{(0)}}$$
$$+ \frac{1}{2} \left\langle 0 \left| \left(\frac{\partial^2 H}{\partial r_A^2} \right)_0 - 2 \left(\frac{\partial^2 H}{\partial r_A \partial r_B} \right)_0 + \left(\frac{\partial^2 H}{\partial r_B^2} \right)_0 \right| 0 \right\rangle. \qquad (241)$$

which resembles the gauge invariance in susceptibility calculations [49].

These relations may be used to simplify expressions for the second order energy. Thus, suppose we now take $c_A = c_B = \frac{1}{2}$ as was done by Berlin [Section D, Eqs. (338)–(339)]. Then (238) becomes

$$E^{(2)} = \frac{1}{4} \sum_{k>0} \frac{|\langle 0 | (\partial H/\partial r_A)_0 - (\partial H/\partial r_B)_0 | k \rangle|^2}{E^{(0)} - E_k^{(0)}}$$
$$+ \frac{1}{8} \left\langle 0 \left| \left(\frac{\partial^2 H}{\partial r_A^2} \right)_0 + 2 \left(\frac{\partial^2 H}{\partial r_A \partial r_B} \right)_0 + \left(\frac{\partial^2 H}{\partial r_B^2} \right)_0 \right| 0 \right\rangle \qquad (242)$$

With the aid of (241), this becomes

$$E^{(2)} = \mathrm{Re}\left[\sum_{k=0} \frac{\langle 0 | (\partial H/\partial r_A)_0 | k \rangle \langle k | (\partial H/\partial r_B)_0 | 0 \rangle}{E^{(0)} - E_k^{(0)}} \right]$$
$$+ \frac{1}{2} \left\langle 0 \left| \left(\frac{\partial^2 H}{\partial r_A \partial r_B} \right)_0 \right| 0 \right\rangle. \qquad (243)$$

The second term in (243) can be thought of as the contribution to $E^{(2)}$ in the absence of relaxation of the charge distribution to the new molecular configuration, and the first term as the effect of relaxation. Equation (243) is quite the same as Eq. (360) in the discussion of the Hellmann–Feynman theorem, except that the change of wavefunction is written as a first-order correction of perturbation theory [$\Psi^{(1)}$, Eq. (233)], rather than as $(\partial \Psi / \partial R)\, \delta R$.

The contribution of the internuclear repulsion term in H to $(\partial H / \partial r_A)_0$ is $-Z_A Z_B / R_0^2$ and, to $(\partial^2 H / \partial r_A\, \partial r_B)_0$, $2 Z_A Z_B / R_0^3$. Because of the orthogonality of the electronic eigenfunctions, the internuclear repulsion makes no contribution to the first term of (243). The other terms in H which depend on nuclear positions are the electron–nuclear attraction operators. We have

$$\frac{\partial}{\partial r_A}\left(\sum_i \frac{-Z_A}{r_{Ai}}\right) = Z_A \sum_i \frac{\cos\theta_{Ai}}{r_{Ai}^2} \tag{244}$$

(see "Notes on Notation and Coordinate Systems"). In differentiating a second time, we must deal with the singularity at $r_{Ai} = 0$. From classical electrostatics one has

$$\nabla^2(r_{Ai}^{-1}) = -4\pi\,\delta(\mathbf{r}_{Ai}) \tag{245}$$

so that

$$\frac{\partial^2}{\partial r_A^2}\left(\sum_i \frac{-Z_A}{r_{Ai}}\right) = -Z_A \sum_i \left(\frac{-4\pi}{3}\,\delta(\mathbf{r}_{Ai}) + \frac{3\cos^2\theta_{Ai} - 1}{r_{Ai}^3}\right) \tag{246}$$

Note that the electron–nuclear attraction terms do not contribute to $(\partial^2 H / \partial r_A\, \partial r_B)_0$. Now (243) becomes

$$E^{(2)} = \frac{Z_A Z_B}{R_e^3} + \mathrm{Re}\!\left[\sum_{k>0} \frac{\langle 0\,|\,(\partial H/\partial r_A)_0\,|\,k\rangle\langle k\,|\,(\partial H/\partial r_B)_0\,|\,0\rangle}{E^{(0)} - E_k^{(0)}}\right] \tag{247}$$

If $\Psi^{(0)}$ could be written as a single determinant, the $\Psi_k^{(0)}$ would consist of determinants differing from $\Psi^{(0)}$ by one spin orbital (singly excited).

Formula (247) was used by Murrell [50] to discuss force constants in diatomic molecules. Letting R_0 refer to the equilibrium configuration, we have $k_e = \frac{1}{2}E^{(2)}$. We might expect only the valence electrons to contribute to the force constant, the core electrons following the nuclei. The nuclei plus core electrons are then replaced by effective nuclear charges, say $\bar{Z}_A$ and $\bar{Z}_B$, where $\bar{Z}_A$ is Z_A minus the number of core electrons. Then (247) would be

$$\frac{k_e}{2\bar{Z}_A \bar{Z}_B} = \frac{1}{R_e^3} + \mathrm{Re}\!\left[\sum_{k>0} \frac{\langle 0\,|\,\sum_i (\cos\theta_{Ai}/r_{Ai}^2)\,|\,k\rangle\langle k\,|\,\sum_i (\cos\theta_{Bi}/r_{Bi}^2)\,|\,0\rangle}{E_0 - E_k}\right] \tag{248}$$

The sums over i are over valence electrons only. If we now plot [50] the observed force constants divided by $2\bar{Z}_A\bar{Z}_B$ against R_e^{-3} (R_e, observed equilibrium internuclear distance) for diatomics, it is found that, except for the alkali metal diatomics, all molecules fall to the right of the line $k_e/(2\bar{Z}_A\bar{Z}_B)$ $= R_e^{-3}$, so that the "relaxation" term is negative. This energy lowering might be expected. For Li_2, Na_2, and K_2, the points fall closely on the line. Isoelectronic molecules fall along straight lines, while molecules for which both atoms belong to the same group of the periodic table fall on smooth curves. Murrell [50] gives some discussion of these results.

Expressions such as (247) are not useful for actual calculation because of the appearance of all the excited states. If we can argue that many of these do not contribute and that one or a small number give most of the contribution, there is a hope of estimating force constants. This approach was investigated by Bader [51] with reference to potential interaction constants in certain triatomic and tetraatomic molecules. He was able to predict their signs correctly.

The use of a variational approximation to $\Psi^{(1)}$ in the framework of variation–perturbation theory [Eq. (167)] can be used to avoid the infinite expansion. Then $\Psi^{(1)}$ of Eq. (233) may be approximated by minimizing, with respect to variations in a trial function, Φ, a functional which becomes an approximation to the electronic term in (247). Furthermore, the third-order energy, which is the coefficient of $(\delta R)^3$ in the expansion of E about $E(0)$, and equivalent to the anharmonic force constant, may then also be computed. This approach was applied to H_2 by Benston and Kirtman [52]. Starting with several different approximations to $\Psi^{(0)}$, each at the internuclear distance for which the force (first-order energy) vanished, they showed that quite good agreement with the experimental values could be obtained by this method. (In at least one case, however, the result was so bad that k_e was negative.) It seemed that the quality of Φ was more important than the quality of $\Psi^{(0)}$.

Since one knows that Hartree–Fock functions give quite good approximations to the ground state wave functions of many molecules, and since the perturbation is one-electron, it is suggested that the Hartree–Fock perturbation theory, discussed in Subsection 2, be used. This means the version appropriate to a basis expansion in most cases. But there is an additional complication now. The basis orbitals we would use in general would be atomic-like functions centered on the nuclei, so the subspace spanned by the basis is actually changing when we change R. The dependence of the *basis functions* on the perturbation parameter must be considered. The necessary extension of the Stevens–Pitzer–Lipscomb equations was given by Gerratt and Mills [53]. The Stevens–Pitzer–Lipscomb equations appear as a special case.

We consider, with Gerratt and Mills [53], the Roothaan equations [Eqs. (75)–(76)]

$$\mathbf{F}\mathbf{c}_i = \varepsilon \mathbf{S}\mathbf{c}_i \tag{249}$$

for p basis functions. The matrix operator $\mathbf{F}$ consists of the matrix $\mathbf{h}$ of the one-electron part of the Hamiltonian and the Fock matrix $\mathbf{G}$, where

$$
\begin{aligned}
G_{jk} &= \sum_{l=1}^{N} \sum_{m=1}^{p} \sum_{n=1}^{p} c_{ml} c_{nl} (\langle \chi_j \chi_m \mid g \mid \chi_k \chi_n \rangle - \langle \chi_j \chi_m \mid g \mid \chi_n \chi_k \rangle) \\
&= \sum_{l=1}^{N} \sum_{m=1}^{p} \sum_{n=1}^{p} c_{ml} c_{nl} g_{jm,\,kn}.
\end{aligned}
\tag{250}
$$

We take all functions as real, and use superscripts to indicate orders of perturbation theory. Suppose (249) has been solved for the one-electron operator $h^{(0)}$ and we wish to solve it for $h^{(0)} + \gamma h^{(1)}$. The perturbation being one-electron, only $\mathbf{h}$ changes explicitly, but the self-consistency requirement means $\mathbf{G}$ also changes. In addition, we have the complication that the basis orbitals change as the one-electron Hamiltonian changes. We must write

$$\chi_k(\gamma) = \chi_k^{(0)} + \gamma \chi_k^{(1)} + \cdots, \qquad k = 1, \ldots, p \tag{251}$$

where $\chi_k^{(0)}$, $\chi_k^{(1)}$, ... are known. Thus the spin orbitals

$$\lambda_i = \sum_{k=1}^{p} c_{ki} \chi_k, \qquad i = 1, \ldots, N \tag{252}$$

depend on γ in two ways: through a change in the expansion coefficients with γ,

$$c_{ki} = c_{ki}^{(0)} + \gamma c_{ki}^{(1)} + \cdots \tag{253}$$

and through the changes in the χ_k themselves. Furthermore, all the matrix elements appearing in (249) must be expanded in powers of γ. For the one-electron part,

$$\langle \chi_j \mid h^{(0)} + \gamma h^{(1)} \mid \chi_k \rangle = h_{jk}^{(0)} + \gamma h_{jk}^{(1)} + \cdots \tag{254}$$

where $h_{jk}^{(0)} = \langle \chi_j^{(0)} \mid h^{(0)} \mid \chi_k^{(0)} \rangle$ and

$$h_{kj}^{(1)} = \langle \chi_j^{(0)} \mid h^{(1)} \mid \chi_k^{(0)} \rangle + \langle \chi_j^{(0)} \mid h^{(0)} \mid \chi_k^{(1)} \rangle + \langle \chi_j^{(1)} \mid h^{(0)} \mid \chi_k^{(0)} \rangle. \tag{255}$$

Similarly,

$$g_{jm,kn} = g_{jm,kn}^{(0)} + \gamma g_{jm,kn}^{(1)} + \cdots \tag{256}$$

and

$$S_{jk} = \langle \chi_j \mid \chi_k \rangle = S_{jk}^{(0)} + \gamma S_{jk}^{(1)} + \cdots \tag{257}$$

We now expand the orbital energies

$$\varepsilon_i = \varepsilon_i^{(0)} + \gamma \varepsilon_i^{(0)} + \cdots, \qquad i = 1, \ldots, N \tag{258}$$

and substitute all expansions into (249).

We will abbreviate

$$G_{jk}^{(0)} = \sum_{l=1}^{N} \sum_{m=1}^{p} \sum_{n=1}^{p} c_{ml}^{(0)} c_{nl}^{(0)} g_{jm,kn}^{(0)} \tag{259}$$

$$G_{jk}^{(1)} = \sum_{l=1}^{N} \sum_{m=1}^{p} \sum_{n=1}^{p} (c_{ml}^{(0)} c_{nl}^{(0)} g_{jm,kn}^{(1)} + c_{ml}^{(0)} c_{nl}^{(1)} g_{jm,kn}^{(0)} + c_{ml}^{(1)} c_{nl}^{(0)} g_{jm,kn}^{(0)}) \tag{260}$$

and use matrix notation. The set of coefficients corresponding to the ith molecular orbital is denoted

$$\mathbf{c}_i = \mathbf{c}_i^{(0)} + \gamma \mathbf{c}_i^{(1)} + \cdots \tag{261}$$

(we emphasize the difference in notation from the discussion of Subsection 2). Grouping terms by powers of γ, we have

$$(\mathbf{h}^{(0)} + \mathbf{G}^{(0)})\mathbf{c}_i^{(0)} + \gamma[(\mathbf{h}^{(0)} + \mathbf{G}^{(0)})\mathbf{c}_i^{(1)} + (\mathbf{h}^{(1)} + \mathbf{G}^{(1)})\mathbf{c}_i^{(0)}] + \cdots$$
$$= \varepsilon_i^{(0)}\mathbf{S}^{(0)}\mathbf{c}_i^{(0)} + \gamma[\varepsilon_i^{(1)}\mathbf{S}^{(0)}\mathbf{c}_i^{(0)} + \varepsilon_i^{(0)}\mathbf{S}^{(1)}\mathbf{c}_i^{(0)} + \varepsilon_i^{(0)}\mathbf{S}^{(0)}\mathbf{c}_i^{(1)}] + \cdots \tag{262}$$

The zero-order terms give the Roothaan equations from which we start:

$$(\mathbf{h}^{(0)} + \mathbf{G}^{(0)})\mathbf{c}_i^{(0)} = \varepsilon_i^{(0)}\mathbf{S}^{(0)}\mathbf{c}_i^{(0)}. \tag{263}$$

The rather complicated first-order equations are

$$(\mathbf{h}^{(0)} + \mathbf{G}^{(0)})\mathbf{c}_i^{(1)} + (\mathbf{h}^{(1)} + \mathbf{G}^{(1)})\mathbf{c}_i^{(0)} = \varepsilon_i^{(1)}\mathbf{S}^{(0)}\mathbf{c}_i^{(0)} + \varepsilon_i^{(0)}\mathbf{S}^{(1)}\mathbf{c}_i^{(0)} + \varepsilon_i^{(0)}\mathbf{S}^{(0)}\mathbf{c}_i^{(1)}.$$

They reduce to those of Stevens et al. [18] when we put $\mathbf{S}^{(1)} = 0$, $h_{jk}^{(1)} = \langle \chi_j^{(0)} \mid h^{(1)} \mid \chi_k^{(0)} \rangle$, and all $g_{jm,kn}^{(1)} = 0$. The first-order energy can be obtained by enumeration of terms in the expression for the energy [Eqs. (62) and (74)]. Then

$$E^{(1)} = \sum_{i=1}^{N} (\mathbf{c}_i^{(0)\dagger}\mathbf{h}^{(1)}\mathbf{c}_i^{(0)} + \mathbf{c}_i^{(1)\dagger}\mathbf{h}^{(0)}\mathbf{c}_i^{(0)} + \mathbf{c}_i^{(0)\dagger}\mathbf{h}^{(0)}\mathbf{c}_i^{(1)})$$
$$+ \tfrac{1}{2} \sum_{i=1}^{N} (\mathbf{c}_i^{(0)\dagger}\mathbf{G}^{(1)}\mathbf{c}_i^{(0)} + \mathbf{c}_i^{(1)\dagger}\mathbf{G}^{(0)}\mathbf{c}_i^{(0)} + \mathbf{c}_i^{(0)\dagger}\mathbf{G}^{(0)}\mathbf{c}_i^{(1)})$$

This may be simplified using (263) and the orthonormality conditions: $\mathbf{c}_i^{(0)\dagger}\mathbf{S}^{(0)}\mathbf{c}_j^{(0)} = \delta_{ij}$ (zero order) and

$$\mathbf{c}_i^{(1)\dagger}\mathbf{S}^{(0)}\mathbf{c}_i^{(0)} + \mathbf{c}_i^{(0)\dagger}\mathbf{S}^{(1)}\mathbf{c}_j^{(0)} + \mathbf{c}_i^{(0)\dagger}\mathbf{S}^{(0)}\mathbf{c}_j^{(1)} = 0$$

(first order). The result, using the definition of $\mathbf{G}^{(1)}$, is

$$E^{(1)} = \sum_{i=1}^{N}\left(\mathbf{c}_i^{(0)\dagger}\mathbf{h}^{(1)}\mathbf{c}_i^{(0)} - \varepsilon_i^{(0)}\mathbf{c}_i^{(0)\dagger}\mathbf{S}^{(1)}\mathbf{c}_i^{(0)} + \tfrac{1}{2}\sum_{j=1}^{N}\sum_{k,l,m,n=1}^{p} c_{ki}^{(0)}c_{lj}^{(0)}c_{mi}^{(0)}c_{nj}^{(0)}g_{kl,mn}^{(1)}\right).$$

(264)

This means that it may be obtained from the unperturbed coefficients and the changes in the basis functions with γ.

Note that $g_{kl,mn}^{(1)}$ consists of four terms like

$$\langle \chi_k^{(0)}(1)\chi_l^{(0)}(2) \mid g_{12}(1 - \mathscr{T}_{12}) \mid \chi_m^{(0)}(1)\chi_n^{(1)}(2)\rangle$$

and, according to (251), $\chi_n^{(1)}(2) = \partial\chi_n(2)/\partial\gamma$. If γ is a displacement of nucleus B along the internuclear direction, and χ_n is an atomic orbital centered on A, the derivative vanishes. But if χ_n is centered on B, for example a $1s$ orbital with exponential parameter ζ, we have explicitly

$$\frac{\partial\chi_n(2)}{\partial\gamma} = \frac{\partial}{\partial z_B}[\exp(-\zeta r_{2B})] = \zeta\exp(-\zeta r_{2B})\left(\frac{z_2 - z_B}{r_{2B}}\right) \quad (265)$$

since $r_{2B}^2 = (x_2 - x_B)^2 + (y_2 - y_B)^2 + (z_2 - z_B)^2$ and the z axis is the internuclear axis. The function (265) may be thought of as a $1p$ atomic orbital and the integrals involving it are different from those generally encountered. The amount of extra work we have to do to compute the $g_{kl,mn}^{(1)}$ depends on how we choose the perturbation. Gerratt and Mills [54] considered diatomic hydrides and changed R by moving the H only. Then the only $\partial\chi_n/\partial\gamma$ which were nonzero were those for which χ_n was the hydrogen 1s orbital.

We refer to Benston and Kirtman [52] for the reduction of the expression for the second-order energy. Here, we obtain it from (234), noting that $\Psi^{(1)}$ consists of a sum of singly excited determinants and is orthogonal to $\Psi^{(0)}$, the Hartree–Fock function, while H_2 is one-electron. The electronic contribution is

$$\sum_{i=1}^{N}\langle \lambda_i^{(0)} \mid h^{(1)} \mid \lambda_i^{(1)}\rangle + \sum_{i=1}^{N}\langle \lambda_i^{(0)} \mid h^{(2)} \mid \lambda_i^{(0)}\rangle.$$

Including the nuclear–nuclear repulsion term and taking $c_A = 0$ in (229) and (230), we have

$$E^{(2)} = \frac{2z_A z_B}{R_e^3} + \sum_{i=1}^{N} \left\langle \lambda_i^{(0)}(1) \left| \frac{Z_B \cos \theta_{1B}}{r_{1B}^2} \right| \lambda_i^{(1)}(1) \right\rangle$$

$$- \sum_{i=1}^{N} \left\langle \lambda_i^{(0)}(1) \left| \frac{3 \cos^2 \theta_{1B} - 1}{r_{1B}^3} - \frac{4\pi}{3} \delta(\mathbf{r}_{1B}) \right| \lambda_i^{(0)}(1) \right\rangle. \quad (266)$$

For each of the three molecules considered (LiH, BH and HF) the force constant was high by 50% or so (although it should be mentioned that the value for HF is ten times that for LiH) [54]. A problem is that the basis set was not extensive enough, since the forces calculated by the Hellmann–Feynman theorem did not agree with the energy derivatives, as they should for a true Hartree–Fock function. Also, the calculation was made at the experimental R_e and not at R_e defined for this function by the Hellmann–Feynman force or the energy minimum.

In confocal ellipsoidal coordinates (see "Notes on Notation") the R-dependence of the Hamiltonian can be made explicit, since

$$H = R^{-2}t + R^{-1}v$$

with t and v dimensionless (see Sections III.C and III.D). If $H(R) - H(R_e)$ is taken as a perturbation, we can derive an explicit R-dependence for the difference of the corresponding energies. Parr and White [55] studied such a scheme, taking R_e as the equilibrium distance and using, as the perturbation parameter,

$$\gamma = 1 - R_e/R.$$

The unperturbed equation is

$$H(R_e)\Psi_e = (R_e^{-2}t + R_e^{-1}v)\Psi_e = E(R_e)\Psi_e$$

and the perturbed equation, $H(R)\Psi = E(R)\Psi$, was rearranged to

$$[(1 - \gamma)R_e^{-2}t + R_e^{-1}v]\Psi = E(1 - \gamma)^{-1}\Psi. \quad (267)$$

Then Ψ and $(1 - \gamma)^{-1}E$ were expanded as power series in γ, and the usual Rayleigh–Schrödinger formulas applied, with the perturbation being $-\gamma R_e^{-2}t$ [55]. Then $E(1 - \gamma)^{-1}$ was produced in the form $\sum \gamma^i w_i$, with $w_0 = E(R_e)$ and w_1 the negative of the kinetic energy for $R = R_e$. According to the virial theorem (Section C) $w_1 = E(R_e)$. In terms of the w_i,

$$E - E(R_e) = \sum_{i=2}^{\infty} (w_i - w_{i-1})\gamma^i. \quad (268)$$

If w_i vanishes for $i \geq 3$, or if w_i approaches a constant for $i \geq 3$, this is the Fues potential (Section D of Chapter II). Tests with experimental potential curves showed that $w_3 - w_2$ is indeed small compared to $w_2 - w_1$ for most molecules. This formalism has been implemented [55] for H_2.

Even without explicit numerical evaluation, the form of equations (267) and (268) helps to give understanding of potential curves and suggests simple models for their calculation (see Vol. 2, Chap. III Sect. B). More experience is needed before we can judge their practical utility for calculation.

5. *Double Perturbation Theory*

Double perturbation theory treats the changes in energy and wavefunction of a system under the influence of two simultaneous perturbations. In applications of interest to us, one will generally be an electrostatic perturbation, such as in the calculation of polarizabilities, and the other a perturbation giving corrections to the approximate Hamiltonian used. For example, in Section A.3 we considered the error in an expectation value, calculated with an approximate wavefunction, by introducing a (possibly fictitious) Hamiltonian of which the approximate wave function was an eigenfunction. The difference between the exact Hamiltonian and this one would be the perturbation. Then corrections to the approximate wavefunction could be calculated by perturbation theory, as in Section B.3. If we want to study the effect of the approximate nature of the wave function on the calculated second-order energy due to an electric field, we have two perturbations to consider simultaneously. Since the expectation value of any operator can be considered as the first-order energy in a perturbation scheme in which that operator is the perturbation, corrections to an expectation value can also be considered from the point of view of double perturbation theory. Further applications of the theory arise whenever, for any reason, we want to consider a perturbation as the sum of two parts. For a detailed review of the formalism, see the article of Hirschfelder *et al.* [3, Chapter IV].

We consider the unperturbed Hamiltonian H_0 and the perturbations V_1 and V_2. Analogously to the treatment of a single perturbation, in equations (138) through (151), we consider

$$H(\gamma_1, \gamma_2) = H_0 + \gamma_1 V_1 + \gamma_2 V_2 \qquad (269)$$

The eigenfunctions and eigenvalues are functions of both γ_1 and γ_2, and we assume the dependence is analytic:

$$\Psi_k = \sum_{i,j} \gamma_1{}^i \gamma_2{}^j \Psi_k^{(i,j)} \tag{270a}$$

$$E_k = \sum_{i,j} \gamma_1{}^i \gamma_2{}^j E_k^{(i,j)}. \tag{270b}$$

The expansions are substituted into the eigenvalue equation for H and the coefficient of $\gamma_1{}^m \gamma_2{}^n$ for each choice of m and n is equated to zero. This gives the equations of double perturbation theory. In general,

$$H_0\Psi_k^{(i,j)} + V_1\Psi_k^{(i-1,j)} + V_2\Psi_k^{(i,j-1)} = \sum_{m=0}^{i} \sum_{n=0}^{j} E_k^{(m,n)}\Psi_k^{(i-m,j-n)} \tag{271}$$

where it is understood that functions with negative superscripts are to be ignored. We demand that Ψ_k be normalized to all orders of γ_1 and γ_2, which requires that $\Psi_k^{(0,0)}$ be normalized and that

$$\sum_{m=0}^{i} \sum_{n=0}^{j} \langle \Psi_k^{(m,n)} \mid \Psi_k^{(i-m,j-n)} \rangle = 0 \tag{272}$$

whenever either i or j is greater than zero.

The equations (271), when either i or j is zero, are just the equations of ordinary perturbation theory. For instance, when $i = 1$ and $j = 0$ we obtain

$$H_0\Psi_k^{(1,0)} + V_1\Psi_k^{(0,0)} = E_k^{(0,0)}\Psi_k^{(1,0)} + E_k^{(1,0)}\Psi_k^{(0,0)} \tag{273}$$

which is just Eq. (143). For $i = 1$ and $j = 1$, equation (271) reads

$$H_0\Psi^{(1,1)} + V_1\Psi^{(0,1)} + V_2\Psi^{(1,0)}$$
$$= E^{(0,0)}\Psi^{(1,1)} + E^{(1,0)}\Psi^{(0,1)} + E^{(0,1)}\Psi^{(1,0)} + E^{(1,1)}\Psi^{(0,0)}.$$

If we multiply by $\Psi^{(1,0)*}$ and integrate over configuration space, we obtain on rearrangement

$$\langle \Psi^{(0,0)} \mid V_1 - E^{(1,0)} \mid \Psi^{(0,1)} \rangle + \langle \Psi^{(0,0)} \mid V_2 - E^{(0,1)} \mid \Psi^{(1,0)} \rangle$$
$$= \langle \Psi^{(0,0)} \mid E^{(1,1)} \mid \Psi^{(0,0)} \rangle.$$

We may invoke the complex conjugate of (273) and the normalization of

$\Psi^{(0,0)}$ to obtain

$$E^{(1,1)} = \langle \Psi^{(1,0)} \mid E^{(0,0)} - H_0 \mid \Psi^{(0,1)} \rangle + \langle \Psi^{(0,0)} \mid V_2 - E^{(0,1)} \mid \Psi^{(1,0)} \rangle.$$

If we use in the first term the equation corresponding to (273) for the perturbation V_2, this becomes

$$E^{(1,1)} = \langle \Psi^{(1,0)} \mid V_2 \mid \Psi^{(0,0)} \rangle + \langle \Psi^{(0,0)} \mid V_2 \mid \Psi^{(1,0)} \rangle. \tag{274a}$$

Of course we can also derive

$$E^{(1,1)} = \langle \Psi^{(0,1)} \mid V_1 \mid \Psi^{(0,0)} \rangle + \langle \Psi^{(0,0)} \mid V_1 \mid \Psi^{(0,1)} \rangle. \tag{274b}$$

The result of Eq. (274) is an example of Dalgarno's interchange theorem [3, Sect. IVE]. The theorem states, in general, that all the energies $E^{(1,j)}$ may be calculated using the functions $\Psi^{(0,j)}$, which arise from the perturbation equations involving V_2 only; while all $E^{(i,1)}$ may be calculated from the functions $\Psi^{(i,0)}$, which arise from the perturbation equations involving V_1. For calculation of $E^{(1,1)}$, one can either solve Eq. (273) for $\Psi^{(1,0)}$ and calculate as in (274a), or solve the corresponding equation for $\Psi^{(0,1)}$ and calculate as in (274b). If V_1 is simpler than V_2, the former is the way to proceed. Note that all three energies which are first order, $E^{(1,0)}$, $E^{(1,1)}$, and $E^{(0,1)}$, may be obtained without solving an equation involving V_2. In the presence of degeneracy, however, there are no interchange theorems [3, p. 294].

In Section A.3, the interchange theorem was used, when we calculated the first correction to the expectation value of the operator Q [Eqs. (30) *et seq.*]. Here, Q may be taken as V_1, and the difference between the exact Hamiltonian H and the approximate Hamiltonian H_Φ may be taken as V_2. Then $E^{(1,0)}$ is the expectation value calculated with the approximate function Φ, while $E^{(0,0)} + E^{(0,1)}$ is the expectation value of the exact Hamiltonian over the approximate function. $E^{(1,1)}$ is the first order correction to the expectation value (first order in V_2, the correction Hamiltonian). The interchange theorem states that this quantity may be calculated from the first-order correction to the wave function when Q is taken as the perturbation. This is the first equality in the equation after (31).

In connection with long-range induction and dispersion forces (Vol. 2, Ch. I Sect. A), the electric polarizabilities of atoms are of interest. The dipole polarizability is proportional to the second-order term in the energy when the perturbation is the interaction of the atom with an electric field. If an approximate wave function is used for the unperturbed function (atom in the absence of the field) and the difference between the exact and approximate Hamiltonians is taken as V_2, the leading correction to the polarizability is obtained from $E^{(2,1)}$ in a double perturbation theory expansion. The approximate polarizability is $E^{(2,0)}$. According to the interchange theorem, $E^{(2,1)}$ is calculable from the function $\Psi^{(2\ 0)}$, which is obtained as the second order correction when V_1 is the perturbation. If the van der Waals constant itself is computed by perturbation theory (Section A.6 of Chapter I, Volume 2), the errors due to the use of approximate atomic wave functions may be treated by similar relations [3, p. 294].

An example of the use of double perturbation theory, in which the perturbation is divided into two parts for convenience in the calculation, is the work on the ground state of H_2 by Liu $et\ al.$ [56]. The zero-order spatial wave function for the ground state was taken as the simple product $\phi(1)\phi(2)$, where

$$\phi = Ne^{-sR\xi}$$

in ellipsoidal coordinates. Here, N is a normalizing factor and s, which is a scaling parameter, may be chosen variationally to minimize the sum of zero and first-order energies, that is, the expectation value of the full Hamiltonian. The spatial function is to be multiplied by a spin singlet. The product $\phi(1)\phi(2)$ is an eigenfunction of

$$H_0' = \sum_{i=1}^{2} -\tfrac{1}{2}\nabla_i^2 - \frac{4s\xi_i}{R(\xi_i^2 - \eta_i^2)} - \frac{2s^2(1 - \eta_i^2)}{(\xi_i^2 - \eta_i^2)}$$

with eigenvalue $E^{(0,0)} = -8s^2$. The difference between H_0 and the true Hamiltonian was written as $V_1 + V_2$ with

$$V_1 = \sum_{i=1}^{2} \frac{4s}{R(\xi_i^2 - \eta_i^2)} - \frac{4(1 - s)\xi_i}{R(\xi_i^2 - \eta_i^2)}$$

and

$$V_2 = 1/r_{12}.$$

Then $E^{(1,0)}$ and $E^{(0,1)}$ could be written in closed form. Here, V_1, being one-electron, is simpler to handle than V_2. Indeed, the first-order equation in V_1 was solved in closed form to give $\Psi^{(1,0)}$, while the first-order equation in V_2 was solved approximately only, using a five-term variation–perturbation expansion. The value $E^{(1,1)}$ was computed from $\Psi^{(1,0)}$. The approximation to $\Psi^{(0,1)}$ was used only for the computation of $E^{(0,2)}$.

The importance of double perturbation theory for this kind of application will depend on whether accurate calculations of molecular electronic energies by perturbation theory are successful. The previously discussed application, to the errors in properties calculated from approximate wave functions, should grow in importance, since only when calculated values are accompanied by error estimates can quantum chemistry really be said to have reached full maturity.

REFERENCES

1. J. O. Hirschfelder, *J. Chem. Phys.* **43**, S199 (1965).
2. P. Claverie, *Int. J. Quantum Chem.* **5**, 273 (1971).
3. J. O. Hirschfelder, W. Byers Brown, and S. T. Epstein, *Advan. Quantum Chem.* **1**, 255 (1964).
4. J. O. Hirschfelder, *in* "Perturbation Theory and Its Application in Quantum Mechanics" (C. H. Wilcox, ed.), p. 3. Wiley, New York, 1966.
5. A. Dalgarno, *Rev. Mod. Phys.* **35**, 522 (1963).
6. E. A. Hylleraas, *Z. Phys.* **65**, 209 (1930); E. A. Hylleraas and J. Midtdal, *Phys. Rev.* **103**, 829 (1956); **109**, 1013 (1958).
7. S. Prager and J. O. Hirschfelder, *J. Chem. Phys.* **39**, 3289 (1963).
8. R. E. Knight and C. W. Scherr, *J. Chem. Phys.* **37**, 2503 (1962).
9. R. E. Knight and C. W. Scherr, *Phys. Rev.* **128**, 2675 (1962); *Rev. Mod. Phys.* **35**, 436 (1963).
10. J. I. Musher, *Ann. Phys.* (*New York*) **32**, 416 (1965).
11. A. Dalgarno, *in* "Quantum Theory," Pt. I, "Elements" (D. R. Bates, ed.), Sect. 5.7. Academic Press, New York, 1961.
12. W. J. Meath and J. O. Hirschfelder, *J. Chem. Phys.* **41**, 1628 (1964); and references therein.
13. J. N. Silverman and J. C. van Leuven, *Phys. Rev.* **162**, 1175 (1967); and subsequent papers.
14. P.-O. Löwdin, *J. Mol. Spectrosc.* **10**, 12 (1963); *J. Math. Phys.* **3**, 969 (1962); **6**, 1341 (1965); *Phys. Rev. A* **139**, 357 (1965); *J. Chem. Phys.* **43**, S175 (1965); J. O. Hirschfelder, *J. Chem. Phys.* **44**, 273 (1966), derives the bracketing theorem without formal operator algebra.
15. R. McWeeny and B. T. Sutcliffe, "Methods of Molecular Quantum Mechanics." Sects. 2.4, 8.4. Academic Press, London and New York, 1969.
16. A. Dalgarno, *Proc. Roy. Soc. Ser. A* **251**, 282 (1959).
17. P. W. Langhoff, M. Karplus, and R. P. Hurst, *J. Chem. Phys.* **44**, 505 (1966).
18. R. M. Stevens, R. Pitzer, and W. N. Lipscomb, *J. Chem. Phys.* **38**, 550 (1963).
19. O. Sinanoğlu, *Phys. Rev.* **122**, 493 (1961).
20. S. Seung and E. B. Wilson, Jr., *J. Chem. Phys.* **47**, 5343 (1967).
21. R. L. Matcha and W. Byers-Brown, *J. Chem. Phys.* **48**, 74 (1968).
22. J. Midtdal, *Phys. Rev. A* **138**, 1010 (1965); J. Linderberg, *J. Mol. Spectrosc.* **9**, 95 (1962).
23. C. W. Scherr, F. C. Sanders, and R. E. Knight, *in* "Perturbation Theory and Its Applications in Quantum Mechanics" (C. H. Wilcox, ed.). Wiley, New York, 1966.
24. J. Goodisman, *J. Chem. Phys.* **48**, 2981 (1968).
25. J. Goodisman, *J. Chem. Phys.* **50**, 903 (1969).

26. O. Sinanoğlu, *J. Chem. Phys.* **33**, 1212 (1960).

27. O. Sinanoğlu, *Proc. Roy. Soc. Ser. A* **260**, 379 (1961).

28. J. Goodisman, *J. Chem. Phys.* **51**, 3540 (1969).

29. M. Geller, H. S. Taylor, and H. B. Levine, *J. Chem. Phys.* **43**, 1727 (1965); B. Kirtman and D. R. Decious, *Ibid.* **48**, 3133 (1968).

30. F. W. Byron, Jr., and C. F. Joachain, *Phys. Rev.* **157**, 7 (1967).

31. F. W. Byron, Jr., and C. F. Joachain, *Phys. Rev.* **157**, 1 (1967); K. J. Miller and K. Ruedenberg, *J. Chem. Phys.* **48**, 3414 (1968).

32. H. E. Montgomery, B. L. Bruner, and R. E. Knight, *J. Chem. Phys.* **52**, 470 (1970).

33. B. Kirtman, *Chem. Phys. Lett.* **1**, 631 (1968); B. Kirtman and R. L. Mowery, *J. Chem. Phys.* **55**, 1447 (1961); B. Kirtman, *Ibid.* **55**, 1457 (1971).

34. C. Møller and M. S. Plesset, *Phys. Rev.* **46**, 618 (1934).

35. R. E. Stanton, *J. Chem. Phys.* **42**, 2353 (1965).

36. H. J. Silverstone and O. Sinanoğlu, *J. Chem. Phys.* **46**, 854 (1967).

37. O. Sinanoğlu, *Advan. Chem. Phys.* **6**, 315 (1964).

38. K. A. Brueckner, *Phys. Rev.* **97**, 1353 (1955).

39. K. A. Brueckner, *Phys. Rev.* **100**, 36 (1955).

40. K. F. Freed, *Annu. Rev. Phys. Chem.* **22**, 313 (1971).

41. H. P. Kelly, *Phys. Rev. Lett.* **23**, 455 (1969); T. Lee, N. C. Dutta, and T. P. Das, *Ibid.* **25**, 204 (1970).

42. J. M. Schulman and D. N. Kaufman, *J. Chem. Phys.* **53**, 477 (1970); J. I. Musher and J. M. Schulman, *Phys. Rev.* **173**, 93 (1968).

43. H. P. Kelly, *Int. J. Quantum Chem. Symp.* **3**, 349 (1970); N. C. Dutta, T. Isihara, C. Matsubara, and T. P. Das, *Ibid.* **3**, 367 (1970).

44. K. Kumar, "Perturbation Theory and the Nuclear Many-Body Problem." North-Holland Publ., Amsterdam, 1962; A. L. Fetter and J. D. Wallecka, "Quantum Theory of Many-Particle Systems." McGraw-Hill, New York, 1971; D. A. Kirzhnits, "Field Theoretical Methods in Many-Body Systems." Pergamon, Oxford, 1967; N. H. March, W. H. Young, and S. Sampanthar, "The Many-Body Problem in Quantum Mechanics." Cambridge Univ. Press, London and New York, 1967; and many others.

45. H. P. Kelly, *Int. J. Quantum Chem. Symp.* **1**, 25 (1967); *in* "Perturbation Theory and Its Applications in Quantum Mechanics" (C. H. Wilcox, ed.), p. 215. Wiley, New York, 1966.

46. K. F. Freed, *Phys. Rev.* **173**, 1 (1968); *Chem. Phys. Lett.* **4**, 496 (1970); S. Diner, J.-P. Malrieu, and P. Claverie, *Theor. Chim. Acta* **13**, 1 (1969).

47. W. Byers Brown, *Proc. Cambridge Phil. Soc.* **54**, 250 (1958).

48. V. V. Rossikhin and V. P. Morozov, *Theor. Exp. Chem. (USSR)* **2**, 396 (1966); *Teor. Eksp. Khim.* **2**, 528 (1966).

49. J. H. van Vleck, "The Theory of Electric and Magnetic Susceptibilities," p. 276. Oxford Univ. Press, London and New York, 1932.

50. J. N. Murrell, *J. Mol. Spectrosc.* **4**, 446 (1960).

51. R. F. W. Bader, *Mol. Phys.* **3**, 137 (1960).

52. M. L. Benston and B. Kirtman, *J. Chem. Phys.* **44**, 126 (1966).

53. J. Gerratt and I. M. Mills, *J. Chem. Phys.* **49**, 1719 (1968).

54. J. Gerratt and I. M. Mills, *J. Chem. Phys.* **49**, 1730 (1968).

55. R. G. Parr and R. J. White, *J. Chem. Phys.* **49**, 1059 (1968).

56. B. Liu, W. D. Lyon, and W. Byers Brown, *J. Chem. Phys.* **44**, 562 (1966).

C. Virial Theorem

1. *Scaling and the Virial Theorem*

The virial theorem expresses the balance between kinetic and potential energy for atomic and molecular systems. In principle, it allows us to derive $U(R)$ from kinetic or potential energy alone. For atoms, the expectation value of the potential energy, when only Coulombic interactions are considered, is of opposite sign and twice the magnitude of the expectation value of the kinetic energy. For diatomic molecules in the fixed-nucleus approximation, where the kinetic energy is electronic only, this holds whenever $dU/dR = 0$, which obtains in the limit of dissociation, where one has separated atoms, and also at the equilibrium internuclear distance. We shall derive the theorem for fixed-nucleus electronic wave functions by a method given by Hirschfelder and Kincaid [1]. This is a generalization of Fock's proof [2] for atomic systems, which we give first, and demonstrates simultaneously the conditions under which an approximate wave function will satisfy the virial theorem. Another proof, more closely connected with classical mechanics, is given by Slater [3]. A related derivation by Cottrell and Patterson [4] is suitable for particle-in-box-type potentials.

Consider an atom described by a normalized wavefunction $\psi_1(\mathbf{r}_1, \mathbf{r}_2, \ldots, \mathbf{r}_n)$. Here, $\mathbf{r}_i$ is the coordinate of the ith electron relative to the nucleus. Now imagine a wave function ψ_s obtained from ψ_1 by scaling, i.e., multiplying, all electronic coordinates by a factor s. To normalize ψ_s, we must multiply the function by $s^{3n/2}$. Then

$$\int \cdots \int \psi_s^* \psi_s \, d\mathbf{r}_1 \cdots d\mathbf{r}_n = s^{3n} \int \cdots \int |\psi_1(s\mathbf{r}_1, \ldots, s\mathbf{r}_n)|^2 \, d\mathbf{r}_1 \cdots d\mathbf{r}_n$$

$$= s^{3n} \int \cdots \int |\psi_1(\boldsymbol{\rho}_1, \ldots, \boldsymbol{\rho}_n)|^2 \, s^{-3} \, d\boldsymbol{\rho}_1 \cdots s^{-3} \, d\boldsymbol{\rho}_n \tag{275}$$

which is unity because ψ_1 is normalized. The expectation value of the kinetic energy over ψ_s is

$$s^{3n} \int \cdots \int \psi_1^*(s\mathbf{r}_1 \cdots s\mathbf{r}_n) T_r \psi_1(s\mathbf{r}_1 \cdots s\mathbf{r}_n) \, d\mathbf{r}_1 \cdots d\mathbf{r}_n$$

$$= s^{3n} \int \cdots \int \psi_1^*(\boldsymbol{\rho}_1 \cdots \boldsymbol{\rho}_n) s^2 T_\varrho \psi_1(\boldsymbol{\rho}_1 \cdots \boldsymbol{\rho}_n) s^{-3} \, d\boldsymbol{\rho}_1 \cdots s^{-3} \, d\boldsymbol{\rho}_n. \tag{276}$$

Here, T_r is the kinetic energy operator in terms of r and T_ϱ the operator in terms of the scaled coordinates ϱ. The expression $T_r = s^2 T_\varrho$ follows from the form of the operator, which behaves here like a homogeneous function of degree -2 in the electronic coordinates. From (2), $\langle T \rangle_s = s^2 \langle T \rangle_1$, the subscripts on the expectation values indicating the functions used in the calculation. If the potential energy is of Coulombic form, it is homogeneous of degree -1 in the electronic coordinates. By manipulations like those of Eq. (276), we find that $\langle V \rangle_s = s \langle V \rangle_1$. The energy with the scaled function is

$$E_s = s^2 \langle T \rangle_1 + s \langle V \rangle_1. \tag{277}$$

Treating s as a variational parameter, we set $dE_s/ds = 0$, which gives for the optimum value of s

$$s_0 = -\tfrac{1}{2} \langle V \rangle_1 / \langle T \rangle_1 \tag{278}$$

and for the optimum energy

$$E_{s=s_0} = -\tfrac{1}{4} \langle V \rangle_1{}^2 / \langle T \rangle_1. \tag{279}$$

If ψ_1 had been the exact wave function, it could not have been improved by scaling, so s_0 must have come out to be unity. Then, according to (278), $\langle T \rangle = -\tfrac{1}{2} \langle V \rangle$ for the exact function. Since $\langle T \rangle + \langle V \rangle = E$, the virial theorem implies $\langle T \rangle = -E$, $\langle V \rangle = \tfrac{1}{2} E$. If ψ_1 had not been exact, s_0 might have different from unity. But for the scaled function,

$$\langle V \rangle_s = -\tfrac{1}{2} \langle V \rangle_1{}^2 / \langle T \rangle_1 \tag{280a}$$

$$\langle T \rangle_s = \tfrac{1}{4} \langle V \rangle_1{}^2 / \langle T \rangle_1 \tag{280b}$$

so the virial theorem is satisfied [2]:

$$\langle T \rangle_s = -\tfrac{1}{2} \langle V \rangle_s. \tag{281}$$

This proof shows that a variational wave function for which a scaling is an allowed variation will satisfy the virial theorem. This will occur when all parameters which multiply distances are chosen variationally, or if a scaling parameter is among the parameters varied.

The argument we have given is valid for a molecular wave function which describes all particles, nuclei and electrons. For the fixed-nucleus electronic problem, the situation is changed, since the nuclear configuration enters as a parameter and not a coordinate. Let $\psi_1(\mathbf{r}_1, \mathbf{r}_2, \ldots, \mathbf{r}_n; \mathbf{P})$ be an electronic wave function for a nuclear configuration specified by $\mathbf{P}$. The scaling of the wave function multiplies only electronic coordinates by s.

$$\psi_s = s^{3n/2}\psi_1(s\mathbf{r}_1, s\mathbf{r}_2, \ldots, s\mathbf{r}_n; \mathbf{P}). \tag{282}$$

The normalization factor of $s^{3n/2}$ is still valid, as is the manipulation of Eq. (276). Thus $\langle T\rangle_s = s^2\langle T\rangle_1$ again. For the interelectronic repulsion, we have similarly $\langle V_{ee}\rangle_s = s\langle V_{ee}\rangle_1$. But the operator V_{ne} for electron–nuclear attraction is not homogeneous in the electronic coordinates alone. We have instead

$$V_{ne}(s\mathbf{r}_1 \cdots s\mathbf{r}_n; s\mathbf{R}) = s^{-1}V_{ne}(\mathbf{r}_1, \mathbf{r}_2, \ldots, \mathbf{r}_n; \mathbf{R}).$$

Thus if we use ψ_s for a calculation at a scaled nuclear configuration $\mathbf{P}/s$, we obtain

$$\langle V_{ne}\rangle_{s;\mathbf{P}/s} = s^{3n}\int \cdots \int |\psi_1(s\mathbf{r}_1 \cdots s\mathbf{r}_n; \mathbf{P})|^2 V_{ne}(\mathbf{r}_1 \cdots \mathbf{r}_n; \mathbf{P}/s)\, d\mathbf{r}_1 \cdots d\mathbf{r}_1$$

$$= s\int \cdots \int |\psi_1(\boldsymbol{\rho}_1 \cdots \boldsymbol{\rho}_n; \mathbf{P})|^2 V_{ne}(\boldsymbol{\rho}_1 \cdots \boldsymbol{\rho}_n; \mathbf{P})\, d\boldsymbol{\rho}_1 \cdots d\boldsymbol{\rho}_n$$

$$= s\langle V_{ne}\rangle_{1;\mathbf{P}}$$

The second subscript on the expectation values gives the configuration for which the calculation is performed. We now put $\mathbf{P} = s\mathbf{R}$ to get

$$\langle V_{ne}\rangle_{s;\mathbf{R}} = s\langle V_{ne}\rangle_{1;s\mathbf{R}}. \tag{283}$$

We now specialize to a diatomic molecule, so $\mathbf{R}$ is simply the internuclear distance R. Since the internuclear repulsion is Z_AZ_B/R^2, $V_{nn}(R) = sV_{nn}(sR)$ and the scaled energy is

$$E_s(R) = s\langle V\rangle_{1;sR} + s^2\langle T\rangle_{1;sR} \tag{284}$$

The subscript sR on the expectation value of kinetic energy is, of course, unnecessary.

Following Hirschfelder and Kincaid [1] we first differentiate (284) with R fixed to get

$$\left(\frac{\partial E_s(R)}{\partial s}\right)_R = \langle V\rangle_{1;sR} + 2s\langle T\rangle_{1;sR}$$
$$+ \left(\frac{\partial(sR)}{\partial s}\right)_R\left[s\frac{d\langle V\rangle_{1;sR}}{d(sR)} + s^2\frac{d\langle T\rangle_{1;sR}}{d(sR)}\right]. \tag{285}$$

We again set this equal to zero to find the optimum value of s. For a diatomic molecule, the square bracket in Eq. (285) is equal to $(\partial E_s(R)/\partial R)_s$

divided by s. (For a polyatomic, a sum over the independent nuclear co-ordinates would appear in its place.) Then we have

$$\left(\frac{\partial E_s(R)}{\partial s}\right)_R = \langle V\rangle_{1;sR} + 2s\langle T\rangle_{1;sR} + \frac{R}{s}\left(\frac{\partial E_s(R)}{\partial R}\right)_s = 0. \quad (286)$$

Suppose first that ψ_1 is the exact electronic function and R the equilibrium internuclear distance. Then the last term in (286) vanishes and $s = 1$, which gives $\langle V\rangle = -2\langle T\rangle$. If R is not the equilibrium configuration, we recover

$$2\langle T\rangle + \langle V\rangle + R(dE/dR) = 0, \quad (287)$$

the virial theorem for diatomic molecules in the fixed-nucleus approxima-tion. This appears as a special case of the Hellmann Feynman theorem in Section D, Eq. (350). Next suppose that ψ_1 is not exact, but that R is the equilibrium configuration for the scaled function. Then the last term in (286) vanishes and

$$s_0 = -\tfrac{1}{2}\langle V\rangle_{1;sR}/\langle T\rangle_{1;sR}$$

and the scaled function will have $\langle T\rangle = -\tfrac{1}{2}\langle V\rangle$. Finally, let R be arbitrary. Multiply (286) by s to get [4a]

$$0 = \langle V\rangle_{s;R} + 2\langle T\rangle_{s;R} + R(dE_s/dR). \quad (288)$$

Therefore, when a fixed-nucleus molecular wave function has been optim-ized with respect to a scaling parameter on the electronic coordinates, the virial theorem, Eq. (287), will be satisfied. Note that the introduction of a scaling factor means a change in the internuclear distance. For a wave function of a given functional form, there will be a different scaling factor at each R.

If the initial wavefunction is written in ellipsoidal coordinates (or, for a polyatomic, in a system of coordinates such that, for given values of the coordinates, all actual interparticle distances are proportional to some inter-nuclear distance), the criterion for satisfaction of the virial theorem is somewhat different [5]. In this case, some sort of scaling automatically occurs as R is changed. It suffices that all parameters be chosen to satisfy the variational theorem. Let $\langle H\rangle$ be the energy calculated with such a function. Then

$$\frac{dE}{dR} = \sum_i \frac{\partial\langle H\rangle}{\partial\lambda_i}\frac{d\lambda_i}{dR} + \frac{\partial}{\partial R}\langle T + V\rangle \quad (289)$$

where $\{\lambda_i\}$ are parameters entering the function. If they are chosen variationally, $\partial\langle H\rangle/\partial\lambda_i = 0$. The second term in (289) is the explicit dependence on R for fixed $\{\lambda_i\}$. If $\langle T\rangle$ is calculated in ellipsoidal coordinates, it will be R^{-2} times some integral which is independent of R. Then $\partial\langle T\rangle/\partial R = -2\langle T\rangle/R$. The potential energy being Coulombic, $\langle V\rangle$ will be R^{-1} times some integral independent of R, so $\partial\langle V\rangle/\partial R = -\langle V\rangle/R$. Therefore,

$$dE/dR = -(2/R)\langle T\rangle - (1/R)\langle V\rangle$$

which is equation (287). Note also that the virial theorem will still hold if some of the parameters are not chosen variationally, but held constant with R. As will be seen in the next chapter, there is a close connection with the Hellmann–Feynman theorem here.

It is possible to go in a different way [6] from Eq. (281), which holds for a molecular wavefunction which treats nuclei and electrons equivalently, to Eq. (287), which applies to the fixed-nucleus case. Also, we may note here that (287) holds whether or not the internuclear repulsion V_{nn} is included in the energy. This follows because

$$R(dV_{nn}/dR) = -V_{nn}$$

Thus the virial theorem holds for the interaction potential $U(R)$ or for the function $U(R) - V_{nn}$.

If one has a Hamiltonian which is divided into a zero-order and per-turbing part,

$$H = (T + V_0) + \gamma V_1 \tag{290}$$

with γ independent of R and only Coulombic interactions present, the virial theorem holds for all values of γ. Therefore, if perturbation theory is used, the theorem holds in each order. Thus,

$$2\langle T\rangle^{(0)} + \langle V_0\rangle^{(0)} + R\,dE^{(0)}/dR = 0,$$

$$2\langle T\rangle^{(m)} + \langle V_0\rangle^{(m)} + \langle V_1\rangle^{m-1} + R\,dE^{(m)}/dR = 0.$$

Here, the superscript on an expectation value refers to terms of a particular order, e.g.,

$$\langle V_1\rangle^{(1)} = \langle\Psi^{(1)}\mid V_1\mid\Psi^{(0)}\rangle + \langle\Psi^{(0)}\mid V_1\mid\Psi^{(1)}\rangle.$$

For the remainder of this section, we are concerned with procedures for scaling molecular wave functions. The equation for the optimum scaling factor may be derived from Eqs. (285) and (286).

$$s_0 = - \left[\frac{\langle V \rangle_{1;sR} + sR \langle V \rangle'_{1;sR}}{2\langle T \rangle_{1;sR} + sR \langle T \rangle'_{1;sR}} \right] \qquad (291)$$

where the primes mean derivatives of the expectation values with respect to sR. Now sR is the original (unscaled) internuclear distance, so all the quantities in (291) are obtainable from the unscaled molecular wave function. The derivatives of the expectation values may be obtained by doing several computations for internuclear distances close to sR [7].

Now consider a linear variation calculation (Section A.4), where the trial function is

$$\Psi = \sum_i c_i \Phi_i \qquad (292)$$

and the $\{c_i\}$ are determined from a secular equation. The resulting wave function may not satisfy the virial theorem, but of course may be made to satisfy it by scaling. But then the coefficients will no longer be optimum, since Ψ will be expanded in scaled basis functions. The process may be repeated; at each step, redetermination of the $\{c_i\}$ or scaling, the energy is lowered. This has been discussed by Löwdin [8]. Some simplification is possible. When one goes over to scaled basis functions,

$$\bar{\Phi}_i(\mathbf{r}_1, \ldots, \mathbf{r}_n) = s^{3n/2} \Phi_i(s\mathbf{r}_1, \ldots, s\mathbf{r}_n)$$

the overlap matrix $\mathbf{S}$ is unchanged, while the elements of the kinetic energy matrix $\mathbf{T}$ are multiplied by s^2 and the elements of the potential energy matrix $\mathbf{V}$ by s. (There will also be a scaling of the nuclear configuration in the case of a fixed-nucleus electronic wavefunction.) Solution of the original secular equation ($s = 1$) yields the kinetic and potential energies for the best function of form (292). Invoking the Hellmann–Feynman theorem [Eq. (335)] allows us to calculate dE/ds for this function from the derivative of the Hamiltonian matrix with respect to s, which is

$$[(d/ds)(s^2\mathbf{T} + s\mathbf{V})]_{s=1} = 2\mathbf{T} + \mathbf{V}.$$

Knowing, in addition to the eigenvalue, its derivative with respect to s simplifies the problem of finding the wave function which is optimum with respect to both scaling and linear variation. If we use basis functions of the form

$$\Phi(\mathbf{k}) = k_1^{3/2} k_2^{3/2} \cdots k_n^{3/2} \sum_P (-1)^P \phi(k_1\mathbf{r}_1, \ldots, k_n\mathbf{r}_n),$$

i.e., different scaling factors for different electrons, the determination of the optimum nonlinear parameters $\{k_i\}$ can be simplified by using the virial theorem [9].

A scaling parameter may generally be added to the parameters varied in a calculation with little additional work. However, after varying the scaling parameter, other parameters must subsequently be reoptimized. This includes R itself. If we start with a wave function for some nuclear configuration and, if the parameter s is chosen variationally, we have no control over the final configuration. One actually sets

$$\left(\frac{\partial E_s(R)}{\partial s}\right)_{sR} = \langle V\rangle_{1;sR} + 2s\langle T\rangle_{1;sR}$$

equal to zero [contrast Eq. (286)]. The result is to give a lower energy than the original one, but for a configuration with internuclear distances s^{-1} times the original ones. The virial theorem will hold for the optimally scaled function in the form $\langle V\rangle = -2\langle T\rangle$. In fact, the new R will be an equilibrium distance in the following sense: As s varies with sR constant, we generate energies for a series of new distances. Letting P be the initial configuration, the sequence of energies can be thought of as a potential curve $E_s(P/s)$. The best s gives the minimum point on the curve.

McLean [10] has considered in detail the use of scaling to improve molecular electronic wave functions. Suppose we have variational fixed-nucleus wave functions of the same degree of approximation for different internuclear distances P_i, $i = 1 \cdots N$. By the same degree of approximation, we mean that they all have the same form and number of variational parameters, whose values are chosen in each case to minimize the energy at the corresponding P_i. By scaling the wave function for P_i with a scaling factor s and allowing s to vary, we generate a function $E^{(i)}$ according to (284), which we now rewrite as

$$E^{(i)}(P_i/s) = s\langle V\rangle_{1;P_i} + s^2\langle T\rangle_{1;P_i}. \tag{293}$$

Each such function corresponds to $E_s(P/s)$ of the preceding paragraph. McLean [10] shows such curves for a few diatomic systems. If we had a set of $E^{(i)}$ for various P_i, we could seek, for a desired R, the minimum energy obtainable by scaling a variational function of this degree of approximation. This would give a curve of $E(R)$ which would be the best obtainable from such a variational function with scaling. It is clear that no $E^{(i)}$ can go below this optimally scaled curve. But for each $E^{(i)}$ there is an R such that $E^{(i)}(R)$ is lower than any $E^{(j)}(R)$ ($j \neq i$), so that the optimally scaled curve cannot lie wholly below this $E^{(i)}$ for this value of R. The optimally scaled curve thus is formed as the envelope of the $E^{(i)}(R)$, being tangent to each of them at one point, from below.

Suppose there are N curves available. Let the optimal curve be written as $E^{(0)}(R)$. It should be possible to determine N points on $E^{(0)}(R)$. Specifically, let the N points of contact with the $E^{(i)}$ be denoted by R_i, the values of the optimal curve there being $E^{(0)}(R_i)$. From (293) we have

$$E^{(0)}(R_i) = (P_i/R_i)\langle V \rangle_{1;P_i} + (P_i/R_i)^2 \langle T \rangle_{1;P_i} \qquad (294a)$$

and the condition of tangency gives

$$E^{(0)'}(R_i) = (P_i/R_i^2)\langle V \rangle_{1;P_i} - 2(P_i^2/R_i^3)\langle T \rangle_{1;P_i}. \qquad (294b)$$

These are $2N$ equations ($i = 1 \cdots N$) for the N R_i and the N parameters in $E^{(0)}(R)$, since all other quantities in (294a, b) are known. Equations (294a, b) are nonlinear and best solved iteratively, starting from a set of guesses for the R_i. Rearranging,

$$(P_i/R_i)^2 \langle T \rangle_{1;P_i} = -E^{(0)}(R_i) - R_i E^{(0)'}(R_i), \qquad (295a)$$

$$(P_i/R_i)\langle V \rangle_{1;P_i} = 2E^{(0)}(R_i) + R_i E^{(0)'}(R_i). \qquad (295b)$$

The quantities on the left of (295a) and (295b) are the scaled kinetic and potential energies. Adding twice (295a) to (295b), we see that the optimally scaled curve satisfies the virial theorem, equation (288), at each of the points of contact. The functional forms used for $E^{(0)}$ by McLean [10] were a truncated series in powers of R and such a series together with a term in R^{-1}. The latter, as expected, seemed preferable, in that it produced curves which more closely resembled accurately calculated potential curves. It must be noted that the improvement due to scaling tends to be small when reasonably good approximate functions are used. This is because the scaling mainly corrects the size of the inner-shell charge distributions, which most affect $\langle T \rangle$ and $\langle V \rangle$, and these are ordinarily well calculated.

In addition to scaling a set of wave functions to get a "best" $U(R)$ curve for a given molecule, we can in some cases scale wavefunctions for one molecule to get $U(R)$ for a second [11]. The procedure is quite analogous. Very little additional calculation is needed. It is only required that the two molecules be isoelectronic. The expectation values for kinetic energy and interelectronic repulsion energy (before scaling) are unchanged in going from molecule A to molecule B; those for the electron–nuclear attraction energies for B may often be obtained from those for A in a trivial way (multiplication by ratios of nuclear charges). Then the scaling may be carried out to produce a wave function for molecule B at a new internuclear distance R_A/s. Optimal $U(R)$ curves for B may be obtained by essentially McLean's procedure. Here, the optimal scaling factors can be much different from unity, which rarely occurs for scaling within one molecule.

Löwdin [8] and Kohn [12] have also discussed scaling where the Hamiltonian is written as $H_0 + \gamma W$, and perturbation theory is used [cf. Eq. (290)]. Suppose $H_0 = T + V$, with V and W homogeneous of degree -1. Then we can scale the zero-order function $\Psi^{(0)} (H_0 \Psi^{(0)} = E^{(0)} \Psi^{(0)})$ to minimize the sum of the zero- and first-order energies

$$(\partial/\partial s)(E^{(0)} + E^{(1)}) = (\partial/\partial s)(s^2 \langle T \rangle_0 + s \langle V + \gamma W \rangle_0) = 0. \qquad (296)$$

The scale factor is given by

$$s = -(2\langle T \rangle_0)^{-1}(\langle V + \gamma W \rangle_0) = 1 + \gamma \langle W \rangle_0/(2E^{(0)}) \qquad (297)$$

where we have used the fact that $\Psi^{(0)}$, being an exact eigenfunction of H_0, obeys the virial theorem. The energy is now

$$s^2 \langle T \rangle_0 + s \langle V + \gamma W \rangle_0 = E^{(0)} + E^{(1)} + \gamma^2 \langle W \rangle_0^2/(4E^{(0)}).$$

The scaled wave function may be thought of as $\Psi^{(0)} + \gamma \Phi$, Φ becoming an approximation to the first-order wave function, so the term following $E^{(1)}$ is an upper bound to the exact second-order energy.

We can think of the scaling as an adjustment of the size of a wave function with no change in shape. It is clear that proper scaling is a necessary condition for a good approximate wave function. As we have pointed out, it is easy to carry out the scaling, which will also ensure that the virial theorem is satisfied. On the other hand, if we have an approximate wave function in which no explicit overall scaling parameter has been introduced, the degree to which the virial theorem is satisfied is an indication of accuracy. Here, we have a theoretical relation between expectation values that we may use to check approximate wave functions. It is nice not always to have to have recourse to experimentally measured properties. Hurley [13] was able to show that only variational functions of optimal scale satisfy the virial theorem.

Where the only restriction on a variational procedure is one of functional form, the virial theorem will always be satisfied. It is satisfied for the Hartree–Fock function, for instance. Equation (287) then holds between $\langle T \rangle$,

$\langle V \rangle$, and $E = \langle H \rangle$ calculated with the variationally determined function. This does not imply that E or the other expectation values are close to the exact values.

2. Consequences of the Virial Theorem

Since, for a diatomic system, it suffices to know T or V as a function of R to obtain $U(R)$, the virial theorem seems to simplify calculation of $U(R)$. The kinetic energy being a one-electron operator, it should be easier to calculate than the total energy. (The danger is that an approximate wave function not determined variationally may not satisfy the virial theorem, so that its kinetic energy may bear no relation to its total energy.) In addition, it may be possible to formulate simple models for evaluating the kinetic energy or potential energy alone. The potential energy operator involves Coulombic interactions only, so that, if the electron density is known, all forces may be calculated by classical electrostatics. No additional "quantum mechanical" forces need be invoked; quantum mechanics is needed only for the determination of the wave function which yields the electron density. This point will be discussed in more detail in the next section.

In this subsection, we first mention some consequences of the virial theorem which relate to $U(R)$, and then discuss attempts at using the virial theorem for calculation of force constants and related quantities. We close with a brief mention of the hypervirial theorems, of which the virial theorem is a special case.

At the equilibrium internuclear distance, the point of minimum energy on the potential curve,

$$-\langle T \rangle = E = \tfrac{1}{2}\langle V \rangle. \tag{298}$$

If we have a wave function for which the virial theorem holds (either because scaling has been explicitly carried out or because a large number of parameters have been varied), Eq. (298) may be used to define the equilibrium internuclear distance R_e rather than

$$(dE/dR)_{R_e} = 0.$$

If R_e is to be found by interpolation and high accuracy is required, it is better to look for the zero on a plot of $\langle T \rangle + E$ as a function of R, rather than the minimum on a plot of E as a function of R. The latter is flat near R_e, which leads to inaccuracies.

Using Eq. (287) together with

$$E = \langle T \rangle + \langle V \rangle,$$

we can get the potential curve in terms of $\langle T \rangle$ or $\langle V \rangle$ alone:

$$2E + R(dE/dR) = \langle V \rangle, \qquad (299a)$$

$$-d(ER)/dR = -R(dE/dR) - E = \langle T \rangle. \qquad (299b)$$

Again, V_{nn} may or may not be included in V. In what follows, we assume it is included, so E is identical with $U(R)$. McLean [10] noted that, for an unscaled wave function, for which $\langle T \rangle$ from (299a, b) would not agree with the directly computed expectation values, (299a, b) should be used. The effect of scaling is much greater on the expectation values than on $d(ER)/dR$.

Given the dependence of E on R, the curves of $\langle T \rangle$ and $\langle V \rangle$ versus R follow rigorously. In particular, considering the equilibrium internuclear distance and the separated atoms ($dE/dR = 0$ for both cases), we have

$$2(E^{at} - E^{eq}) = (\langle V \rangle^{at} - \langle V \rangle^{eq}) = 2(\langle T \rangle^{eq} - \langle T \rangle^{at}).$$

If the molecule is stable, so the first member is positive, this says that the potential energy *must* decrease and the kinetic energy *must* increase on molecule formation. Equations (299) also imply $2E > \langle V \rangle$ and $\langle T \rangle > -E$ for $dE/dR < 0$ (which for the usual situation is for $R < R_e$), and $2E < \langle V \rangle$ and $\langle T \rangle < -E$ for $dE/dR > 0$ (which usually means $R > R_e$). Differentiating (299b) yields

$$d\langle T \rangle/dR = -R(d^2E/dR^2) - 2(dE/dR).$$

This means that a positive force constant (d^2E/dR^2 positive, where $dE/dR = 0$) requires that the kinetic energy be decreasing with R at R_e. If $U = -CR^{-6}$ (van der Waals), $\langle T \rangle = -5CR^{-6}$ and $\langle V \rangle = 4CR^{-6}$, so the latter increases and the former decreases with decreasing R (relative to their values in the separated atoms).

A discussion of potential curves of one-electron molecules by an extension of (299a) has recently appeared [14]. It is obvious from the form of V that for the one-electron case,

$$V = Z_A(\partial V/\partial Z_A) + Z_B(\partial V/\partial Z_B).$$

Taking expectation values and invoking the Hellmann–Feynman theorem [see Section D, particularly the discussion of Eqs. (327) and (328)] yields, according to (299a),

$$2E + R(\partial E/\partial R) = Z_A(\partial E/\partial Z_A) + Z_B(\partial E/\partial Z_B).$$

The solution to this equation is

$$E(Z_A, Z_B, R) = Z_A Z_B f(Z_A R, Z_B R)$$

with f arbitrary. For the homonuclear case, $E = Z^2 g(ZR)$ with g arbitrary, and E is determined for all Z and R once sufficient boundary conditions are imposed. In particular, if $E = U(R)$ for $Z = 1$, then in general $E = Z^2 U(ZR)$. This is evident on applying scaling to the differential equation. For the heteronuclear case, it suffices to have E for $Z_A = 1$ and arbitrary Z_B and R.

Equation (287), differentiated with respect to R, gives an equation for the force constant:

$$R(d^2E/dR^2) - R^{-1}(2\langle T \rangle + \langle V \rangle) + (d/dR)(2\langle T \rangle + \langle V \rangle) = 0. \quad (300)$$

When (300) is evaluated at $R = R_e$, $2\langle T \rangle + \langle V \rangle$ vanishes, according to the virial theorem. The last term may be written $2(dE/dR - d\langle V \rangle/dR)$, and $dE/dR = 0$ at R_e. Thus we obtain

$$R_e k_e = 2d\langle V \rangle/dR$$

since the force constant k_e is (d^2E/dR^2) evaluated at $R = R_e$. The anharmonic force constant l_e may be obtained [15] in terms of derivatives of $\langle V \rangle$ with respect to R in a similar way. By differentiation of Eq. (300),

$$R(d^3E/dR^3) = -(d^2/dR^2)(2\langle T \rangle + \langle V \rangle + 2E) = -4(d^2E/dR^2) + d^2\langle V \rangle/dR^2$$

when $R = R_e$. Since l_e is (d^3E/dR^3) evaluated at $R = R_e$, we obtain

$$R_e l_e = (d^2\langle V \rangle/dR^2)_e - 4R_e^{-1}(d\langle V \rangle/dR)_e.$$

Of course, these and higher force constants may likewise be written in terms of the derivatives of $\langle T \rangle$ [15]. Thus, experimental measurement of the vibrational potential constants leads to information on the variation of kinetic and potential energy with internuclear distance. Schwendeman [16] used the four measured spectroscopic parameters of CO to obtain the derivatives of total energy, kinetic energy, and potential energy. These derived values were compared to theoretical values, obtained by fitting power series expansions to calculated energies for different values of R. A discussion of problems related to curve fitting was given.

Equation (300) actually implies a number of different relations. In the last term, we can differentiate an integral like $\int \Psi^* V \Psi \, d\tau$ in several different ways, depending on the coordinate system in which the electronic coordinates are kept fixed. If the system is like ellipsoidal coordinates, where interparticle distances are proportional to R for fixed values of the coordinates, we have the operator equations

$$\partial T/\partial R = -2T/R, \qquad \partial V/\partial R = -V/R. \tag{301}$$

We write $(\partial \Psi/\partial R)_e$ for the derivative of the normalized eigenfunction with R, the electronic coordinates being held fixed in such a system. In Section D, this kind of differentiation is used to give the virial theorem as a special case of the Hellmann–Feynman theorem. Now (300) becomes

$$\left(\frac{d^2E}{dR^2}\right)_e = R^{-2}(6\langle T \rangle + 2\langle V \rangle) - 2R^{-1} \operatorname{Re} \int \left(\frac{\partial \Psi}{\partial R}\right)_e^* (2T + V)\Psi \, d\tau, \tag{302}$$

valid for a function satisfying the virial theorem. The first term on the left side of (302) is the expectation value of $(\partial^2 H/\partial R^2)_e$, as may be seen by differentiation of (301):

$$\left(\frac{\partial^2 H}{\partial R^2}\right)_e = \left(\frac{\partial}{\partial R}\left[\frac{-2T}{R}\right]\right)_e + \left(\frac{\partial}{\partial R}\left[\frac{-V}{R}\right]\right)_e = R^{-2}[6T + 2V]. \tag{303}$$

For a change in R, Ψ changes in size and in shape; $(\partial \Psi/\partial R)_e$ roughly corresponds to the change of shape only. If the last term in (302) is neglected, we get a simple approximate expression for the second derivative. At the equilibrium internuclear distance we may use (298), and d^2E/dR^2 becomes $-2E/R^2$. If $(\partial \Psi/\partial R)_e$ is calculated by perturbation theory (Section B.4), it may be written as

$$\left(\frac{\partial \Psi}{\partial R}\right)_e = \sum_{i>0} \frac{\Psi_i \langle \Psi_i | (\partial H/\partial R)_e | \Psi \rangle}{E - E_i} \tag{304}$$

where i runs over excited states. Recognizing that $2T + V$ is $-R(\partial H/\partial R)_e$ and remembering that we are in the ground state so $E - E_i$ is negative, we see that the second term in (302) is negative. Thus

$$d^2E/dR^2 < \langle(\partial^2 H/\partial R^2)_e\rangle \tag{305}$$

Davidson [17] has proved that (305) holds for any function where all

R-dependent parameters are chosen variationally. Now (301) implies

$$-T = R(\partial H/\partial R)_e + H \qquad (306a)$$

$$V = R(\partial H/\partial R)_e + 2H \qquad (306b)$$

which, with (302), now yields

$$d^2E/dR^2 = R^{-2}[-2E - 4R\langle(\partial H/\partial R)_e\rangle] - 2R^{-1}\,\text{Re}\langle(\partial\Psi/\partial R)_e\,|(2T+V)|\Psi\rangle \qquad (307a)$$

$$< R^{-2}[-2E - 4R\,dE/dR] \qquad (307b)$$

Replacement of $(\partial H/\partial R)_e$ by dE/dR is justified by the Hellmann–Feynman theorem. Inequality (307b) means that $d^2(R^2E)/dR^2$ is everywhere negative. At the equilibrium distance R_e, this gives a weak condition relating the force constant, R_e, and the energy. This and related inequalities are discussed by Davidson [17]. They have implications for the forms allowable for a potential curve $U(R)$.

Another exact expression is obtained by differentiating (299a) but, in the integrals, keeping electronic coordinates fixed in a *space-fixed* system. We obtain

$$\int (\partial/\partial R)(|\Psi|^2)V\,d\tau + \langle\partial V/\partial R\rangle = 3\,dE/dR + R(d^2E/dR^2). \qquad (308)$$

Now $\partial V/\partial R = \partial H/\partial R$ in such a coordinate system, and the expectation value is dE/dR by the Hellmann–Feynman theorem. If the first term in (308) is neglected, then the equation may be solved explicitly, as pointed out by Clinton [6]. The solution is

$$E = a + b/R \qquad (309)$$

and this clearly represents an unbound molecular state. Thus the change in $|\Psi|^2$ with R is responsible for the *existence* of a force constant and cannot be neglected. The situation is to be contrasted with Eq. (302); there, neglecting $(\partial\Psi/\partial R)_e$ corresponds to letting the charge distribution change with R by changing its size.

At the equilibrium distance, Eq. (308) may be written either as

$$(d^2E/dR^2)_{R_e} = R_e^{-1}(\partial\langle V\rangle/\partial R)_{R_e} \qquad (310)$$

or, since the preceding discussion shows that $\langle\partial V/\partial R\rangle$ vanishes here, as

$$(d^2E/dR^2)_{R_e} = R_e^{-1}\left[\int (\partial/\partial R)(|\Psi|^2)V\,d\tau\right]_{R_e}. \qquad (311)$$

These two expressions for the force constant are discussed by Salem [17a]. Using first-order perturbation theory (Section B), $\partial\Psi/\partial R$ at R_e may be written in terms of the complete set of eigenfunctions of the electronic Hamiltonian at R_e. Then the force constant $k_e = (d^2U/dR^2)_{R_e}$ is expressed as an infinite sum resembling a second-order energy of perturbation theory. An average-energy approximation then gives the force constant as an expectation value of $\partial(V^2)/\partial R$, but this is not likely to prove useful [17a].

The virial theorem leads to several methods for calculation of $U(R)$. By differentiating (299b) we find

$$-R^{-1}\,d(R^2\langle T\rangle)/dR = R^2(d^2E/dR^2) + 4R(dE/dR) + 2E. \quad (312)$$

Other second-order differential equations for E may be derived similarly. A formal solution to (312) is

$$E = R^{-1}(R_e E_e - \int_{R_e}^{R} \langle T\rangle_{R'}\,dR') \quad (313)$$

where the constants of integration have been chosen such that, for $R = R_e$ (equilibrium internuclear distance), $E = E_e$ and dE/dR vanishes (since $E = -\langle T\rangle$ here). Equation (313) again shows that all properties of the potential curve are calculable when the variation of $\langle T\rangle$ with R is known (together with two constants of integration). This equation has been used by Borkman and Parr [18] and is discussed in Volume 2, Chapter III, Section B.

Denoting the left side of (312) by $2S$, we discuss some properties of the function

$$S(R) = -\langle T\rangle - \tfrac{1}{2}R\,d\langle T\rangle/dR. \quad (314)$$

One might expect the second term to be less important than the first, particularly for small R. The internuclear repulsion terms on the right of (314) cancel out; for the remainder of this paragraph, they are not included in E. Since E approaches a finite value (united atom energy) as R goes to 0, $R^2(d^2E/dR^2)$ and $R(dE/dR)$ vanish here, and $S(0) = E(0)$. By the virial theorem, this is minus the kinetic energy of the united atom. Furthermore, Clinton [19] points out that if $U\ (= E + V_{nn})$ goes to zero for $T \to \infty$ as some inverse power of R higher than the first, $R^2(d^2U/dR^2)$ and $R(dU/dR)$ approach zero as $R \to \infty$ and S goes over to the negative of the kinetic energy of the separated atoms. It may also be shown [19] that S is always negative. Putting $\alpha = -S/\langle T\rangle$ and using (299b) for $\langle T\rangle$, we obtain the differential equation

$$R^2(d^2U/dR^2) + (4 - 2\alpha)R(dU/dR) + (2 - 2\alpha)U = 0.$$

If α may be approximated as unity for all R, the solution to the equation is $U = cR^{-1}$. Allowing for some variation of α, Clinton [19, 20] derived

$$U = \operatorname{Re}(AR^{-N})$$

where A and N are complex, a generalization of the Sutherland potential to complex exponents (see Section D of Chapter II). There are four real parameters in U. Relations between potential constants predicted from this function proved satisfactory, and it compared favorably with the more common functional forms for $U(R)$ for a number of molecules.

Bishop [21] has suggested calculating force constants from the quantity $\langle T \rangle / E$. Differentiating $\langle T \rangle / E$ from Eq. (299b),

$$d(\langle T \rangle / E)/dR = (R/E)(d^2E/dR^2) - (1/E)(dE/dR). \tag{315}$$

Then for $R = R_e$ we get the expression for the force constant:

$$(d^2E/dR^2)_{R_e} = -(E_{R_e}/R_e)(d(\langle T \rangle / E)\, dR)_{R_e}$$

Let $(\langle T \rangle / E)$ be calculated for a series of R-values. The results are to be fitted to a polynomial, R_e is to be found as the value of R which makes $\langle T \rangle / E$ equal unity, and the force constant calculated from the slope of the curve at this value of R.

A generalization of (313) is

$$RE(R) - R_0E(R_0) = -\int_{R_0}^{R} \langle T \rangle_{R'} \, dR' \tag{316}$$

where R_0 need not be R_e. The kinetic energy operator is one-electron, and easier to calculate or perhaps estimate than the potential energy. Equation (316) then suggests avoiding some work by calculating $\langle T \rangle$ as a function of R and performing a numerical integration to get $E(R)$. The danger, analogous to that in use of the Hellmann–Feynman theorem to avoid work, is that E and $\langle T \rangle$ satisfy (316) if the function satisfies the virial theorem. Otherwise, what we calculate from (316) may be unrelated to the energy obtained by direct calculation. It could be better or worse. In some tests on H_2 with various wave functions, Benston and Kirtman [22] found that force constants were poor if the virial theorem expressions were used with unscaled wave functions, but fairly reliable otherwise. A similar conclusion was reached by Thorhallson and Chong [7], who computed the "virial force," $-(2\langle T \rangle + \langle V \rangle)/R$, for valence bond configuration interaction wave functions for LiH at three different internuclear distances. By fitting the three points to a parabola, they derived values for R_e, k_e, and l_e. When the wave functions were scaled, there was a great improvement in the values of k_e and l_e (now yielding results in substantial agreement with experiment) and also some improvement in the value of R_e. It was noted [7] that scaling changes the energy only slightly, but the forces very markedly. An early discussion of the dangers in using the virial theorem with approximate functions to get $U(R)$ was given by Coulson and Bell [4a].

A convenient choice of R_0 in (316), used by Hurley [23] in his important discussion, is infinite nuclear separation. Here, the virial theorem is known to hold in the form

$$\langle T \rangle_\infty = -E(\infty). \tag{317}$$

To use this, we put $R_0 = sR$ and $R' = s'R$ in (316) to give

$$E(R) - sE(sR) = -\int_{R_0/R}^{1} \langle T \rangle_{s'R} \, ds'.$$

Adding $-s\langle T \rangle_{sR}$ to both sides yields

$$E(R) - s[E(sR) + \langle T \rangle_{sR}] = \int_{1}^{s} [\langle T \rangle_{s'R} - \langle T \rangle_{sR}] \, ds' - \langle T \rangle_{sR} \tag{318}$$

Now let $s \to \infty$. For very large internuclear distance, the sum of the energy and the kinetic energy [see (299b)] will fall off to zero exponentially, so the second term on the left vanishes. Then, using (317), this becomes

$$E(R) = \int_{1}^{\infty} (\langle T \rangle_{s'R} - \langle T \rangle_\infty) \, ds' + E(\infty). \tag{319}$$

When R is the equilibrium internuclear distance, (319) gives the binding energy as

$$\int_{1}^{\infty} (\langle T \rangle_\infty - \langle T \rangle_{s'R}) \, ds'.$$

Baumann *et al.* [24] suggest taking advantage of the simplicity of calculation of $\langle T \rangle$ compared to E by using the condition $\langle T \rangle = -E$, which obtains at the equilibrium R [see (299b)], to determine a wave function. An experimental value is used for E, and a parameter in the trial function is chosen so that $\langle T \rangle$ calculated with the function at the equilibrium R is the negative of this. Since only one parameter can be so chosen, the interest is for systems other than many-electron diatomics.

Since T is one-electron, its expectation value for a single-determinant wavefunction is a sum of contributions of the occupied spin orbitals. Equation (312), for $R = R_e$, becomes

$$(d^2E/dR^2)_{R_e} = -(2/R) \sum_i d\langle \lambda_i | T | \lambda_i \rangle/dR \tag{320}$$

where λ_i are the occupied spin orbitals. Rossinkhin et al. [25] considered the calculation of force constants for a Hartree–Fock function, which satisfies the virial theorem, in which each λ_i was expanded in atomic orbitals. They gave formulas for the harmonic and higher force constants in terms of integrals over the basis functions. As is seen from (320), the integrals involve the derivative of an atomic orbital with respect to R, as in the work of Gerratt and Mills [Section B, Eqs. (249) et seq.], as well as formulas for the change in the expansion coefficients with R. Rossikhin et al. [25] calculated the second, third, and fourth derivatives of $U(R)$ for the first row hydrides, obtaining results correct to a factor of 2 in all cases.

We conclude with a word about hypervirial theorems, which point up connections between a number of subjects discussed in this book. As has been emphasized recently by Hirschfelder and co-workers [26–28], the virial theorem is really one of a class of theorems which may be written

$$\langle \Psi | [H, W] | \Psi \rangle = 0. \tag{321}$$

Here, Ψ is a bound-state eigenfunction of H, W is an arbitrary Hermitian time-independent operator, and $[H, W]$ is the commutator

$$[H, W] = HW - WH$$

Since $H\Psi = E\Psi$ with E real, the validity of (321) is evident. Relations of the form (321) are known as hypervirial theorems. By taking

$$W = \frac{1}{2} \left(x_i \frac{d}{dx_i} + \frac{d}{dx_i} x_i \right)$$

where i runs over Cartesian coordinates of all particles, and considering that the potential energy of the particles is homogeneous of degree -1 (Coulombic) in the coordinates of all the particles, so that

$$\sum_i x_i(dV/dx_i) = -V,$$

we can derive the virial theorem. Suppose [26, 27] that an approximate wave function Φ satisfies

$$\partial\Phi/\partial a = (i/\hbar)W \tag{322}$$

with a a parameter in the wavefunction and W an operator. The parameter a may also appear in the Hamiltonian. Differentiating the expectation value of the Hamiltonian with respect to a [cf. Eq. (325)] and using (322) we can show

$$\frac{\partial E}{\partial a} \int \Phi^* \Phi \, d\tau - \int \Phi^* \frac{\partial H}{\partial a} \Phi \, d\tau = \frac{i}{\hbar} \int \Phi^* [(H - E), W] \Phi \, d\tau \quad (323)$$

The right side of (323) is like Eq. (321). This shows that, if Φ satisfies the hypervirial theorem for the W of Eq. (322), the function satisfies the Hellmann–Feynman theorem (Section D) with respect to a and vice versa. Some special choices of W and their implications are discussed by Epstein and Hirschfelder [27] and Hirschfelder and Coulson [29].

Since the various hypervirial relations, Eq. (321), must be satisfied by the exact wave function and do not necessarily hold for an approximate wavefunction, they can be used as constraints in a constrained variation procedure (Section A.3). A desired region of configuration space may be emphasized by choosing a hypervirial theorem for which the operator W is large in the region of interest. This was emphasized by Coulson [29] in a discussion of the uses of the hypervirial and related theorems. He was able to show that a wave function for which

$$\int \Phi^* [H, [H, f(x)]] \Phi \, d\tau = 0$$

for all functions $f(x)$ must be an eigenfunction of H. This is the hypervirial relation of (321) with $W = [H, f(x)]$. Coulson also showed that it suffices to have

$$\int \Phi^* [H, f(x) \, d/dx] \Phi \, d\tau = 0$$

for all $f(x)$, to assure that Φ is an eigenfunction. The proofs involve straightforward algebraic manipulations and integration by parts, which lead, in both cases, to

$$(d/dx)\{T\Phi/\Phi + V\} = 0.$$

It is assumed that the Hamiltonian involves only the usual kinetic energy operators and a multiplicative potential energy operator. The quantity in braces is the local energy, and a wave function for which the local energy is a constant over configuration space must be an eigenfunction of the Hamiltonian. This is the basis for local energy methods (see Section E).

REFERENCES

1. J. O. Hirschfelder and J. F. Kincaid, *Phys. Rev.* **52**, 658 (1937).
2. V. Fock, *Z. Phys.* **63**, 855 (1930).
3. J. C. Slater, "Quantum Theory of Molecules and Solids," Vol. 1, Sect. 2.4 and Appendix 3. McGraw-Hill, New York, 1963.
4. T. L. Cottrell and S. Patterson, *Phil. Mag.* **42**, 391 (1951).
4a. C. A. Coulson and R. P. Bell, *Trans. Faraday Soc.* **41**, 141 (1945).
5. T. L. Cottrell and S. Patterson, *Trans. Faraday Soc.* **47**, 233 (1951).
6. W. L. Clinton, *J. Chem. Phys.* **33**, 1603 (1960).
7. J. Thorhallson and D. P. Chong, *Chem. Phys. Lett.* **4**, 405 (1969).
8. P.-O. Löwdin, *J. Mol. Spectrosc.* **3**, 46 (1959).
9. R. T. Brown, *J. Chem. Phys.* **48**, 4698 (1968).
10. A. D. McLean, *J. Chem. Phys.* **40**, 2774 (1964).
11. J. Goodisman, *J. Phys. Chem.* **69**, 2520 (1965); **70**, 1675 (1966).
12. W. Kohn, *Phys. Rev.* **71**, 635 (1967).
13. A. C. Hurley, *J. Chem. Phys.* **37**, 449 (1962).
14. B. J. Laurenzi, *Theor. Chim. Acta* **13**, 106 (1969); B. J. Laurenzi and A. F. Saturno, *J. Chem. Phys.* **53**, 579 (1970).
15. R. H. Schwendeman, *J. Chem. Phys.* **44**, 556 (1966).
16. R. H. Schwendeman, *J. Chem. Phys.* **44**, 2115 (1966).
17. E. R. Davidson, *J. Chem. Phys.* **36**, 2527 (1962).
17a. L. Salem, *J. Chem. Phys.* **38**, 1227 (1963).
18. R. F. Borkman and R. G. Parr, *J. Chem. Phys.* **48**, 1116 (1968); **50**, 58 (1969).
19. W. L. Clinton, *J. Chem. Phys.* **38**, 2339 (1963).
20. W. L. Clinton and S. D. Frattali, *J. Chem. Phys.* **39**, 3316 (1963).
21. D. M. Bishop, *Mol. Phys.* **6**, 305 (1963).
22. M. L. Benston and B. Kirtman, *J. Chem. Phys.* **44**, 119 (1966).
23. A. C. Hurley, *Proc. Roy. Soc. Ser. A* **226**, 170 (1954).
24. H. Baumann, E. Heilbronner, and J. N. Murrell, *Theor. Chim. Acta* **5**, 87 (1966).
25. V. V. Rossikhin, V. P. Morozov, and L. I. Bezzub, *Theor. Exp. Chem. (USSR)* **4**, 22 (1968); *Teor. Eksp. Khim.* **4**, 37 (1968).
26. J. O. Hirschfelder, *J. Chem. Phys.* **33**, 1462 (1960).
27. S. T. Epstein and J. O. Hirschfelder, *Phys. Rev.* **123**, 1495 (1961).
28. J. O. Hirschfelder and C. A. Coulson, *J. Chem. Phys.* **36**, 941 (1962).
29. C. A. Coulson, *Quart. J. Math. Oxford Ser.* **16**, 279 (1965).

D. Hellmann–Feynman Theorem

1. *Hellmann–Feynman Theorem and Electrostatic Forces*

The Hellmann–Feynman theorem [1, 2] is, at first sight, both simple and powerful. Suppose one wants to calculate the change of the energy of a system with change of the value of some parameter appearing in the

Hamiltonian. One can write

$$\frac{dE}{dQ} = \frac{d}{dQ}\left(\frac{\int \Psi^* H\Psi\, d\tau}{\int \Psi^*\Psi\, d\tau}\right)$$

$$= \frac{\int \Psi^*\Psi\, d\tau(d/dQ)(\int \Psi^* H\Psi\, d\tau) - \int \Psi^* H\Psi\, d\tau(d/dQ)(\int \Psi^*\Psi\, d\tau)}{(\int \Psi^*\Psi\, d\tau)^2}$$

$$(324)$$

where Ψ is the eigenfunction of H with eigenvalue E. The values of Ψ and E will change with Q as well as H, so that

$$\frac{d}{dQ}\int \Psi^* H\Psi\, d\tau = \int \frac{d\Psi^*}{dQ}H\Psi\, d\tau + \int \Psi^*\frac{dH}{dQ}\Psi\, d\tau + \int \Psi^* H\frac{d\Psi}{dQ}\, d\tau$$

$$= \int \Psi^*\frac{dH}{dQ}\Psi\, d\tau + E\frac{d}{dQ}\int \Psi^*\Psi\, d\tau. \qquad (325)$$

We have used $H\Psi = E\Psi$ in the second step. Using (325), together with $E\int \Psi^*\Psi\, d\tau = \int \Psi^* H\Psi\, d\tau$, in equation (324), we see that it reduces to

$$dE/dQ = \int \Psi^*(dH/dQ)\Psi\, d\tau \Big/ \int \Psi^*\Psi\, d\tau, \qquad (326)$$

that is, the derivative can be calculated as the expectation value of the operator dH/dQ. This is a great simplification. To get the change in the energy with Q when $Q = Q_0$, we do not need several computations to give values of E for different values of Q near Q_0. It suffices to calculate the expectation value of dH/dQ with the single wave function for $Q = Q_0$.

The most important case, for our purposes, is where $Q = R$, since we are concerned with the forces on the nuclei. Since, in most coordinate systems, the kinetic energy operator and the operator for interelectronic repulsion are R-independent, only the derivative with R of the nuclear–electronic attraction operator contributes to the operator in (326). An important consequence of this, as recognized by Hellmann [2] and Feynman [1], is that the forces can be considered completely classically. We require the expectation value of a multiplicative one-electron operator, which is calculable from the one-electron density or charge density. Quantum mechanics enters only in the determination of the density. Once this is known, forces and therefore the shapes of potential curves can be understood in terms of classical electrostatics. The theorem, used in connection with forces, is often referred to as the electrostatic theorem.

The sufficiency of classical electrostatics makes it possible to construct simple models for calculation of $U(R)$ and to give simple interpretations of force constants and related properties. We shall mention such work in this section. The simple models have been more successful than calculations based solely on the Hellmann–Feynman theorem. The danger in using the theorem for calculations is that one does not know whether a given wave function satisfies it. This is a feature shared by the virial theorem, to which the Hellmann–Feynman theorem is related, as appeared in Section C. The necessary conditions for a wave function to satisfy the theorem, and their implications, are discussed in Subsection 2. In Subsection 3, application of the theorem to calculation of $U(R)$ is discussed, and some related theorems which have been used are mentioned in Subsection 4.

If Eq. (326) is applied to the fixed-nucleus electronic Hamiltonian of a molecule with Q being Z_A, the charge on a nucleus A, the only terms in H which depend on Q are the electron–nuclear attraction terms corresponding to nucleus A,

$$V_A = - \sum_i Z_A e^2 \, | \, \mathbf{r}_i - \mathbf{r}_A \, |^{-1}, \tag{327}$$

and the internuclear repulsion terms involving A,

$$V_{NA} = \sum_B^{(B \neq A)} Z_A Z_B e^2 \, | \, \mathbf{r}_A - \mathbf{r}_B \, |^{-1}. \tag{328}$$

The sum over i in Eq. (327) is over electrons, $\mathbf{r}_i$ is the position of electron i, and $\mathbf{r}_A$ is the position of nucleus A. Obviously, dH/dQ is just $Z_A^{-1}(V_A + V_{NA})$ in this case. Then the theorem states that the change in energy with change in Z_A is calculable from V_{NA} and the expectation value of V_A. Thus, if the electron–nuclear attraction energy is known as a function of Z_A, the dependence of the energy on Z_A is known. Now let each nuclear charge be written as QZ_B, defining a "charge-scaling" parameter Q such that $Q = 1$ corresponds to the actual molecule. Integrating (326) with respect to Q from $Q = 0$ to $Q = 1$,

$$E = E(Q \to 0) + \int_0^1 dQ \langle \partial V_{ne}/\partial Q \rangle.$$

The expectation value of $\partial V_{ne}/\partial Q$ is a sum of expectation values like $| \, \mathbf{r} - \mathbf{r}_A \, |^{-1}$. The energy E vanishes for $Q \to 0$. Wilson [3] pointed out that this allows calculation of E from the one-electron density as a function of Q, which is a function of only four variables (three spatial coordinates plus Q).

The relation (326) depended, for its derivation, on Ψ being the exact eigenfunction of H. It is not necessarily true when Ψ is an approximation to the exact eigenfunction (see Subsection 2), but the theorem will hold where Ψ is the exact eigenfunction of H within a subspace. Let Ψ be written as a linear combination of the same basis functions for all values of Q:

$$\Psi = \sum_{k=1}^{N} c_k \Phi_k \tag{329}$$

The coefficients $\{c_k\}$ and the energy E are obtained from the equations of linear variation:

$$\sum_{j=1}^{N} (H_{ij} - ES_{ij})c_j = 0, \qquad i = 1, \ldots, N \tag{330}$$

Here, $H_{ij} = \langle \Phi_i \mid H \mid \Phi_j \rangle$ and $S_{ij} = \langle \Phi_i \mid \Phi_j \rangle$. The energy is

$$E = \sum_{i,j} c_i^* c_j H_{ij} \Big/ \sum_{i,j} c_i^* c_j S_{ij}. \tag{331}$$

H_{ij} changes with Q, so the $\{c_i\}$ and E are Q-dependent. Differentiating (331) with respect to Q, and paralleling the manipulations of Eqs. (324)–(327), we find

$$\frac{dE}{dQ} = \frac{\sum_{i,j} c_i^* c_j H_{ij}'}{\sum_{i,j} c_i^* c_j S_{ij}} \tag{332}$$

where $H_{ij}' = \langle \Phi_i \mid dH/dQ \mid \Phi_j \rangle$. This is just the expectation value of dH/dQ over Ψ. Hellmann [2] proved the theorem for the exact wavefunction by considering the expression

$$dE/dQ = \langle dH/dQ \rangle + 2 \, \mathrm{Re} \int (d\Psi^*/dQ) H\Psi \, d\tau, \tag{333}$$

which is valid when Ψ is normalized independently of Q. The second term vanishes because $\langle H \rangle$ is an extremum with respect to any variation in Ψ which preserves normalization, and $\delta\Psi = \delta Q(d\Psi/dQ)$ is such a variation. Thus the theorem is proved. By the same token, if Ψ is obtained by extremizing $\langle H \rangle$ within a subspace, and Ψ changes with Q but always remains within the subspace, $dE/dQ = \langle dH/dQ \rangle$. The linear variation case is such a situation.

If we use, in an expression like (329), a basis set in which individual 'functions depend on the parameter Q, the set nevertheless remaining complete for all Q, a generalization of the Hellmann–Feynman theorem may be proved [4]. Since S_{ij} is now Q-dependent, we have

$$\frac{dE}{dQ} = \frac{\mathbf{c}^{\dagger\prime}\mathbf{Hc} + \mathbf{c}^{\dagger}\mathbf{H}'\mathbf{c} + \mathbf{c}^{\dagger}\mathbf{Hc}'}{\mathbf{c}^{\dagger}\mathbf{Sc}} - \frac{E}{\mathbf{c}^{\dagger}\mathbf{Sc}}(\mathbf{c}^{\dagger\prime}\mathbf{Sc} + \mathbf{c}^{\dagger}\mathbf{S}'\mathbf{c} + \mathbf{c}^{\dagger}\mathbf{Sc}'). \quad (334)$$

Here, we have introduced an obvious matrix notation and used primes for differentiation with respect to Q. Equation (330) now takes the form

$$(\mathbf{H} - E\mathbf{S})\mathbf{c} = 0 \qquad\qquad (335)$$

Using (335) and its adjoint, we reduce Eq. (334) to

$$\frac{dE}{dQ} = \frac{\mathbf{c}^{\dagger}(\mathbf{H}' - E\mathbf{S}')\mathbf{c}}{\mathbf{c}^{\dagger}\mathbf{Sc}}$$

As previously, the derivatives of the wave function with respect to Q have dropped out.

If the exact wave function and energy is calculated as a Rayleigh–Schrö-dinger perturbation expansion, the Hellmann–Feynman theorem holds for all values of the perturbation parameter (γ in Section B). Thus it holds for each order. If γ is identical to the parameter Q, dH/dQ is just the perturbing Hamiltonian, and it may be shown [4, 5], that the mth-order term in $\langle H \rangle$ (contributions from functions of $0, \ldots, m$th order) can be computed in terms of the $(m-1)$-order terms in the expectation value of dH/dQ:

$$\langle dH/dQ \rangle^{(m-1)} = m\langle H \rangle^{(m)} \qquad\qquad (336)$$

This and related expressions have been discussed by several authors [4, 5].

For the remainder of this section, we will be concerned with the Hell-mann–Feynman theorem where Q is the coordinate of a nucleus in a problem in which we are interested in the electronic energy as a function of nuclear positions. The change in electronic energy due to a small displacement of nucleus A in the z direction is the negative of the force on A in this direction. According to Eq. (326)–(328), it can be computed from dV_{NA}/dz_A and the expectation value of

$$dV_A/dz_A = \sum_i Z_A e^2(z_A - z_i)\,|\,\mathbf{r}_i - \mathbf{r}_A\,|^{-3} \qquad\qquad (337)$$

since V_A and V_{NA} are the only operators in the Hamiltonian which depend on the position of A. The operator dV_A/dz_A represents the negative of the electrostatic force on the nucleus due to the electrons. Thus the electronic force on A may be defined as the derivative of the electronic energy with respect to a displacement of A or as the expectation value of the force operator of Eq. (337). For the exact function, these are exactly equal; for one that is not exact, they may not be (see Subsection 2) but one has no reason to choose one over the other as the definition of the force. The use of the force concept in quantum chemistry has recently been reviewed [5a].

Since the force may be considered purely from the point of view of electrostatics, visualization of the bonding process is possible. For instance, an attractive interaction between two atoms must be associated with a buildup of electronic charge between them. Building on this, Berlin [6] defined binding and antibinding regions, such that electronic charge located in the former tends to attract the nuclei of a diatomic to each other, and vice versa. We may calculate the force on nucleus A by considering the change in electronic energy for a displacement of A, the other nucleus, B, remaining fixed. Thus, $|\mathbf{r}_i - \mathbf{r}_A|$ changes, but not $|\mathbf{r}_i - \mathbf{r}_B|$. The force F opposing an increase of R was defined by Berlin as the average of the force on nucleus A and the force opposing a displacement of B when A is kept fixed. This is just one of many possible definitions (see Section B.4) and corresponds to changing R by δR by moving each nucleus outward by $\frac{1}{2}\delta R$. Then we have

$$F = Z_A Z_B e^2 / R^2 - \tfrac{1}{2}e^2 \int f\varrho \, d\tau \qquad (338)$$

where ϱ is the electron density (taken as positive) and

$$f = Z_A \cos \theta_A / r_A{}^2 + Z_B \cos \theta_B / r_B{}^2. \qquad (339)$$

The first term in (338) is antibinding. Binding is due to a sufficient buildup of electron density in regions where f is positive to offset the internuclear repulsion. These regions of space are referred to as binding. The boundaries between binding and antibinding regions are given by the surfaces $f = 0$. We can speak correspondingly of binding and antibinding molecular orbitals.

Bader and Jones [7] extended Berlin's ideas within the molecular orbital framework. If the molecular orbitals are expanded in atomic orbitals, the expectation value for the force on nucleus A includes contributions from the orbitals of atom A alone, from those of atom B alone, and cross terms. The different contributions were discussed by Bader and Jones for a number of molecules [7]. The simplicity of interpretation when the electrostatic theorem was invoked was emphasized. Ransil and Sinai [8] have continued analyses of this kind, considering charge densities from approximate and accurate Hartree–Fock wavefunctions for Li_2, F_2, HF, and LiF. The He–He repulsion was discussed in terms of the electrostatic theorem by Salem [9].

The transfer of electron density from between the nuclei to the antibinding regions is a consequence of the exclusion principle when two closed-shell atoms interact.

The ready physical interpretability of the one-electron density suggests that it should be easier to construct an approximate density than an approximate wave function. Then it should be easier to calculate an approximate force than an approximate energy. Since the force is the derivative of the energy with respect to a nuclear displacement, and since one knows the energy for certain nuclear configurations (e.g., united atom limit, separated atom limit), the Hellmann–Feynman theorem provides a simple route for the calculation of potential energy curves, in which only one-electron integrals need be evaluated. Hurley [10, 11] has investigated the situation and given some very important analysis (see Subsection 2).

For a diatomic molecule, the force on nucleus A in the direction from B to A is given by

$$F_A = \frac{Z_A Z_B e^2}{R^2} - Z_A e^2 \int \frac{\cos \theta_{Ai}}{|\mathbf{r}_i - \mathbf{r}_A|^2} \varrho(\mathbf{r}_i) \, d\tau_i. \tag{340}$$

Hurley [10] computed the force curves from such an expression for several states of H_2^+ and H_2, using simple molecular orbital and valence bond wavefunctions to construct ϱ. The results were compared to the force curves obtained by (a) differentiating the energies obtained from these and more complicated wave functions and (b) differentiating the experimental potential energy curve. Results were good in some cases, extremely poor in others. It appeared that certain approximate wave functions could give good potential energy curves by calculation of energy as a function of R, but poor curves if one proceeded by way of force calculations. The two need not be identical, since the Hellmann–Feynman theorem does not hold for all functions.

A vivid illustration of the danger in naïvely applying the Hellmann–Feynman theorem was given by Salem and Wilson [12]. Consider an atom in an electric field. Taking the unperturbed electron density as an approximation to the density in the presence of the field, we see that the electrons exert no force on the nucleus, so that the force on the nucleus is that due to the field itself, which then tends to move it out of the atom! What makes this result look even worse is that we get the right force, namely zero, by calculating it from the energy. It is easy to see that the energy of the atom due to the field, using the same approximate wave function, will be independent of nuclear position. Then, differentiating with respect to a nuclear displacement, the correct result is obtained.

2. *Validity of the Theorem*

Hurley [11] was able to derive conditions under which the theorem holds for an approximate wave function. He imagined such a function to be expanded in a set of basis functions which do not themselves depend on position, as in Eq. (329). Here, Ψ depends on the space and spin coordinates of the electrons, denoted by $\mathbf{r}$, and on the N parameters $\{c_r\}$, which in turn depend on the nuclear configuration. This is specified by $\{X_\alpha, Y_\alpha, Z_\alpha\}$ for $\alpha = 1, \ldots, M$ (M nuclei). If Ψ is normalized, the electronic energy for some nuclear configuration is

$$E(X_\alpha, Y_\alpha, Z_\alpha) \equiv \mathscr{E}(X_\alpha, Y_\alpha, Z_\alpha, c_r)$$

$$= \int \Psi^*(\mathbf{r}, \mathbf{c}) H(X_\alpha, Y_\alpha, Z_\alpha) \Psi(\mathbf{r}, \mathbf{c})\, d\tau. \qquad (341)$$

The force in the x direction on nucleus α, as obtained by differentiation of the energy, is

$$F_{\alpha x} = -\frac{dE}{dX_\alpha} = -\frac{\partial \mathscr{E}}{\partial X_\alpha} - \sum_{r=1}^{N} \frac{\partial \mathscr{E}}{\partial c_r} \frac{\partial c_r}{\partial X_\alpha}. \qquad (342)$$

The term $\partial \mathscr{E}/\partial X_\alpha$ takes into account the direct dependence of E on X_α, i.e., the occurrence of X_α in the Hamiltonian; the second term takes into account the dependence of Ψ on X_α by way of the parameters $\{c_r\}$. Invoking the Hellmann–Feynman theorem means calculating only the first term. Thus the Hellmann–Feynman theorem holds when the sum in (342) vanishes. This requires that, for each r, either c_r be independent of $\{X_\alpha\}$, or $\partial \mathscr{E}/\partial c_r = 0$. The latter holds when the value of c_r is chosen variationally. The exact wave function (an infinite number of parameters chosen variationally) and the eigenfunction within a fixed subspace obviously satisfy the theorem, as we have shown. In general, the Hellmann–Feynman theorem holds for an approximate function in which *all parameters which change with nuclear displacement* have their values chosen variationally.

Hurley pointed out that, in the usual approximate wavefunctions, constructed from atomic orbitals centered on the nuclei, there are hidden parameters which are dependent on R and are almost never determined variationally. This becomes evident when the wave functions are written out in a space-fixed system of Cartesian coordinates. The definition of the atomic orbitals are in terms of distances from a nucleus, whose position may change with R. Thus, the 1s function on nucleus A is explicitly

$$\phi = (Z^3/\pi)^{1/2} \exp(-Z \mid \mathbf{r}_i - \mathbf{r}_A \mid).$$

For the Hellmann–Feynman theorem to hold, $\mathbf{r}_A$ and the corresponding quantities in all basis functions used must be replaced by origins which either are held fixed as the nuclei move or chosen variationally for each nuclear configuration. The first case is exemplified by one-center calculations. The second leads to "floating" orbitals, i.e., orbitals centered at points which may be off the nuclei.

For the cases he considered, Hurley [11] found that treating the origins of the atomic orbitals as variational parameters led to great improvement in the calculated values for the equilibrium internuclear distance and the depth of the potential well. In particular, floating functions are important for molecules dissociating into ions if reasonable results are to be obtained at large internuclear distances [12]. Suppose the charge clouds of the ions are nonoverlapping. The charge cloud centered on a nucleus A does not contribute to the force on that nucleus; the Hellmann–Feynman theorem then predicts a force on A equal to the net charge (nuclear charge minus number of electrons) on the other center (B), times the charge on nucleus A, divided by the square of the internuclear distance. On the other hand, one might predict a force equal to the *net* charge on A, times the charge on B, divided by the square of the internuclear distance. The results may differ from each other and from the correct classical result. Floating charge distributions correct the situation, as would hybridization or anything leading to an asymmetry in the atomic charge distribution. Similarly, for the atom in an electric field (end of Subsection 1), if we allow the electron density to polarize in response to the field—for instance by floating—it will exert a force on the nucleus which will cancel off that of the field.

Subsequent work with floating orbitals for H_2^+ and H_2 by Shull and Ebbing [13] produced less encouraging results, as errors appeared to have entered certain computations. Floating an orbital moves the cusp in the electron density from the nucleus, which is unfavorable energetically, so the energy obtained cannot be much better than that obtained without floating. From the point of view of a variational calculation, introduction of additional parameters could be more fruitful elsewhere.

Since the number of parameters is not a well-defined concept, Hurley [14] has stated the validity criterion more generally. A related discussion, given by Hall [15], applies to many situations in which we use an approximate function to calculate a particular property. Hall introduced the notion of "stability." Consider the eigenfunction ψ of the perturbed Hamiltonian

$H + \lambda P$ with eigenvalue E; here λ is a parameter. As in the discussion at the beginning of this section, we have

$$dE/d\lambda = \langle \partial H/\partial \lambda \rangle = \langle P \rangle \tag{343}$$

provided that

$$\int (\partial \psi^*/\partial \lambda)(H - E)\psi \, d\tau + \int \psi^*(H - E)(\partial \psi/\partial \lambda) \, d\tau = 0. \tag{344}$$

An approximate function obeying condition (344) for $\lambda = \lambda_0$ is referred to as being stable at λ_0 with respect to the perturbation P. Evidently, the exact nondegenerate eigenfunction of H is stable for $\lambda_0 = 0$ independently of P. An approximate wave function is stable if it has no way of adjusting to the change of λ. If it contains parameters $\{\mu_i\}$ which depend on λ, it will be stable at λ_0 if the values of the $\{\mu_i\}$ are determined by minimizing the energy with the Hamiltonian $H + \lambda_0 P$. Then the Hellmann–Feynman theorem is obeyed for the parameter λ with the Hamiltonian $H + \lambda P$, $\lambda = \lambda_0$. Hall [15] shows that self-consistent-field wave functions are stable to all one-electron perturbations provided there is no degeneracy, which implies, for parameters entering the one-electron Hamiltonian, Stanton's result (see page 226) that Hartree–Fock functions obey the Hellmann–Feynman theorem. An optimally scaled function is stable with respect to changes of scale in the Hamiltonian. Hall's arguments mean that, for a wave function to give a good estimate of the expectation value of an operator P, it should contain parameters which enable it to adjust to changes in the Hamiltonian $H + \lambda P$ when λ changes near $\lambda = 0$. The perturbation may be thought of as latent. These parameters should be chosen variationally. The significance for Hellmann–Feynman force calculations is obvious. To adjust to an electric field perturbation, parameters allowing for polarization are needed.

The conditions sufficient for satisfaction of the electrostatic Hellmann–Feynman theorem by an approximate wave function are different when the function is given in confocal ellipsoidal coordinates (ξ, η, φ) for a diatomic molecule rather than in space-fixed coordinates. The analysis was given by Phillipson [16]. In this coordinate system, the distances of an electron from the two nuclei are given by

$$r_{Ai} = (R/2)(\xi_i + \eta_i); \qquad r_{Bi} = (R/2)(\xi_i - \eta_i).$$

When $\Phi = \Phi(\xi_i, \eta_i)$, Φ changes with R through the coordinates as well as through any parameters changing with R. We write $d\Phi/dR$ as $d^{(2)}\Phi/dR + d^{(1)}\Phi/dR$ to represent, respectively, the two contributions. Phillipson put, as in Hurley's treatment,

$$\frac{d^{(1)}\Phi}{dR} = \sum_k \frac{\partial \Phi}{\partial \alpha_k} \frac{\partial \alpha_k}{\partial R} \tag{345}$$

As seen in Eqs. (324)–(325), the theorem will hold for an approximate function Φ in the form $dE/dR = \langle \partial H/\partial R \rangle_\Phi$ when

$$S = \int \frac{d\Phi^*}{dR} H\Phi \, d\tau + \int \Phi^* H \frac{d\Phi}{dR} \, d\tau$$

$$- \langle H \rangle_\Phi \left[\int \frac{d\Phi^*}{dR} \Phi \, d\tau + \int \Phi^* \frac{d\Phi}{dR} \, d\tau \right] = 0.$$

The contribution of $d^{(1)}\Phi/dR$ to S will vanish if, for all k,

$$\int \frac{\partial \Phi^*}{\partial \alpha_k} H\Phi \, d\tau + \int \Phi^* H \frac{\partial \Phi}{\partial \alpha_k} \, d\tau$$

$$= \langle H \rangle_\Phi \left[\int \frac{\partial \Phi^*}{\partial \alpha_k} \Phi \, d\tau + \int \Phi^* \frac{\partial \Phi}{\partial \alpha_k} \, d\tau \right] \tag{346}$$

Equation (346) will be obeyed if α_k for each R is chosen variationally.

Now Phillipson considered the change due to the coordinates,

$$\frac{d^{(2)}\Phi}{dR} = \sum_{i=1}^{N} \left[\frac{\partial \Phi}{\partial \xi_i} \frac{\partial \xi_i}{\partial R} + \frac{\partial \Phi}{\partial \eta_i} \frac{\partial \eta_i}{\partial R} \right] = \frac{Q(\xi, \eta)}{R} \Phi \tag{347}$$

where Q is a function of the ξ_i and η_i and derivatives with respect to them. It is calculable from the explicit expressions for ξ_i and η_i in space-fixed coordinates at the midpoint of the molecule. Phillipson gave the explicit expression, and showed that the contribution of $d^{(2)}\Phi/dR$ to S was

$$S^{(2)} = - \int \Phi^* \left(2\frac{T}{R} + \frac{d\tilde{V}}{dR} + \frac{V}{R} \right) \Phi \, d\tau. \tag{348}$$

Here, $d\tilde{V}/dR$ is obtained by differentiating V in ellipsoidal coordinates. Thus it is now shown [16] that a wave function in ellipsoidal coordinates, for which all parameters are chosen variationally, will obey

$$dE/dR = \int \Phi^* (dV/dR)\Phi \, d\tau \Big/ \int \Phi^* \Phi \, d\tau \tag{349}$$

if $S^{(2)}$ vanishes. Otherwise, since $d\tilde{V}/dR = -V/R$,

$$dE/dR = \int \Phi^* (dV/dR)\Phi \, d\tau \Big/ \int \Phi^* \Phi \, d\tau + S$$

$$= (-\langle V \rangle_\Phi - 2\langle T \rangle_\Phi)/R. \tag{350}$$

Equation (349) is the usual electrostatic theorem, and Eq. (350) is the virial theorem.

It must be emphasized that the fact that the Hellmann–Feynman theorem holds for a function Ψ means only that the forces calculated by differentiation of $\langle H \rangle$ and from the expectation value of a force operator will be identical. It does not mean that the forces will be good approximations to those calculated with the exact wave function, although Hall's arguments, as well as those of others, indicate a relation between the two. As we have mentioned, calculation of potential energy curves by way of forces is tempting. The calculations are easy, and it would seem easier to guess good densities, which are all we need with the electrostatic theorem, than good wave functions. However, it seems that forces are generally more sensitive than energies to errors in the wave functions. Approximate wave functions which give reasonable potential curves by way of energy calculations often give ridiculous results if we calculate forces, although one can cite examples [17] of molecular orbital functions which yield very incorrect results for $U(R)$ for large R while giving reasonable one-electron densities, from which forces might be correctly calculated.

The problem is in part the fact that approximate wave functions give more accurate values for the energy than for other properties. [An exception is the case of Hartree–Fock functions, where the error in $\langle dH/dQ \rangle$ will go as ε^2 when dH/dQ is one-electron. The difference is in fact dE^{cor}/dQ, and the error in this quantity is formally of the same size as the error in the energy itself (E^{cor}).] As in Subsection A.3, we can define a Hamiltonian H_0 such that

$$H_0 \Psi^{app} = \langle H \rangle^{app} \Psi^{app}$$

where Ψ^{app} is an approximate wavefunction and compute corrections to Ψ^{app} by perturbation theory, $H - H_0$ being the perturbation [12]. Then we can see that, for a wave function correct to nth order, we can calculate the expectation value of dH/dQ correct to nth order but the energy correct to order $2n + 1$. Thus the derivative

$$d\langle H \rangle / dQ = \langle dH/dQ \rangle \tag{351}$$

(the Hellmann–Feynman theorem has been used here for the exact function) could be calculated to a much higher order of accuracy with Ψ^{app} if we used the expression on the left side of (351) than if we used the conceptually simpler expression on the right side. In some important cases, however, the situation is less bleak, as shown by Yaris [18]. In the preceding argument, the parameter Q is unrelated to the parameter of smallness in the perturbation theory. If Q appears only in the perturbation, and not in the zero-order Hamiltonian, the nth-order wave function determines $\langle dH/dQ \rangle$ to order $(n + 1)$. Furthermore, for this case (351) is satisfied in each order: The derivative of the nth-order energy is equal to the nth-order term in the expectation value of dH/dQ [5, 18] [see Eq. (336)]. If, in addition, the perturbation can be factored into a product of a function of coordinates and a function of Q, the nth-order wave function can be shown to determine the expectation value of dH/dQ to order $(2n + 1)$.

The Hartree–Fock function also gives both sides of (351) to equal accuracy, since it obeys the Hellmann–Feynman theorem, being the best function of a given form. If it is approximated by expansion in some finite basis set, the theorem will hold so long as the basis functions are independent of Q. This is not the case where atomic orbitals centered on the nuclei are used and Q is a nuclear displacement. As the basis set becomes complete, the theorem again holds. On the other hand, if the Hartree–Fock function is improved energetically by an incomplete configuration interaction, the Hellmann–Feynman theorem may not be satisfied.

Stanton's proof [19] of the validity of the theorem for Hartree–Fock functions is perhaps more illuminating. The derivative of the Hartree–Fock energy with respect to Q is given as in Eq. (325), with Ψ the Hartree–Fock function. Assume Ψ is normalized independently of Q, so $(d/dQ) \int \Psi^* \Psi \, d\tau = 0$. The derivative of a determinantal function with respect to a parameter Q may be written as a sum of determinantal functions, in each of which some orbital ϕ_i is replaced by its derivative with respect to Q, ϕ_i'. Since

$$\int \phi_i^* \phi_i \, d\tau = 1$$

for all Q, the real part of $\int \phi_i^* \phi_i' \, d\tau$ vanishes. By inclusion of a phase factor, we may take ϕ_i' orthogonal to ϕ_i. Any contribution of ϕ_j $(j \neq i)$ in ϕ_i' may be set equal to 0, since ϕ_j and ϕ_i' appear in the same determinant. Then $d\Psi/dQ$ consists of determinants which are singly excited from Ψ, and Brillouin's theorem makes $\langle \Psi \mid H \mid d\Psi/dQ \rangle$ vanish. Thus

$$dE^{\mathrm{HF}}/dQ = \int \Psi^{\mathrm{HF}*} H' \Psi^{\mathrm{HF}} \, d\tau.$$

Coulson [20] has extended the proof to show that the Hellmann–Feynman theorem is valid for open-shell functions which are optimized projections of single determinants.

For a Hartree–Fock function the electronic contribution to the force may be written as a sum of orbital contributions:

$$-dE^{e}/dR = \sum_{i} 2f_{i} \qquad (352)$$

Here, f_{i} is the expectation value of the force operator for the ith spatial orbital. This is equal to

$$-dE^{e}/dR = -\sum_{i} d(\varepsilon_{i} + h_{i})/dR \qquad (353)$$

where ε_{i} is the orbital energy and h_{i} is the expectation value of the one-electron part of the Hamiltonian over the ith spatial orbital [see Eq. (71)]. The virial theorem, which is satisfied by a Hartree–Fock function, allows us to write the energy derivative in terms of orbital contributions in still another way [see Eq. (287)]:

$$-dE^{e}/dR = R^{-1} \sum_{i} (2T_{i} + \varepsilon_{i} + h_{i}) \qquad (354)$$

where T_{i} is the expectation value of kinetic energy for the ith orbital. Furthermore, according to Eq. (299b),

$$-d(RE^{e})/dR = \sum_{i} 2T_{i}. \qquad (355)$$

The expressions (352)–(355) were given by Hurley [14, p. 176]. He pointed out that, while they are equal for a Hartree–Fock function, the contributions of a given orbital to each sum are not equal. Indeed, (353) and (354) are only formally sums of contributions of individual orbitals, because ε_{i} itself involves a sum of orbital contributions. As a result, only (352) and (355) can be generalized to open-shell Hartree–Fock functions (sums of determinants) by substituting occupation numbers other than two: The others require further modification.

3. Applications

Since the Hellmann–Feynman theorem applies to both the exact and Hartree–Fock functions, we may write

$$dE^{cor}/dQ = \int \Psi^{HF*} H' \Psi^{HF} \, d\tau - \int \Psi^{ex*} H' \Psi^{ex} \, d\tau \qquad (356)$$

Ψ^{ex} being the exact eigenfunction and E^{cor} the correlation energy as usually defined. Suppose Q enters only the one-electron part of the Hamiltonian. Then H' is one-electron, and its expectation value for the Hartree–Fock should be close to the exact expectation value. This means that dE^{cor}/dQ is small. In particular, the correlation energy should change little with change in internuclear distance or nuclear charge. Hartree–Fock and exact potential curves should be roughly parallel over short distances, which is encouraging for the calculation of potential constants. One can argue in particular [21, 22] that molecular geometries are well predicted by the Hartree–Fock function; that is, they are correct to second order. Let the exact energy be expanded through quadratic terms in a power series about R_e. The slope at R_e', the Hartree–Fock equilibrium internuclear distance (where $dE^{HF}/dR = 0$), is $k(R_e' - R_e)$ plus higher terms, k being the force constant. Invoking the Hellmann–Feynman theorem for exact and Hartree–Fock functions, we see that the slope at R_e' is the error in the electronic contribution to the force at this point, which is second order ($\varepsilon^2 \langle \partial H/\partial R \rangle_X$, where X is the correction to the Hartree–Fock function [cf. Eq. (9) of Sect. A]).

Schwendeman [21], who gave this argument first, also noted that it is misleading to some extent. The explicit expression for the difference in equilibrium internuclear distances is

$$R_e' - R_e = \frac{\varepsilon^2 \langle \partial H/\partial R \rangle_X}{(d^2/dR^2)(E^{ex} + V_{nn})}.$$

It turns out that, typically, the two terms in the denominator cancel to about 90%. Since ε is usually about 0.1, the difference in equilibrium internuclear distances is pseudo-first-order and not second order. This leads to first-order errors in force constants. Schwendeman [21] suggested that, if one calculates force constants from theoretical energy or force results, one should evaluate energy derivatives at the experimental equilibrium internuclear distance, rather than the equilibrium internuclear distance predicted from the calculations.

Even if the quantity of Eq. (356) is small, the correlation energies for two widely different configurations need not differ by a small quantity. For example, dissociation energies computed as differences of Hartree–Fock energies for molecule and separated atoms are inaccurate in some cases (see Chapter II, Volume 2). However, Stanton [19] gives some examples of how the smallness of dE^{cor}/dQ may be used to correct Hartree–Fock values for equilibrium distance and dissociation energies. The correlation energy for $R = R_e$ should not differ much from that of the united atom, for instance. This conjecture was borne out in calculations [23] on the first row diatomic hydrides, as well as by some results given by Stanton [19].

Force calculations using something close to a Hartree–Fock function were first carried out for H_2 [24]. Potential constants from the force and energy curves were similar, but not always identical, and agreed fairly well with experiment. Experiments with other, simpler H_2 wave functions were carried out by Benston and Kirtman [25]. The forces were generally in error by quantities of the same magnitude as their values, but it was noted that the force curves were roughly parallel to the correct one near the equilibrium R, so derivatives of forces could still be correct. On the other hand, reasonable equilibrium distances could be obtained from the energy curves. Force curves have also been calculated [26] from near-Hartree–Fock functions for N_2, CO, BeO, BF, and LiF, and potential constants derived.

Force calculations, invoking the electrostatic theorem, were performed by Hurley [11] for several diatomic systems. Atomic orbitals were combined to form inner-shell, lone pair, and bond orbitals for the molecule, and these were used to generate pair functions. The resulting wave functions included both valence bond and molecular-orbital-type functions as special cases. While energy calculations for such functions became quite difficult, calculation of the force was tractable. The parameters (not including floating of the atomic orbitals) were chosen from other work on related systems, and it was shown that reasonable results were obtained in all cases. On the other hand, changing certain parameters sometimes changed calculated forces markedly, and one must be able to estimate their values correctly for this procedure to be useful.

Electrostatic force calculations are really useful only if we can get the required approximate wave function without performing a variational calculation; that is, we need a recipe for determining the parameters. Bader [17] has experimented with such methods. The values for the exponential parameters in simple molecular orbital functions for H_2 and He_2 were determined by interpolation between the united atom and the separated atom values. The potential curves were in very good agreement with experiment for both problems.

Note that we define the equilibrium distance, from the Hellmann–Feynman theorem, as the value of R for which there is no force on either nucleus. For a function not satisfying the theorem, the expectation values of the forces on the two nuclei need not be identical, nor need the forces vanish at the internuclear distance corresponding to minimum energy. In fact, even for what are considered good approximate functions one sometimes gets forces which are seriously in error. Thus, Bader [27], considering the best minimal basis LCAO–MO functions for first-row homonuclear diatomic molecules, found net forces at the equilibrium R between $0.076e^2/a_0^2$ (attractive) for Li_2 to $-1.814e^2/a_0^2$ (repulsive) for F_2. An analysis was given for the origin of the errors. This means satisfaction of the Hellmann–Feynman theorem is a highly sensitive measure of the accuracy of a wave function, one which can be checked without recourse to experimental data. As mentioned, a true Hartree–Fock function must satisfy the theorem, so this is also a test for how closely we have approached the Hartree–Fock function by an approximation technique.

On the other hand, because the expectation value of the force operator seems so sensitive to small errors in the wavefunction, it has been suggested [28] that the constrained variation method be used with expectation values of force operators as constraints. For example, one can demand equality of the forces on the two nuclei of a heteronuclear diatomic, or equality of any two expressions which are equal to dE/dR for the exact wave function (see page 235). These constraints can be invoked without outside (i.e., experimental) information. For some simple cases, demanding zero forces on the nuclei suffices to determine the parameters in a trial wave function. For example, if the ground state H_2 wave function is taken in the simple molecular orbital form, with the molecular orbitals a linear combination of 1s orbitals, the exponential parameter is fixed by demanding a zero force at the known equilibrium internuclear distance. For systems with more electrons, and thus more parameters, additional information is needed. Bader and Jones [29] have determined an electronic wave function for HF by requiring that the forces on the two nuclei vanish and that the correct dipole moment be obtained. Here, the orbital exponents for atomic orbitals were taken from other calculations, and only three mixing parameters remained.

Expressions for the force constant can be obtained by differentiating the expression for the force given by the Hellmann–Feynman theorem. They are valid only for a function which satisfies the theorem in the first place. Assuming dV/dR is real,

$$\frac{d^2E}{dR^2} = \frac{d}{dR}\left[\frac{\int \Phi^*(dV/dR)\Phi\, d\tau}{\int \Phi^*\Phi\, d\tau}\right]$$

$$= \left\langle\frac{d^2V}{dR^2}\right\rangle_\Phi + 2\,\mathrm{Re}\int \Phi^*\left(\frac{dV}{dR}-\frac{dE}{dR}\right)\frac{d\Phi}{dR}\,d\tau \Big/ \int \Phi^*\Phi\, d\tau \quad (357)$$

The second term in (357) represents the effect of the change in Φ with R. The force constant is d^2E/dR^2 evaluated at the equilibrium R (R_e) where dE/dR vanishes.

If Φ and V are written in ellipsoidal coordinates, Phillipson [16] derived

$$\left(\frac{d^2E}{dR^2}\right)_e = \frac{d^{(1)}}{dR}\left\langle\frac{dV}{dR}\right\rangle + \left\langle\frac{dV}{dR}\right\rangle_e \frac{(d^{(1)}/dR)[\int \Phi^*\Phi \, d\tau]}{[\int \Phi^*\Phi \, d\tau]^2} \tag{358}$$

where

$$\frac{d\Phi}{dR} = \frac{d^{(1)}\Phi}{dR} + \frac{Q\Phi}{R} \tag{359}$$

[see Eqs. (345)–(347)]. In Eq. (358) the second term vanishes. This shows that writing a wave function in ellipsoidal coordinates takes care of some of the dependence on R, so that the force constant may be expressed as the change of the Hellmann–Feynman force with R due to changes in the variational parameters [16].

If the wave function is expressed in space-fixed coordinates, so that it changes with R only due to changes in variational parameters, the force expression takes on a simpler form [Eq. (349)]: the changes in these parameters do not enter. However, an additional term appears in the force constant expression [16]. Differentiating (349) and using the present notation,

$$\left(\frac{d^2E}{dR^2}\right)_e = \frac{d}{dR}\left\langle\frac{dV}{dR}\right\rangle = \frac{d^{(1)}}{dR}\left\langle\frac{dV}{dR}\right\rangle + \left\langle\frac{d^2V}{dR^2}\right\rangle. \tag{360}$$

Salem [30] has worked this out in detail. Using Berlin's convention that R is increased by moving both nuclei outward by equal distances, and atomic units,

$$\frac{dE}{dR} = \frac{-Z_A Z_B}{R^2} + \int |\Phi|^2 \sum_i \left(Z_A \frac{\cos\theta_{Ai}}{2r_{Ai}^2} + Z_B \frac{\cos\theta_{Bi}}{2r_{Bi}^2}\right) d\tau$$

where the definition of r_A, θ_A, r_B, and θ_B is given in the Notes on Notation and Coordinate Systems. In computing the second derivative, it was noted that [Eq. (244)]

$$\frac{\partial}{\partial x_A}\left(\frac{\cos\theta_{Ai}}{r_{Ai}^2}\right) = \frac{3\cos^2\theta_{Ai} - 1}{r_{Ai}^3} - \frac{4\pi}{3}\delta(\mathbf{r}_i - \mathbf{r}_A). \tag{361}$$

The internuclear axis is taken as the z axis and $\mathbf{r}_A$ is the position of A. Explicitly, (360) becomes

$$\left(\frac{d^2E}{dR^2}\right)_e = \frac{1}{2}\int\left(\frac{\partial|\Phi|^2}{\partial Z_B} + \frac{\partial|\Phi|^2}{\partial Z_A}\right)\left(\sum_i \frac{Z_A\cos\theta_{Ai}}{2r_{Ai}^2} + \sum_i \frac{Z_B\cos\theta_{Bi}}{2r_{Bi}^2}\right)d\tau$$

$$+ \frac{1}{4}\left[Z_A q_A + Z_B q_B + \frac{4\pi}{3}(Z_A\varrho(A) + Z_B\varrho(B)) + \frac{4Z_A Z_B}{R^3}\right]. \tag{362}$$

Here, $\varrho(A)$ is the electron density at the position of nucleus A; q_A is the field gradient, which is the expectation value of the operator q_A,

$$q_A = \frac{2Z_B}{R^3} - \frac{3\cos^2\theta_{Ai} - 1}{r_{Ai}^3} \tag{363}$$

and similarly for $\varrho(B)$ and q_B. The sums in (362) and (363) are over electrons. Related expressions for the force constant were given when one nucleus or the other is fixed as R changes. Thus, when B is fixed we obtain

$$\left(\frac{d^2E}{dR^2}\right)_e = Z_A\left[q_A + \frac{4\pi}{3}\varrho(A) - \int \frac{\partial|\Phi|^2}{\partial z_A}\frac{\cos\theta_{Ai}}{r_{Ai}^2}\,d\tau\right]. \tag{364}$$

Averaging (364) with the corresponding expression for moving nucleus B gives another expression for the force constant; combination with (362) gives still another.

These are equal for an exact wave function. The various formulas for the second derivatives of the energy, and the relations between them, were discussed by Byers Brown [31]. The relations between the second derivatives express the invariance of the energy to translations (and, for a nonlinear molecule, to rotations) of the molecule as a whole. The remarks surrounding Eqs. (224), (225), (240), and (241) give examples of such relations. Byers Brown also divided the second-order energy derivatives into classical electrostatic and relaxation (referred to by him as quantal) contributions, as other authors have done (see page 230). He noted that, despite the fact that they complicate calculations, one could not always neglect the quantal terms, since the invariance of the energy to overall translations requires a cancellation between classical and quantal terms.

The terms in (364) are (1) the expectation value of the second derivative of the electrostatic potential at A due to charges outside the nucleus, (2) the expectation value of the second derivative of the electrostatic potential at A due to (electronic) charges right at the nucleus and (3) the effect of readjustment of the electron density as R changes. Term (1) is measured as the nuclear quadrupole coupling, term (2) is related to the magnetic hyperfine interaction for molecules with unpaired electrons. The three terms have a one-to-one correspondence with those obtained by perturbation theory. The last term is large and negative and for H_2 it is close in value to the second one, which is positive. If (2) and (3) approximately canceled, the force constant would roughly equal Z_Aq_A. This is true for diatomic hydrides (A = H). Salem [30] discussed this fact. If all changes in the electron density as the nuclei move are neglected, only term (2) appears [32]. A further discussion of the terms in Eq. (364) has been given by Bader and Band-rauk [26]. They considered, in particular, the last term. The change in $\varrho = |\varPhi|^2$ comes primarily from the part of the charge density which follows the movement of A (B is fixed in this formula). This term is always negative, since it makes it easier to move the nucleus than if the electronic charge distribution remained rigid. It was shown that for a charge density which follows A rigidly there is exact cancellation between this term and the other electronic contributions, so the force constant is just $2Z_A Z_B/R_e^3$. According to the Hellmann–Feynman theorem, the change in energy due to a displacement of A is given by the expectation value of the force operator plus the nuclear contribution. If the electron density rigidly follows A, the expectation value is unchanged.

For the molecules, N_2, CO, BF, BeO, and LiF, the electronic contribution to k_e was analyzed and interpreted [26]. Contour plots of densities and density differences were used to show the relaxation process and elucidate differences between the different systems. The changes in the polarizations of the charge densities, relative to separated-atom charge densities, were of central importance. Other work along these lines was performed for the homonuclear molecules and the diatomic hydrides [27, 33]. The relaxation of the charge distribution on motion of the proton was investigated. In addition, forces for seven values of R were calculated from near-Hartree–Fock functions, and quadratic and cubic force constants were determined by fitting the force curves to Dunham expansions. Accuracy of a few percent was obtained.

Although the use of the Hellmann–Feynman theorem in conjunction with approximate densities is dangerous, Salem [34] obtained reasonable results for the harmonic and anharmonic force constants with crude densities for the valence electrons of H_2, Li_2, and Na_2. He did not consider inner-shell electrons explicitly, but used an effective nuclear charge for the core, chosen so that the force vanished at the equilibrium distance. Further calculation pointed up the extreme sensitivity of calculated forces to the wavefunctions used. This is of paramount importance for the contribution of "atomic" forces, which in a molecular orbital method are nonvanishing matrix elements of the force operator for nucleus A between atomic orbitals centered on A.

For an approximate function of the type of Eqs. (341)–(342) satisfying the Hellmann–Feynman theorem, the force constant is

$$\frac{-d}{dR} \int |\varPsi(x_i; c_r)|^2 \left(\frac{\partial V}{\partial R}\right) d\tau = -\sum_r \frac{\partial \langle \partial V/\partial R \rangle}{\partial c_r} \frac{\partial c_r}{\partial R} - \left\langle \frac{\partial^2 V}{\partial R^2}\right\rangle_{\varPsi}$$

evaluated at $R = R_e$. Assuming the virial theorem is satisfied, another expression in terms of the changes in parameters with R may be derived [16]. Tests of these with approximate H_2^+ and H_2 functions [35] show a wide variety of force constants are obtainable, again reminding us that assumption of the validity of the Hellmann–Feynman theorem for an approximate function is dangerous.

More recently, Anderson and Parr [36] obtained force constants by combining physical intuition with a variant of equation (364). They considered $\nabla_A^2 W$ (W = electronic energy), in which the field gradient terms sum to zero:

$$\nabla_A^2 W = 4\pi Z_A \varrho(A) - Z_A \int \nabla_A \varrho \cdot \nabla_A \left[\sum_i r_{Ai}^{-1} \right] d\tau.$$

Anderson and Parr wrote the density as a sum of three terms, $\varrho = \varrho_A + \varrho_B + \varrho_{NPF}$, where ϱ_A is supposed to follow perfectly the motion of nucleus A, ϱ_B to follow the motion of nucleus B, and ϱ_{NPF} is what remains. The contributions of ϱ_A to the two parts of the preceding expression cancel, as already mentioned, whereas ϱ_B does not contribute to the integral. It was further argued [36] that the contribution of ϱ_{NPF} to both terms should be small, so

$$\nabla_A^2 W \approx 4\pi Z_A \varrho_B(A).$$

$\varrho_B(A)$ was estimated by finding the electron density at a distance R from B in the direction opposite to A. Derivatives of $\varrho_B(A)$ with respect to R, for higher force constants, were obtained by considering different distances from B in the direction opposite to A. For a large number of molecules, qualitative agreement of calculated k_e, l_e, and m_e with experiment was found [36].

By differentiation with respect to R of the expressions already given for the quadratic force constant, we may obtain a variety of expressions for higher force constants. Such expressions have been discussed by Schwendeman [37], by Rossikhin and Morozov [38] (for Hartree–Fock functions), and by Anderson and Parr [36].

4. Generalizations

In the derivations of the Hellmann–Feynman theorem just given, it is implicit that the electronic coordinates are defined relative to an invariant, space-fixed coordinate system. One way of expressing this is to write dH/dQ, where Q is a nuclear coordinate, as $(\partial H/\partial Q)_r$; that is, the electrons remain fixed while the nuclei are moved. It is also legitimate to speak of $(\partial H/\partial Q)_\rho$, where ρ refers to the coordinates of the electrons in any coordinate system. The coordinate system itself may depend on nuclear positions. The meaning of dE/dQ is unchanged, but the differentiations inside the integrals over electronic coordinates have different meanings. For example, in treating the van der Waals interaction at large distances, Salem and Wilson [12] used space-fixed electronic coordinates, whereas Yaris [18] used coordinates for each electron relative to one nucleus or the other. Formally different results were obtained. The simple electrostatic Hellmann–Feynman theorem, which leads to the classically interpretable forces, is only one of several theorems which can be obtained by differentiating with respect to a nuclear displacement while holding electronic coordinates fixed in various coordinate systems [39]. This has been investigated by Benston and Kirtman [25] for diatomic molecules. The important case, in which the Hellmann–Feynman theorem is shown to give the virial theorem, was previously worked out by Philipson [16], and already mentioned (pages 224–225). By differentiating the virial theorem or any of the expressions for dE/dR we get relations for the second and higher energy derivatives. In each differentiation one can hold the electronic coordinates fixed in any coordinate system. A variety of expressions can thus be generated, as discussed by Benston and Kirtman [25]. One has been considered in Section C after Eq. (301). Another [35] is obtained by holding the x and y coordinates constant in a space-fixed system, but scaling z with R (the molecule lies along the z axis), i.e., the z coordinate of an electron, which remains fixed as R changes, is a distance along the z axis divided by R.

Hurley [14] has referred to

$$dE/dQ = \langle(\partial H/\partial Q)_\rho\rangle \tag{365}$$

where the electronic coordinates ρ are held constant in the differentiation, as the generalized Hellmann–Feynman theorem, and showed under what circumstances it is satisfied for approximate functions. He considered the approximate functions $\Psi(Q')$, defined for a range of values Q' including Q, the value of interest. The energy depends on Q through the wave function as well as through the Hamiltonian. To separate out the Q-dependences in H and Ψ, suppose H depends on Q and Ψ on Q'. Now if (365) is satisfied at Q,

$$(\partial E(Q', Q)/\partial Q')_Q = 0 \tag{366}$$

when $Q' = Q$. For a ground-state wave function, (366) means E is minimized with respect to Q', so $E(Q, Q) < E(Q', Q)$ for any $Q' \neq Q$. Suppose (366) is satisfied for some $Q'' \neq Q$, i.e., the best value of the parameter to use in Ψ is not $Q' = Q$ but $Q' = Q''$. Of course Q'' would depend on Q (the value used in the Hamiltonian), so that we can speak of a function $Q''(Q)$. Then the Hellmann–Feynman theorem will be satisfied if we calculate E using $\Psi(Q'')$ instead of $\Psi(Q)$. This means

$$E(Q', Q) = \int \Psi^*(Q''(Q))H(Q)\Psi(Q''(Q))\, d\tau$$

with Ψ normalized. By definition, $\partial E/\partial Q' = (\partial E/\partial Q'')(\partial Q''/\partial Q')$ vanishes. Then Eq. (366) and thus Eq. (365) are satisfied. Now suppose Ψ is restricted in some way, such that the only permitted variations are through a set of parameters c_r. Then $(\partial E(Q', Q)/\partial Q')_Q$ can be written out by the chain rule. If the parameters are chosen variationally, this vanishes, and the generalized Hellmann–Feynman theorem is satisfied. [See the discussion of Eqs. (343)–(344).] In general, (365) will hold if (1) $\Psi(Q')$ for $Q' \neq Q$ is obtainable for a permitted variation and (2) $\Psi(Q')$ for $Q' = Q$ is selected variationally [14]. This includes the exact wave function of course, but also Hartree–Fock functions for both closed- and open-shell cases, and various extended Hartree–Fock theories. Epstein [40] has proved that a wave function that satisfies the Hellmann–Feynman theorem (365) for two different coordinate systems ρ and σ satisfies a hypervirial theorem (Section C).

Some extensions of the Hellmann–Feynman theorem for linear variation functions, Eq. (332), were given by Levine [41]. Let $\Delta(\lambda) = \det| \lambda - \mathbf{H} |$ be the secular determinant, i.e., E is determined as the value of λ for which $\Delta(\lambda)$ vanishes. For a change in λ by $\delta\lambda$, with $\delta\lambda$ small, Δ changes by $\delta\Delta$, where Levine showed by expansion of the determinant that

$$\delta\Delta/\Delta = \delta(\ln \Delta) = \delta\lambda \operatorname{tr}[(\lambda - \mathbf{H})^{-1}]. \tag{367}$$

If the Hamiltonian depends on Q, then with λ fixed the same arguments give for small δQ

$$\delta(\ln \Delta) = -\delta Q \operatorname{tr}[(\partial\mathbf{H}/\partial Q)(\lambda - \mathbf{H})^{-1}]. \tag{368}$$

The change in λ needed to keep $\ln \Delta$ and hence Δ constant when Q changes by dQ is then

$$d\lambda = dQ \operatorname{tr}[(\partial\mathbf{H}/\partial Q)(\lambda - \mathbf{H})^{-1}]/\operatorname{tr}[(\lambda - \mathbf{H})^{-1}].$$

If we evaluate the traces by going over the basis of eigenvectors of $\mathbf{H}$ with eigenvalues ε_i,

$$d\lambda/dQ = \sum_i (\lambda - \varepsilon_i)^{-1}(\partial H/\partial Q)_{ii} \Big/ \sum_i (\lambda - \varepsilon_i)^{-1}. \tag{369}$$

This reduces to the Hellmann–Feynman theorem when we let λ approach an eigenvalue ε_j. In this case the term for $i = j$ will dominate each sum in (369). The right side will become $(\partial H/\partial Q)_{jj}$, i.e., the expectation value of $\partial H/\partial Q$ over the jth eigenvector. Since the left side of (369) is $d\varepsilon_j/dQ$, we recover (332). Consider now the change of a function of some eigenvalue with Q. We have, using (369),

$$dF(\lambda)/dQ = (dF/d\lambda) \sum_i (\lambda - \varepsilon_i)^{-1}(\partial H/\partial Q)_{ii} \Big/ \sum_i (\lambda - \varepsilon_i)^{-1}$$

Let λ approach an eigenvalue and repeat the preceding reasoning. The result is

$$dF(\varepsilon_j)/dQ = [dF/d\lambda]_{\lambda=\varepsilon_j}(\partial H/\partial Q)_{jj}. \tag{370}$$

A theorem governing *finite* changes in nuclear configuration or other parameters, as the Hellmann–Feynman theorem governs infinitesimal changes, has been developed by Parr and co-workers, and named the *integral Hellmann–Feynman theorem*. The proof is simple [42]. Consider two Hamiltonians, differing by a change of some parameter, like a nuclear configuration, and their ground state eigenfunctions:

$$H_X\Psi_X = E_X\Psi_X \tag{371}$$

$$H_Y\Psi_Y = E_Y\Psi_Y \tag{372}$$

Ψ_X and Ψ_Y are taken as real. Multiply (371) by Ψ_Y, (372) by Ψ_X, and integrate, using the Hermitian property of H_X. Then on subtraction,

$$\int \Psi_X(H_X - H_Y)\Psi_Y \, d\tau = (E_X - E_Y)S \tag{373}$$

where

$$S = \int \Psi_X\Psi_Y \, d\tau. \tag{374}$$

Usually, $H_X - H_Y$ will be one-electron. This permits a simplification like that in the Hellmann–Feynman theorem: All the necessary quantities for determination of the energy difference $E_X - E_Y$ can be calculated from the one-electron transition density

$$\varrho_{XY}(1) = (n/S) \int \Psi_X(1, 2, \ldots, n)\Psi_Y(1, 2, \ldots, n) \, d\tau_2 \cdots d\tau_n \tag{375}$$

This permits simplified calculations and physical visualizability for such processes as change of internuclear distance or nuclear charge.

Richardson and Pack [43] worked with a closely related formalism for energy differences at about the same time as Parr's original paper. For a diatomic, they considered the Hamiltonian H_M for the N electrons in a molecule, and the Hamiltonians H_A and H_B for the electrons in each atom when they are separated. H_A operates on the coordinates of N_A electrons and H_B on the coordinates of N_B electrons, with $N = N_A + N_B$. Denoting by V the difference $H_M - (H_A + H_B)$ they obtained, for the difference in eigenvalues,

$$E = E_M - (E_A + E_B) = \langle \Phi | V | \Psi \rangle / \langle \Phi | \Psi \rangle$$

Here Ψ is the eigenfunction of H_M with eigenvalue E_M and Φ is the product of eigenfunctions for H_A and H_B: $\Phi = \Phi_A \Phi_B$ where $H_A \Phi_A = E_A \Phi_A$ and $H_B \Phi_B = E_B \Phi_B$. Dissociation energies were calculated by using approximate Hartree–Fock wavefunctions for the atoms and molecules. For the molecules H_2, LiH, BH, HF, and Li_2, the results obtained by using the theorem were closer to the experimental results than the energy differences obtained by direct subtraction of expectation values. Other ways of partitioning the molecular Hamiltonian were suggested. One that was explored was to write H as the Hamiltonian for the united atom plus a correction $\bar{V}$. Then the quantity $\langle \Phi | \bar{V} | \Psi \rangle / \langle \Phi | \Psi \rangle$, computed with Φ and Ψ approximate wave functions for united atom and molecule, respectively, is to be added to the united atom energy to get the energy of the molecule. If an exact united atom energy is used, the derived molecular energy should include some of the effects of correlation.

Energy differences may also be computed by integrating the usual Hellmann–Feynman theorem. Thus

$$E_X - E_Y = \int_{Q(X)}^{Q(Y)} \langle \Psi | \partial H / \partial Q | \Psi \rangle \, dQ \tag{376}$$

Here, Ψ is the eigenfunction of the Hamiltonian H which depends on the parameter Q which specifies the configuration. Here, $Q = Q(X)$ for the configuration or internuclear distance X; $Q = Q(Y)$ corresponds to Y.

Epstein et al. [44] have discussed and compared this "integrated Hellmann–Feynman Theorem" with the integral Hellmann–Feynman theorem and other routes to energy differences. A function satisfying the integral Hellmann–Feynman theorem automatically satisfies the integrated Hellmann–Feynman theorem, but the converse is not true.

The question in general is how good we can expect the results from the integral Hellmann–Feynman theorem to be. If there is a parameter of smallness ε describing the error in the wave function, the energies, and hence the energy differences, have errors of order ε^2 (although there is some question as to the utility of the concept here, since ε is relative to the size of energies, not energy differences, which may be of another order of magnitude). However we can say little along this line for $\langle \Phi_X \mid V \mid \Phi_Y \rangle$ [45]. The energy difference, while calculated directly using the integral Hellmann–Feynman theorem (analogous to the derivative of energy in the Hellmann–Feynman theorem, also calculated directly), obeys no variational principle like the energy itself. We do not know how to choose a wave function to make the result of such a calculation as good as possible. A related problem is: Under what circumstances will an approximate wave function satisfy the theorem?

Silverstone [45] showed that, for a Hartree–Fock function with localized orbitals which are unchanged by the change in conformation (they follow the nuclei), the integral Hellmann–Feynman theorem is obeyed. More generally, it was shown [45] that the theorem will be obeyed when Ψ_X and Ψ_Y are both chosen according to the variational principle as linear combinations of the same set of basis functions.

It is interesting to note [44] that energy differences calculated by the integral theorem are not necessarily additive. This means that, with an approximate function which does not satisfy the theorem, the calculated difference $E_X - E_Y$ may not equal the sum of calculated differences $E_X - E_Z$ and $E_Z - E_Y$ where Z is a third nuclear configuration. This was shown by Hayes and Parr [46] in some one-center calculations on HeH$^+$. The wave functions were centered on He and the H nucleus was moved. For the case where only one parameter in the Hamiltonian is varied, as for change of R in a diatomic molecule, the integrated Hellmann–Feynman theorem yields additive energy differences.

Epstein *et al.* [44] gave detailed numerical comparisons of various ways of computing energy differences (direct subtraction, virial theorem, integrated Hellmann–Feynman theorem, integral Hellmann–Feynman theorem) for H$_2{}^+$, using the simplest molecular orbital wave function. The integrated Hellmann–Feynman theorem predicted repulsion at all R, while the integral Hellmann–Feynman theorem often gave reasonable results, except for small changes in R. The leading $(1/R)$ term in the interaction between ions with spherically symmetric charge distributions centered at the nuclei was also considered by both the integral theorem and the integrated theorem. As we have noted, the integrated Hellmann–Feynman theorem gives incorrect answers unless the charge distributions are allowed to float off the nuclear centers. (The virial theorem gives correct results here.) The integral Hellmann–Feynman theorem also gave incorrect results unless some correction to the ionic charge density was invoked, but making such a correction is less straightforward than for the integrated theorem.

Like the Hellmann–Feynman theorem itself, the integral Hellmann–Feynman theorem has not fulfilled the best hopes for it, namely that it would allow calculation of $U(R)$ without determination of the wave function via the Schrödinger equation. The future in theorems like these, which are satisfied by the exact wave functions, would seem to lie in reduction of experimental and precise theoretical results [5a] to easily comprehensible terms.

REFERENCES

1. R. P. Feynman, *Phys. Rev.* **56**, 340 (1939).
2. H. Hellmann, "Einführung in die Quantenchemie," pp. 285–286. Deuticke, Leipzig, 1937.
3. E. B. Wilson, Jr., *J. Chem. Phys.* **36**, 2232 (1962).
4. P.-O. Löwdin, *J. Mol. Spectrosc.* **3**, 46 (1959).
5. A. Fröman, *Phys. Rev.* **112**, 870 (1958); C. W. Scherr and R. E. Knight, *Rev. Mod. Phys.* **35**, 436 (1963).
5a. B. M. Deb, *Rev. Mod. Phys.* **45**, 22 (1973).
6. T. Berlin, *J. Chem. Phys.* **19**, 208 (1951).
7. R. F. W. Bader and G. A. Jones, *Can. J. Chem.* **39**, 1253 (1961).
8. B. J. Ransil and J. J. Sinai, *J. Chem. Phys.* **46**, 4050 (1967); A. K. Chandra and R. Sundar, *Mol. Phys.* **22**, 369 (1971).
9. L. Salem, *Proc. Roy. Soc. Ser. A* **264**, 379 (1961).
10. A. C. Hurley, *Proc. Roy. Soc. Ser. A* **226**, 170 (1954).
11. A. C. Hurley, *Proc. Roy. Soc. Ser. A* **226**, 179, 193 (1954).
12. L. Salem and E. B. Wilson, Jr., *J. Chem. Phys.* **36**, 3421 (1962).
13. H. Shull and D. D. Ebbing, *J. Chem. Phys.* **28**, 866 (1958).
14. A. C. Hurley, *in* "Molecular Orbitals in Chemistry, Physics, and Biology" (P.-O. Löwdin and B. Pullman, eds.), p. 161. Academic Press, New York, 1964.
15. G. G. Hall, *Phil. Mag.* **6**, 249 (1961).
16. P. Phillipson, *J. Chem. Phys.* **39**, 3010 (1963).
17. R. F. W. Bader, *Can. J. Chem.* **38**, 2117 (1960).
18. R. Yaris, *J. Chem. Phys.* **39**, 863 (1963).
19. R. E. Stanton, *J. Chem. Phys.* **36**, 1298 (1962).
20. C. A. Coulson, *Mol. Phys.* **20**, 687 (1971).
21. R. H. Schwendeman, *J. Chem. Phys.* **44**, 2115 (1966).
22. K. F. Freed, *Chem. Phys. Lett.* **2**, 255 (1968); *J. Chem. Phys.* **52**, 253 (1970).
23. P. E. Cade and W. M. Huo, *J. Chem. Phys.* **47**, 614 (1967).
24. J. Goodisman, *J. Chem. Phys.* **39**, 2397 (1963).
25. M. L. Benston and B. Kirtman, *J. Chem. Phys.* **44**, 119 (1966).

E. Local Energy Methods 241

26. R. F. W. Bader and A. D. Bandrauk, *J. Chem. Phys.* **49**, 1666 (1968).
27. R. F. W. Bader, *Can. J. Chem.* **41**, 2303 (1963).
28. D. P. Chong, *Theor. Chim. Acta* **11**, 205 (1968); J. Thorhallson, J. F. Larcher, and D. P. Chong, *Theor. Chim. Acta* **16**, 51 (1970).
29. R. F. W. Bader and G. A. Jones, *Can. J. Chem.* **41**, 2251 (1963).
30. L. Salem, *J. Chem. Phys.* **38**, 1227 (1963).
31. W. Byers Brown, *Proc. Cambridge Phil. Soc.* **54**, 250 (1958).
32. W. T. King, *J. Chem. Phys.* **49**, 2866 (1968).
33. R. F. W. Bader and J. L. Ginsburg, *Can. J. Chem.* **47**, 3061 (1969).
34. L. Salem and M. Alexander, *J. Chem. Phys.* **39**, 2994 (1963).
35. M. L. Benston, *J. Chem. Phys.* **44**, 1300 (1966); W. R. Ross and P. Phillipson, *J. Chem. Phys.* **44**, 844 (1966).
36. A. B. Anderson and R. G. Parr, *J. Chem. Phys.* **53**, 3375 (1970).
37. R. H. Schwendeman, *J. Chem. Phys.* **44**, 556 (1966).
38. V. V. Rossikhin and V. P. Morozov, *Theor. Exp. Chem. (USSR)* **2**, 396 (1966); *Teor. Eksp. Khim.* **2**, 528 (1966).
39. C. A. Coulson and A. C. Hurley, *J. Chem. Phys.* **37**, 448 (1962).
40. S. T. Epstein, *J. Chem. Phys.* **42**, 3813 (1965).
41. R. D. Levine, *Proc. Roy. Soc. Ser. A* **294**, 467 (1966).
42. R. G. Parr, *J. Chem. Phys.* **40**, 3726 (1964).
43. J. W. Richardson and A. K. Pack, *J. Chem. Phys.* **41**, 897 (1964).
44. S. T. Epstein, A. C. Hurley, R. E. Wyatt, and R. G. Parr, *J. Chem. Phys.* **47**, 1275 (1967).
45. H. J. Silverstone, *J. Chem. Phys.* **43**, 4537 (1965).
46. E. F. Hayes and R. G. Parr, *J. Chem. Phys.* **44**, 4650 (1966).

E. Local Energy Methods

The effort needed to evaluate the integrals which arise in the variation and perturbation methods increases with the number of electrons and the complexity of the wave function. Since the earliest quantum mechanical calculations of molecular electronic energies, there has been interest in methods which avoid the integral problem. Many of these are related to the local energy principle, that

$$H\Psi = E\Psi \tag{377}$$

may be rewritten trivially as

$$H\Psi/\Psi = E \tag{378}$$

i.e., $H\Psi/\Psi$ is a constant, independent of coordinates. We call $H\Psi/\Psi$ the local energy, and it turns out [1] to be far from constant when Ψ is an approximate function, even one which is good by the variational criterion and gives good expectation values. This is due to the fact that Ψ has nodes and $H\Psi$ singularities, so that $H\Psi/\Psi$ may become infinite for functions other than the exact function.

1. Basic Formulas

We define the local energy for the function Φ at the point P of configuration space as

$$\varepsilon_P = (H\Phi/\Phi)_P \tag{379}$$

the subscript meaning evaluation at point P. Since we have

$$\varepsilon_P = \text{const}$$

for the exact wave function, the constancy of ε_P can be taken as a measure of the goodness of an approximate function. Let us associate positive weighting factors g_P with a set of points P in configuration space. For these points, consider the mean-square deviation of the local energy from the mean:

$$\overline{W}_\Phi = \sum_P (\varepsilon_P - \bar{\varepsilon})^2 g_P \Big/ \sum_P g_P \tag{380}$$

where

$$\bar{\varepsilon} = \sum_P \varepsilon_P g_P \Big/ \sum_P g_P. \tag{381}$$

We can write, using (381),

$$\overline{W}_\Phi = \overline{\varepsilon^2} - \bar{\varepsilon}^2 \tag{382}$$

where $\overline{\varepsilon^2}$ is defined similarly to $\bar{\varepsilon}$ [Eq. (381)]. The value of $\overline{W}_\Phi$ is zero when ε_ϱ is a constant, which occurs when Φ is an exact eigenfunction of H, so the smallness of $\overline{W}_\Phi$ can be considered a measure of the quality of Φ.

The points and weights are completely arbitrary, but if they are chosen in a particular way $\overline{W}_\Phi$ can be shown to become equal to the width W_Φ, discussed in connection with the variation method (Section A.2). First include the reasonable factor of Φ^2 in the weights (Φ is here taken as real) and define w_P as g_P/Φ^2.

$$\overline{W} = \frac{\sum_P w_P (H\Phi)_P (H\Phi)_P}{\sum_P w_P (\Phi^2)_P} - \left\{ \frac{\sum_P w_P (\Phi H\Phi)_P}{\sum_P w_P (\Phi^2)_P} \right\}^2 \tag{383}$$

If there are a sufficient number of points and if the points P and weights w_P are chosen according to some numerical integration formula, the sums approach integrals. The value W_Φ becomes

$$\frac{\langle H\Phi \mid H\Phi \rangle}{\langle \Phi \mid \Phi \rangle} - \left(\frac{\langle \Phi \mid H \mid \Phi \rangle}{\langle \Phi \mid \Phi \rangle} \right)^2 = W_\Phi = \langle H^2 \rangle - \langle H \rangle^2 \qquad (384)$$

and we recover the width W_Φ on invoking the Hermitian property of H in the first term. This supports the idea of the width as a measure of the quality of the wave function. Preuss [2] argued from perturbation theory that minimization of the width should roughly minimize the deviation of the approximate function from the exact and thus produce a better wave function.

This suggests choosing parameters in Φ to minimize (382) rather than $\langle H \rangle_\Phi$. Determination of an approximate wavefunction in this way is the basis of the local energy methods. A review of some local energy calculations for small systems was given by Frost et al. [3]. The principal advantage of the local energy methods is that, by not necessitating evaluation of integrals, they permit the use of trial functions of arbitrarily complex form. A problem with the methods is that the results are unreliable unless the points at which the local energy is evaluated are chosen carefully.

It is not hard to derive [4] an eigenvalue equation from the minimization of W_Φ, by varying the wave function. Since

$$\delta(W_\Phi) = \delta\langle H^2 \rangle_\Phi - 2\langle H \rangle \, \delta\langle H \rangle_\Phi,$$

setting the coefficient of $\delta\Phi^*$ equal to zero yields

$$H^2\Phi - 2\langle H \rangle H\Phi + (2\langle H \rangle^2 - \langle H^2 \rangle)\Phi = 0. \qquad (385)$$

For a given $\langle H \rangle$, Φ must be an eigenfunction of the operator $(H - \langle H \rangle)^2$, the eigenvalue being W_Φ. Note that Φ enters into the operator by way of $\langle H \rangle$ so this is a pseudoeigenvalue equation. Stated another way, (385) is nonlinear in Φ. Clearly exact eigenfunctions of H are eigenfunctions of $H - \langle H \rangle$ with eigenvalue zero and have vanishing W_Φ, so satisfy (385). This holds for excited states as well as the ground states: *Any* eigenfunction makes W_Φ or $\bar{W}_\Phi$ vanish. The problem in using the method would only be the identification of the solutions to $\delta W_\Phi = 0$ once they are found.

However, the nonlinear character of (385) means it has other solutions. Consider

$$\psi = (\Psi_a \pm \Psi_b) \qquad (386)$$

where Ψ_a and Ψ_b are eigenfunctions of H corresponding to *different* eigenvalues E_a and E_b, so ψ is not an eigenfunction. Now

$$\langle H \rangle_\psi = \tfrac{1}{2}(E_a + E_b), \qquad W_\psi = \tfrac{1}{2}(E_a{}^2 + E_b{}^2) - \tfrac{1}{4}(E_a + E_b)^2.$$

Since

$$(H - \langle H \rangle)^2 \psi = (E_a - \langle H \rangle)^2 \Psi_a \pm (E_b - \langle H \rangle^2) \Psi_b$$

ψ is an eigenfunction with eigenvalue $\tfrac{1}{4}(E_a - E_b)^2$, which is equal to W_ψ. Thus ψ satisfies (385). Of course this solution can be distinguished from a true eigenfunction of the Hamiltonian because $W_\psi \neq 0$. But if we vary an approximate function to make W_Φ stationary the width will never be zero, and one would have to decide whether a given solution is an approximation to an eigenfunction or to a linear combination of eigenfunctions like (386).

A more complete analysis for the case where Φ is taken as a linear combination of fixed basis functions (the usual situation in local energy calculations) was given by Stanton and Taylor [5]. They wrote

$$\Phi = \sum_{i=1}^{N} c_i \phi_i \tag{387}$$

and varied c_i. Defining

$$H_{ij} = \int \phi_i{}^* H \phi_j \, d\tau \tag{388a}$$

$$S_{ij} = \int \phi_i{}^* \phi_j \, d\tau \tag{388b}$$

$$G_{ij} = \int \phi_i{}^* H^2 \phi_j \, d\tau - \int (H\phi_i)^*(H\phi_j) \, d\tau \tag{388c}$$

we obtain, for Φ normalized

$$W_\Phi = \sum_{i,j} c_i{}^* c_j G_{ij} - \left(\sum_{i,j} c_i{}^* c_j H_{ij} \right)^2. \tag{389}$$

W_Φ is to be made stationary keeping $\sum_{i,j} c_i{}^* c_j S_{ij}$ constant, which is done using a Lagrangian multiplier λ. The result is the transcription of (385):

$$\sum_{j=1}^{N} (G_{ij} - 2\langle H \rangle_\Phi H_{ij} - \lambda S_{ij}) c_j = 0, \qquad i = 1 \cdots N. \tag{390}$$

It is easily seen that $\lambda = \langle H^2 \rangle - 2\langle H \rangle^2$. Equations (390) are nonlinear in the coefficients, since $\langle H \rangle$ is quadratic in them. Frost and co-workers [3] considered a relaxation method and an iteration method for solving (390). Direct minimization of W_Φ using steepest descent methods is also a possibility [6].

One does not know in general how many solutions there are for a basis set of N elements. Stanton and Taylor [5] treated (390) by perturbation theory, writing the matrix $\mathbf{G}$ as the square of the matrix $\mathbf{H}$ plus a perturbing term. (If the set of ϕ_i becomes complete, $\mathbf{G} = \mathbf{H}^2$.) The unperturbed equations are obtained from (390) by replacing G_{ij} with $H_{ik}H_{kj}$. The eigenvectors of H are clearly solutions with $\lambda = -\langle H \rangle^2$. But there are other solutions to the unperturbed equations, corresponding to the $N(N-1)$ linear combinations [Eq. (386)] which satisfy Eq. (385). Unless there are degeneracies in H, these two possibilities exhaust the solutions to the unperturbed equation. Stanton and Taylor's analysis suggested that the Eqs. (390) can also have $N + N(N-1)$ solutions for an N-dimensional basis set, but that only N will correspond to minima in W_Φ. The solution to the $N = 2$ case was worked out completely. It turned out that there might be four or two solutions, depending on the values of the matrix elements in (390).

The complications due to the nonlinear nature of (385) and equations derived from it may be avoided if one knows the energy of the state whose wave function is being sought. Instead of minimizing W_Φ [Eq. (384)] one could consider minimizing $\langle (H - E)^2 \rangle_\Phi$, E being the correct energy. The resulting eigenvalue equation,

$$(H^2 - 2EH + E^2)\Phi = \lambda\Phi \qquad (391)$$

is linear. Conroy's method (see Subsection 2) uses this idea.

An additional problem with use of W_Φ is the evaluation of integrals over the square of the Hamiltonian. These are difficult even for the two-electron problem. Note that for use in one of the lower-bound procedures discussed in Section A it suffices to compute $\langle H^2 \rangle$ roughly with an estimate of the error to get a useful lower bound. If we are to carry out a variational calculation involving $\langle H^2 \rangle$, however, we require high accuracy, which makes the minimization of W_Φ not very promising for future work.

On the other hand, minimization of $\bar{W}_\Phi$ should also lead to an approximation to the eigenfunction, and requires no integrals, only sums over points. If we take a trial function of the form of Eq. (387) and again put $w_P = q_P \Phi_P^2$ as earlier to simplify the algebra, then

$$\bar{W} = \sum_{i,j} c_i^* c_j \bar{G}_{ij} - \left(\sum_{i,j} c_i^* c_j \bar{H}_{ij} \right)^2. \qquad (392)$$

Here

$$\bar{G}_{ij} = \sum_P w_P (H\phi_i)_P^* (H\phi_j)_P \qquad (393\text{a})$$

$$\bar{H}_{ij} = \sum_P w_P (\phi_i^* H\phi_j)_P \qquad (393\text{b})$$

Normalization has been assumed, i.e.,

$$\sum_{i,j} c_i^* c_j \bar{S}_{ij} = 1$$

where

$$\bar{S}_{ij} = \sum_P w_P (\phi_i^* \phi_j)_P. \qquad (393\text{c})$$

Even with the basis functions real, $\bar{H}_{ij}$ is not necessarily symmetrical [7]. However,

$$\sum_{i,j} c_i^* c_j \bar{H}_{ij} = \sum_{i,j} c_i^* c_j \bar{H}_{ji}$$

so one can replace $\bar{H}_{ij}$ by $\frac{1}{2}(\bar{H}_{ij} + \bar{H}_{ji})$, which gives a symmetric matrix. It is now clear that we have the same equation as (389) with S, H, and G replaced by $\bar{\mathbf{S}}$, $\frac{1}{2}(\bar{\mathbf{H}}^\dagger + \bar{\mathbf{H}})$, and $\bar{\mathbf{G}}$. Then the equation for the coefficients is

$$\sum_j (\bar{G}_{ij} - \langle H \rangle_\Phi (\bar{H}_{ij} + \bar{H}_{ji}) - \bar{S}_{ij}) c_j = 0 \qquad (394)$$

and the above discussion of the solutions to (390) applies here as well.

Since the exact wave function makes $\bar{W}_\Phi$ vanish for any choice of points, choice of parameters in Φ to minimize $\bar{W}_\Phi$ is always justifiable, unlike minimization of $\langle \bar{H} \rangle_\Phi$, where

$$\bar{H} = \sum_{i,j} c_i^* c_j \bar{H}_{ij} \Big/ \sum_{i,j} c_i^* c_j \bar{S}_{ij}. \qquad (395)$$

On the other hand, having the function within the subspace which minimized $\bar{W}_\Phi$ would not be particularly interesting unless this function resembled the best approximation to the exact wave function within this subspace. In fact, it turns out [8] that, in order to get good wave functions from the minimization of $\bar{W}_\Phi$ without a prohibitively large number of points, the points and weights must be chosen so the sums approximate integrals. Then $\bar{W}_\Phi$ approaches W_Φ, and one is minimizing W_Φ and doing the integrals numerically.

This raises the question of how well we can do in a conventional variation method when integrals are approximated as sums, i.e., by minimization of $\langle \bar{H} \rangle_\Phi$, Eq. (395). Note that $\bar{H}_{ij}$ and $\bar{H}_{ij}$ may not be equal [7], but, as before, we can replace $\bar{H}_{ij}$ by $\frac{1}{2}(\bar{H}_{ij} + \bar{H}_{ji})$ (for real functions) so that we have a symmetric matrix. The equations to solve are just the equations of linear variation,

$$\sum_{j=1}^{N} (\bar{H}_{ij} - E\bar{S}_{ij})c_j = 0, \tag{396}$$

with sums replacing integrals.

Despite the fact that minimization of $\langle \bar{H} \rangle_\Phi$ is not a valid procedure unless the sums are approximations to integrals, whereas minimization of $\bar{W}_\Phi$ always is, the two methods often give equally good results. In particular, Carlson [9] found that results for H_2^+ using (396) and the local energy methods, using the same basis functions and points (chosen to make sums approximate integrals), were comparable. Other calculations also show [5, 9, 10] that one often obtains good results from (396) even when the number of integration points is insufficient to give accurate approximations to the integrals S_{ij} and H_{ij}. When enough points are used so that the sums in (393) roughly equal the integrals in (388), there seems to be little advantage in dealing with Eqs. (394) rather than with the simpler equations (396). When a very few points are used, or the points are carelessly chosen, neither minimization of $\langle \bar{H} \rangle_\Phi$ nor minimization of $\bar{W}_\Phi$ gives reliable results. Modifications of the local energy method which could give more accurate energies with fewer points have been suggested [6]. It should also be noted that Eqs. (396) can arise in a different way [9]. If the exact eigenfunction could be approximated by a finite sum like (387), we would have, for any point P,

$$\sum_{j=1}^{N} c_i(\phi_k^* H\phi_i)_P = E \sum_{j=1}^{N} c_i(\phi_k^*\phi_i)_P.$$

Summing over points P with convenient weights, we obtain equation (396) with $\bar{H}$ nonsymmetric. There is some relation to Szondy's method of moments (Section 2).

Further comparisons of the approaches we have been discussing have been given by Harriss and Carlson [11], and by Lloyd and Delves [12], who proposed that, in a variational calculation in which integrals are to be approximated by sums, one perform local energy calculations along with variational calculations. The divergence of the estimates of the energy by these two methods, as the basis set increases, indicates that the variational results are becoming unreliable. An analysis by Boys [10] also shows why the errors in Eqs. (396) are not as serious as would be expected. Other modifications of the local energy idea have been suggested [2, 4, 13, 14]; in Subsection 2 we discuss those for which extensive calculations on diatomic molecules have been done.

Actually, James and Coolidge [15] suggested a long time ago that minimization of the width or related quantities could give poorer wave functions than the ordinary variational method. This is because $H\Phi/\Phi$ varies most markedly (becoming infinite) where the wave function has nodes or H has singularities. The quantities W_Φ or $\overline{W}_\Phi$ are extremely sensitive to the behavior in these regions, and minimization is accomplished by making Φ behave correctly in these small regions of configuration space, at the expense of other regions. Only "short range" properties, which depend heavily on the wave function in these very regions, would be improved. Thus, to get useful results from local energy procedures, the trial function must eliminate the singularities on $H\Phi/\Phi$ before variation. We discuss the conditions for this on the next paragraph.

For a ground state wave function, Φ will vanish only where two electrons of the same spin come together. Since $H\Phi$ will also vanish here, the local energy does not become infinite. It could become infinite due to the singularities of the Coulombic terms in the potential. As the position of an electron i approaches that of a nucleus A, a singularity $-Z_A/r_{iA}$ (Z_A = nuclear charge and atomic units are used) enters $H\Phi/\Phi$. If the local energy is to remain finite, this must be canceled by a term arising from the kinetic energy operator:

$$\Phi^{-1}\left(-\tfrac{1}{2}\sum_i \nabla_i{}^2\Phi\right) - Z_A/r_{iA} = \text{finite for } r_{iA} \to 0. \qquad (397)$$

Let the positions of all electrons but i be fixed. Express $\nabla_i{}^2$ in spherical polar coordinates centered at A and put r for r_{iA}. As $r \to 0$, only the part of Φ spherically symmetric about the nucleus contributes, as higher angular momentum contributions vanish. Then we require

$$\lim_{r\to 0}[r^{-1}\,d/dr(r^2\,d\Phi/dr)] = -2Z,$$

which can be satisfied if

$$\lim_{r\to 0}[\Phi^{-1}\,d\Phi/dr] = -Z. \qquad (398)$$

The wave function thus has a cusp [15a] at each nucleus. The point in configuration space where two electrons come together may be considered similarly. For electrons i and j, we must have

$$-\Phi^{-1}(\nabla_i{}^2 + \nabla_j{}^2)\Phi + 2/r_{ij} = \text{finite for } r_{ij} \to 0.$$

The coordinates of the two electrons can be expressed in terms of coordinates of their center of mass and relative coordinates. Only the contribution of the latter is important in cancelling the singularity. Repeating the argument for the nuclear cusp, the result now is [15a]

$$\lim_{r_{ij} \to 0} [\Phi^{-1} \, d\Phi/dr_{ij}] = \tfrac{1}{2}. \tag{399}$$

Conditions (398) and (399) are, of course, satisfied by the exact wave function.

So far, we have not considered [15a] spin. For the two-electron case, the spin function may be factored off the total wave function, so that we can ignore it in this method (except that the spatial functions considered must possess the proper exchange symmetry). For a system of more than two electrons, the wave function cannot be factored. A method for calculation of $H\Phi/\Phi$, which should be spin free, has been suggested by Gimarc and Frost [16].

2. *Avoiding Integral Evaluation*

In this section, we discuss the implementation of the local energy and related formulas mentioned previously. For local energy procedures, the importance of the choice of evaluation points was indicated. For much of the earlier work, they were chosen by the methods of Gauss quadrature [17] for each dimension. Von Mohrenstein [18] pointed out that this becomes impractical if we have an integral over many dimensions, since the number of points is c^d (d = number of dimensions, c = number of points in each dimension). For an N-electron problem d could approach $3n$ if a function involving interelectronic coordinates explicitly were used. Since c could hardly be less than 3, the number for $n = 10$ would stop even a large computer.

Von Mohrenstein suggested [18] that we choose points in the $3n$-dimensional space via a Monte Carlo method, taking the density of points proportional to the absolute value of the function to be integrated. In this method, to estimate $\int F \, d\tau$ by a quadrature involving N points, the region of integration is divided into N regions of volume $\Delta\tau_i$ ($i = 1 \cdots N$). The function is evaluated at the points P_i, one at an arbitrary position in each volume, and the desired integral is approximated as

$$\int F \, d\tau \approx \sum_{i=1}^{N} \Delta\tau_i F(P_i).$$

To have the density of points proportional to the value of F, $\Delta\tau_i$ must be inversely proportional to the average size of F in the region. To this end, we introduce an auxiliary function F^0 which is integrable and approximates F, and choose the volumes $\Delta\tau_i$ so that

$$\int_{\Delta\tau_i} F^0 \, d\tau = N^{-1} \int F^0 \, d\tau \tag{400}$$

for all i. Since

$$\int_{\Delta\tau_i} F^0 \, d\tau \approx F^0(P_i) \, \Delta\tau_i$$

we have

$$\int F \, d\tau \approx \sum_{i=1}^{N} \frac{\int_{\Delta\tau_i} F^0 \, d\tau}{F^0(P_i)} F(P_i) = \frac{\int F^0 \, d\tau}{N} \sum_{i=1}^{N} \frac{F(P_i)}{F^0(P_i)}. \tag{401}$$

It may be seen that $\int F \, d\tau$ will be calculated exactly when F and F^0 are exactly proportional. The errors in this method do not depend on the number of dimensions, but decrease as the square root of the number of evaluation points. For a discussion of the errors, the reader is referred to von Mohrenstein [18].

For the one-dimensional case with finite limits of integration a and b, we can be more specific. Let g be the integral of F^0. Then (401) becomes

$$\int_a^b F \, dx \approx \frac{g(b) - g(a)}{N} \sum_{i=1}^{N} \frac{F(x_i)}{g'(x_i)}. \tag{402}$$

This method of numerical integration was used by Conroy [19–21] in his earlier work, assuming the integrals could be approximated by a product of one-dimensional integrals, so F^0 could be approximated as a product of functions in the various coordinates x_k. Thus

$$F^0 = \prod_k dg_k/dx_k$$

with $d\tau = \prod dx_k$, and

$$\int F \, d\tau \approx N^{-1} \prod_k [g_k(b_k) - g_k(a_k)] \sum_{i=1}^{N} \frac{F(P_i)}{g_k'(x_i)}. \tag{403}$$

The evaluation of G_{ij}, H_{ij}, and S_{ij} is done by using Eq. (403). The problem is the choice of the "point density functions" $g_i'(x_i)$. Conroy wrote $g_i'(x_i)$ as a series expansion, with coefficients chosen to fit the function that was to be integrated.

This method does not use the full advantages of Monte Carlo techniques for multidimensional problems. It was subsequently replaced by a more powerful one based on work of Haselgrove [22]. For f a periodic function (of period 2π) in k variables, Haselgrove considered

$$I = (2\pi)^{-k} \int_{-\pi}^{\pi} dx_1 \cdots \int_{-\pi}^{\pi} dx_k f(x_1 \cdots x_k),$$

and showed that it could be approximated as

$$I \approx (2N + 1)^{-1} \sum_{m=-N}^{N} f(2m\alpha_1, 2m\alpha_2, \ldots, 2m\alpha_k)$$

where $\alpha_1 \cdots \alpha_k$ are a set of irrational numbers. For an arbitrary function F, one can write

$$\int_0^1 F(x_1)\, dx = \frac{-1}{2\pi} \int_{-\pi}^{\pi} F(|z_1|/\pi)\, dz_1$$

where $F(|z_1|/\pi)$ is now periodic with period 2π, and then use the method. The error in the technique decreases as the reciprocal of the number of points. Conroy [23] chose rational values for the α's

$$\alpha_i' = p_i/(2N + 1)$$

(p_i an odd integer) and showed that for a proper choice of p_i the error decreased equally rapidly with N. The choice of these p_i was discussed, and sets were given for various dimensionalities.

In addition, Conroy used a modified version of the local energy method. He defined the energy variance as

$$U^2 = \int (H\Phi - E\Phi)^2\, d\tau \bigg/ \int \Phi^2\, d\tau = \langle H^2 \rangle_\Phi - 2E\langle H \rangle_\Phi + E^2 \quad (404)$$

for the approximate function Φ [20], where E was taken to be the exact eigenvalue of energy [cf. Eq. (391)]. Clearly, U^2 may be thought of as the mean square deviation of the local energy from E (with a Φ^2 weighting factor) rather than from the average energy $\langle H \rangle_\Phi$. For a determinantal function, minimization of U^2 implies conditions [24] on the spin orbitals of which it is composed. For the linear variation function (383), minimization of U^2 leads to

$$\sum_{j=1} (G_{ij} - 2EH_{ij} - \lambda' S_{ij})c_j = 0, \qquad i = 1 \cdots n. \quad (405)$$

The eigenvalue λ' is $U^2 - E^2$. For $2H_{ij}$ Conroy substituted

$$V_{ij} = H_{ij} + H_{ji} \quad (406)$$

and substituted sums over points for integrals.

Starting with a value of E one may obtain the eigenvalue of (405) from the secular determinant and select the solution corresponding to the smallest variance. Using the coefficients of the corresponding wavefunction, one calculates a new energy

$$\mathscr{E} = \sum c_i c_j \bar{H}_{ij} / c_i c_j \bar{S}_{ij}.$$

To obtain the initial value of E, Conroy [20] used an extrapolation procedure, based on relationships which are valid when the sums approximate integrals. First we know the upper bound

$$\mathscr{E} = \langle H \rangle_\Phi \geq E. \tag{407}$$

Temple's lower bound formula [Eq. (18)] may be written

$$E \geq \mathscr{E} + \frac{(\mathscr{E} - E)^2}{E_1 - \mathscr{E}} \cdot \frac{U^2}{}.$$

If $\mathscr{E}$ is less than E_1 (energy of first excited state) this implies

$$E \geq \mathscr{E} - U^2/(E_1 - \mathscr{E}). \tag{408}$$

For any trial function the calculated values for $\mathscr{E}$ and U^2 may be represented as a point on a Cartesian plot with abscissa U and ordinate $\mathscr{E}$. According to (407) the point lies above the horizontal line $\mathscr{E} = E$. According to (408) the point lies to the right of a semicircle with center at $U = 0$, $\mathscr{E} = \frac{1}{2}(E + E_1)$ and radius $\frac{1}{2}(E_1 - E)$. Conroy suggested that the extrapolation be performed as follows: Assuming some E, find the best wave function of a class in the sense of minimizing U^2. Let the results be U_{opt}, $\mathscr{E}_{\text{opt}}$. As the best function is "spoiled" in various ways by mixing in other functions in the subspace, the point on the $\mathscr{E}$–U plot traces out different curves (generally, both $\mathscr{E}$ and U increase), and an average curve may be defined. On this curve, the point corresponding to $U = 2U_{\text{opt}}$ is taken and then a line is drawn through it and the point U_{opt}, $\mathscr{E}_{\text{opt}}$; let the slope be Ω. Then the new estimated energy is

$$E = E_1 - 2 \frac{E_1 - \mathscr{E}_{\text{opt}} + \Omega U_{\text{opt}}}{1 + (1 + \Omega^2)^{1/2}}.$$

The scheme is then repeated, starting with the new E. It appears that the method is not very sensitive to the value of E_1 used. Other extrapolation methods [20] were introduced later. Other authors [6, 25] have considered alternate methods for improving local energy techniques to get reliable energies.

Since the "integrals" are actually evaluated as sums one can take the trial function as complicated in form as desired. In particular, one can ensure that the cusp conditions are satisfied, and also that any other necessary or desired properties hold. For the two-electron problems, the trial functions were composed [21, 26] of sums of products of one-electron functions multiplied onto a correlation function (function of r_{12}), the latter assuring proper cusp behavior where two electrons come together and also containing additional variable parameters. The first factor gives the proper shape in three-dimensional space (near nuclei, at infinite distance, etc.). The construction of trial functions for problems of more than two electrons which include the desirable features mentioned previously has also been discussed [26].

Conroy has reported extremely accurate results for energies and other properties of molecular systems with a few electrons (see Chapter II, Volume 2). Enough calculations have not been made for larger systems to judge whether the method can compete with more conventional techniques for calculation of the energies needed for $U(R)$. There are a number of other methods [27] which can be related to the local energy principle, which we will not discuss here since their application to systems larger than H_2 seems impractical.

An exception is the method of moments of Szondy and collaborators [28, 29], which has been used to calculate properties of $U(R)$ for diatomic systems. In addition, the transcorrelated method of Boys and Handy, which we will discuss afterward, traces [30] its parentage to Szondy's work, which we now consider. (Other schemes for avoiding integral evaluation, which have enjoyed little application to molecular calculations, are sometimes also referred to as methods of moment [27].)

Equation (396) was derived from Eq. (377) (see page 247), assuming Ψ could be written as an expansion in the $\{\phi_k\}$, by multiplying by a function $\phi_k{}^*$, evaluating at some point P, and summing over points P:

$$\sum_P w_P(\phi_k{}^* H\Psi - \phi_k{}^* E\Psi)_P = 0 \qquad (409)$$

If Ψ were exactly equal to a linear combination of the $n + 1$ functions ϕ_k ($k = 0 \cdots n$), the $n + 1$ equations of the form (409) would determine the $n + 1$ parameters, which consist of n expansion coefficients, since normalization is arbitrary, and E. More generally, Φ could be any function containing n parameters α_i ($i = 1 \cdots n$) whose values are to be determined so as to make Φ approximate the exact eigenfunction Ψ. Then using some set of functions ϕ_k ($k = 0 \cdots n$) one could obtain $n + 1$ conditions like (409) to determine the α_i and E. Furthermore, the sums over points could be replaced by integrals, and the parameters determined from the $n + 1$ equations:

$$\langle \phi_k \mid (H - E) \mid \Phi \rangle = 0, \qquad k = 0, 1, \ldots, n \qquad (410)$$

The quantities on the left sides of these equations are the moments of $(H - E)\Phi$ with the weight functions ϕ_k. Alternatively, it is possible to take [28]

$$(1 + \beta_0)\Phi + \sum_{i=1}^{n} \beta_i \phi_i$$

as a trial function for linear variation, Φ still depending on the parameters α_i. If Φ is a good function, the improvement due to admixture of the ϕ_i should be small. The best Φ (in terms of choosing the α_i) would satisfy (410), since this makes all the β_i vanish. In particular, if we have a set of ϕ_i which are large in a particular region of configuration space and (410) holds for these, we can say that Φ is probably a good wave function in this region. This reasoning can be used to justify [29] Preuss's modified local energy method [2], in which one tries to make constant the local energy multiplied by a weighting factor in configuration space.

The ϕ_k in Eq. (410), being arbitrary, can be chosen to simplify the integrations needed to evaluate the moment. Indeed, by using sums of Dirac delta functions for the ϕ_k, we eliminate integrals altogether, since

$$\langle \delta(\mathbf{r} - \mathbf{P}) \mid H - E \mid \Phi \rangle = [(H - E)\Phi]_P$$

and we recover a local energy method. Szondy [29] discussed the relation to other approximation techniques. But, as with the local energy methods, the choice of ϕ_k determines whether one can get good results from this method when n is small. In order to avoid the necessity of knowing E accurately before starting the calculation, one may try to have $\langle \phi_i \mid \Phi \rangle$ small, since this deemphasizes the term in E.

Hegyi *et al.* [28] made a simplification in Eq. (410) by assuming the weight functions ϕ_k to have been subjected to a linear transformation which makes

$$\langle \phi_k \mid \Phi \rangle = 0, \qquad k = 1 \cdots n.$$

Then one must determine the n α_i from

$$\langle \phi_k \mid H \mid \Phi \rangle = 0, \qquad k = 1 \cdots n \qquad (411a)$$

and the energy from

$$E = \langle \phi_0 \mid H \mid \Phi \rangle / \langle \phi_0 \mid \Phi \rangle. \qquad (411b)$$

For ground-state singlet systems, Φ was taken as a single determinant of the $2N$ spin orbitals $\{u_i\}$, each a product of the α or β spin function with a linear combination of fixed basis functions. Another set of spin orbitals, $\{v_i\}$, was similarly constructed from a second set of fixed basis functions. The weight functions were constructed from products of the v_i. The method of moments led to a savings of time in the computation of the interelectronic repulsion integrals when fewer basis functions were used for the $\{v_i\}$ than for the $\{u_i\}$, yet the results were of comparable accuracy. Energies and R_e values were calculated for H_2, Li_2, LiH, Be_2, and HF.

Boys [10] generalized Szondy's method and gave a general discussion of the linear variation problem when the Hamiltonian and/or overlap matrix is not Hermitian. This situation arises from Eq. (411) if Φ is expanded in a set of functions different from $\{\phi_k\}$, say $\Phi = \sum_s c_s g_s$. Then we have to deal with

$$\sum_s \langle \phi_k \mid H - E \mid g_s \rangle c_s = \sum_s (H_{ks} - ES_{ks})c_s = 0. \tag{412}$$

Boys's analysis of the error in E due to the insufficiencies of the basis functions is also relevant to the usual secular equation when integrals are replaced by sums over points [Eq. (396)], in which case it refers to the effect on E of errors in numerical evaluation of H_{ks} and S_{ks}. Here, ϕ_k may be written as Qg_k, where Q is an operator (a weighted sum of δ functions) which transforms the integral into a sum over points. Among Boys's conclusions was that the errors could be made small if $\{g_s\}$ formed a good set of functions, even if $\{\phi_k\}$ did not.

This fact suggested a method for introducing correlation factors into a variation function without introducing complicated integrals. The basis functions would be written as a correlation factor (explicitly including interelectronic distances) times determinantal functions:

$$g_s = \prod_{i,j}^{(i>j)} f_{ij} \Phi_s \tag{413}$$

where i and j refer to electrons. If ϕ_k in (412) were taken as $\prod f_{ij}^{-1} \Phi_k$, the integrals needed would be relatively simple, and good results could be obtained for E even if $\{\phi_k\}$ formed a poor basis set for representation of the exact wavefunction. This idea has been developed by Boys and Handy [30] into a method for calculation of accurate energies for atoms and molecules, named the *transcorrelated method*. It is claimed that, despite the complexity of the wave functions used, no integrals "appreciably more complicated than six dimensions" need be computed.

In recent versions of the method, the wave function is written in the form

$$\Psi = \prod_{i,j}^{(i>j)} f_{ij}\Phi = C\Phi$$

where Φ is a determinant, and substituted into the Schrödinger equation. On multiplication by $\prod f_{ij}^{-1}$, one obtains the transcorrelated wave equation,

$$\left[\left(\prod_{i,j}^{(i>j)} f_{ij}^{-1} \right) H \left(\prod_{i,j}^{(i>j)} f_{ij} \right) - E \right]\Phi = 0.$$

Conditions which make the correlation factor uniquely defined have been given [30]. The variational principle, applied to the transcorrelated Hamiltonian $C^{-1}H\,C$, leads to equations for the determination of parameters in the orbitals of Φ and in the correlation factor. Since they all involve matrix elements of the transcorrelated Hamiltonian between determinants, rather than matrix elements of H between correlated functions (i.e., of $C\,H\,C$ between determinants), they are relatively simple to deal with. The integrands being nonsingular, accurate results are obtainable by numerical techniques using a moderate number of points, and molecular systems are not much more complicated than atomic ones [31]. (Recently, expansions in Gaussian orbitals have been used and integrals explicitly evaluated.) For LiH, the calculated energy at $R = 3\,a_0$ was better than that obtained by any variational calculation ($E_{\text{calc}} = -8.063$ a.u., $E_{\text{expt}} = -8.070$ a.u.).

The work required grows only in proportion to the number of electrons, so there is hope for extremely accurate results by this method for large molecules. At this time the successes of the transcorrelated method, which derives from attempts at implementation of the local energy principle to avoid integral evaluation, raise hopes that the difficulties connected with evaluation of integrals in conventional variational calculations may yet be circumvented.

REFERENCES

1. J. H. Bartlett, *Phys. Rev.* **98**, 1067 (1955).
2. H. Preuss, *Z. Naturforsch.* **16a**, 598 (1961); *Theor. Chim. Acta* **2**, 98 (1964); see also Derflinger [14].
3. A. A. Frost, R. E. Kellogg, and E. C. Curtis, *Rev. Mod. Phys.* **32**, 313 (1960).
4. J. Goodisman, *J. Chem. Phys.* **45**, 3659 (1966); J. Goodisman and D. Secrest, *Ibid.* **45**, 1515 (1966).
5. R. E. Stanton and R. I. Taylor, *J. Chem. Phys.* **45**, 565 (1966).
6. T. A. Rourke and E. T. Stewart, *Can. J. Phys.* **45**, 2755 (1967); **46**, 1603 (1968).
7. J. Goodisman, *J. Chem. Phys.* **41**, 2365 (1964).

8. A. A. Frost, R. E. Kellogg, B. M. Gimarc, and J. D. Scargle, *J. Chem. Phys.* **35**, 827 (1961); D. K. Harriss and A. A. Frost, *Ibid.* **40**, 204 (1964).
9. C. M. Carlson, *J. Chem. Phys.* **47**, 862 (1967).
10. S. F. Boys, *Proc. Roy. Soc. Ser. A* **309**, 195 (1969).
11. D. K. Harriss and C. M. Carlson, *J. Chem. Phys.* **51**, 5458 (1969).
12. M. H. Lloyd and L. M. Delves, *Int. J. Quantum Chem.* **3**, 169 (1969).
13. D. H. Bell and L. M. Delves, *J. Comput. Phys.* **3**, 453 (1969).
14. G. Derflinger, *Chem. Phys. Lett.* **1**, 153 (1967).
15. H. M. James and A. S. Coolidge, *Phys. Rev.* **51**, 860 (1937).
15a.T. Kato, *Commun. Pure Appl. Math.* **10**, 151 (1957). R. T. Pack and W. Byers Brown, *J. Chem. Phys.* **45**, 556 (1966).
16. B. M. Gimarc and A. A. Frost, *J. Chem. Phys.* **39**, 1698 (1963).
17. A. H. Stroud and D. Secrest, "Gaussian Quadrature Formulas." Prentice-Hall, Englewood Cliffs, New Jersey, 1966.
18. A. Von Mohrenstein, *Z. Phys.* **134**, 488 (1953).
19. H. Conroy, *J. Chem. Phys.* **41**, 1331 (1964).
20. H. Conroy, *J. Chem. Phys.* **41**, 1336 (1964).
21. H. Conroy, *J. Chem. Phys.* **41**, 1327, 1341 (1964).
22. C. B. Haselgrove, *Math. Comput.* **15**, 323 (1961).
23. H. Conroy, *J. Chem. Phys.* **47**, 5307 (1967).
24. F. W. Birss and S. Fraga, *J. Chem. Phys.* **38**, 2552 (1963); **40**, 3207, 3212 (1964).
25. H. Conroy, *J. Chem. Phys.* **47**, 930 (1967).
26. H. Conroy, *J. Chem. Phys.* **47**, 912 (1967).
27. J. B. Delos and S. M. Blinder, *J. Chem. Phys.* **47**, 2784 (1967); H. J. Silverstone, M. L. Yin, and R. L. Somorjai, *Ibid.* **47**, 4824 (1967); S. M. Rothstein, J. E. Welsch, and H. J. Silverstone, *Ibid.* **51**, 2932 (1969).
28. M. G. Hegyi, M. Mezei, and T. Szondy, *Theor. Chim. Acta* **15**, 273, 283 (1969).
29. T. Szondy, *Acta Phys. Hung.* **17**, 303 (1964); E. Szondy and T. Szondy, *Ibid.* **20**, 253 (1966).
30. S. F. Boys and N. C. Handy, *Proc. Roy. Soc. Ser. A* **309**, 209 (1969); **310**, 43, 63 (1969).
31. N. C. Handy, *J. Chem. Phys.* **51**, 3205 (1969); S. F. Boys and N. C. Handy, *Proc. Roy. Soc. Ser. A* **311**, 309 (1969).

F. Quantum Statistical Calculations

1. *Thomas–Fermi and Thomas–Fermi–Dirac Theories*

The quantum statistical models, in particular the Thomas–Fermi and Thomas–Fermi–Dirac theories, lead to an attractively simple method of calculating atomic and molecular properties. Early work on atoms demonstrated that quantum statistical calculations could give reasonable results for electron densities and electronic properties, so that attempts are being made to apply them to molecules. At the same time, the models are known to contain serious errors. Thus improvements or corrections to these theories have occupied, and continue to occupy, the attention of theoreticians. Probably because of the successes of the more rigorous quantum mechanical methods, relatively little work has been done with molecules by this method. Some comprehensive reviews of the theory involved and of the earlier work, particularly on atoms, are the book of Gombás [1], and the slightly more recent reviews of Gombás [2], and March [3]. We shall discuss the quantum statistical models in general here, and applications to molecular systems in Sections 2 and 3.

The simplicity of these theories arises because they lead to equations for the electron density directly. The electron density is a function of only three coordinates, no matter how many electrons are present, whereas the difficulty in solving the Schrödinger equation rises rapidly with the number of electrons. Yet, most atomic and molecular properties require for their calculation only a knowledge of the electron density, which is obtained in quantum mechanical calculations by integrating out most of the information in a wave function obtained by approximate solution of the Schrödinger equation. The energy actually requires the two-particle density matrix (for the interelectronic repulsion energy), as well as off-diagonal components of the one-particle density matrix (for the kinetic energy). This is one of the things that keeps us from deriving a rigorous variational equation involving the electron density alone.

If one approximates the wave function by a single determinant, the two-particle density matrix may be expressed in terms of the one-particle density matrix [Eq. (94)]. In the interelectronic repulsion energy, the diagonal elements of $\Gamma^{(1)}$ give the direct or Coulombic contribution, and the off-diagonal elements the exchange contribution. The latter is negative, and arises because of the antisymmetry of the wavefunction. We will drop it from consideration for the moment and return to it later, since it is small in magnitude relative to other quantities entering the energy of a molecule.

Now the Coulombic energy may be thought of strictly as the classical electrostatic interaction of the electronic charge distribution with itself. The potential energy thus becomes

$$\langle V \rangle = -e \int V_{ne}\varrho \, d\tau + \tfrac{1}{2}e^2 \int \varrho(\mathbf{r}_1)\varrho(\mathbf{r}_2)r_{12}^{-1} \, d\tau_1 \, d\tau_2 \qquad (414)$$

where V_{ne} is the potential of the nuclei. The kinetic energy may be written as a sum of orbital contributions,

$$\langle T \rangle = \sum_i n_i \int \lambda_i^*(1)T_1\lambda_i(1) \, d\tau \qquad (415)$$

where T_1 is the kinetic energy operator for electron 1. The spin orbitals λ_i which appear with occupation numbers n_i are the natural orbitals of the wave function used. For the single determinant case, the λ_i are just the occupied orbitals and all n_i are unity. T_1 involves differentiations: These are to be carried out on λ_i before multiplication by λ_i^*, so that we cannot calculate the kinetic energy of an arbitrary wave function from just the electron density. For certain simple situations, however, there is a relation between the kinetic energy and the density.

Consider N noninteracting electrons in a box whose sides are L_x, L_y, and L_z. The volume $V = L_x L_y L_z$ will eventually become infinite (free-electron gas). The wave functions are

$$\psi_k = V^{-1/2} e^{i\mathbf{k}\cdot\mathbf{r}} \tag{416}$$

with $k_x = 2n_x/L_x$, $k_y = 2n_y/L_y$, $k_z = 2n_z/L_z$; n_x, n_y, and n_z integers. All states are filled, putting two particles in each, up to some limiting value of $\mathbf{k}$ such that

$$N = \int \sum 2 \, |\psi_k|^2 \, d\tau = \sum_{k_x, k_y, k_z} 2.$$

If the box is large so that the values of k are closely spaced, sums may be replaced by integrals:

$$\sum_{k_x} \to (L_x/2\pi) \int dk_x$$

Then k_F, the maximum value of $|\mathbf{k}|$ or Fermi momentum, is obtained from

$$N = \frac{2V}{8\pi^3} \int dk_x \int dk_y \int dk_z = \frac{V}{4\pi^3} \int_0^{k_F} 4\pi k^2 \, dk = \frac{V}{3\pi^2} k_F^3. \tag{417}$$

The kinetic energy is the integral over all space of

$$2 \sum \psi_{\mathbf{k}}(-\hbar^2/2m) \nabla^2 \psi_{\mathbf{k}} = \sum_{k_x, k_y, k_z} 2(\hbar^2/2mV)(k_x^2 + k_y^2 + k_z^2).$$

We refer to this as the kinetic energy density ϱ_k. Going over to integrals,

$$\varrho_k = (\hbar^2/mV)(V/8\pi^3) \int_0^{k_F} 4\pi k^4 \, dk = (\hbar^2/10m\pi^2)k_F^5.$$

This may be expressed in terms of the density $\varrho = N/V$ by using (417):

$$\varrho_k = (3\hbar^2/10m)(3\pi^2)^{2/3}\varrho^{5/3} = \varkappa_k \varrho^{5/3}. \tag{418}$$

These results would hold for particles moving in a uniform constant potential. For a nonuniform system, we could imagine applying (418) to some region of space over which the potential was roughly constant. The sum over such regions becomes an integral: The total kinetic energy is expressed in terms of the density as $\int \varkappa_k \varrho^{5/3} \, d\tau$. Trouble can be anticipated where the potential varies rapidly, such as near a center of Coulombic force. Indeed, densities from the quantum statistical calculations are very bad near nuclei in atomic and molecular systems. Another case where the assumptions leading to (418) break down is where the density starts to approach zero, at the periphery of the molecule. The number of occupied states is not large enough to replace the sum over states by an integral.

We may now introduce the approximate expression for the kinetic energy and attempt to minimize the total energy

$$E = -e \int V_{\mathrm{ne}}\varrho \, d\tau + \tfrac{1}{2}e^2 \iint \varrho(\mathbf{r}_1)\varrho(\mathbf{r}_2)r_{12}^{-1} \, d\tau_1 \, d\tau_2 + \varkappa_k \int \varrho^{5/3} \, d\tau \quad (419)$$

with the normalization condition

$$\int \varrho \, d\tau = N = \text{const.} \quad (420)$$

N is the total number of electrons in the system. A Lagrange multiplier λ will be employed to maintain the condition (420). Then $\delta E = 0$ leads to an integral equation for ϱ:

$$-eV_{\mathrm{ne}} + e^2 \int \varrho(\mathbf{r}') \, | \, \mathbf{r} - \mathbf{r}' \, |^{-1} \, d\tau' + \tfrac{5}{3}\varkappa_k\varrho^{2/3} = \lambda \quad (421)$$

The first two terms represent the electronic charge times the electrostatic potential seen by an electron (as in SCF theories, the potential depends on the electron density). Calling it V, we rewrite (421) as

$$\varrho = [(eV + \lambda)(3/5\varkappa_k)]^{3/2}$$

Evidently, V must satisfy the Poisson equation,

$$\nabla^2 V = 4\pi e\varrho,$$

wherever the nuclear charge density vanishes. Combining these two equations leads to the Thomas–Fermi equation

$$\nabla^2(\bar{V}) = 4\pi e^2[3\bar{V}/5\varkappa_k]^{3/2} \quad (422)$$

where $\bar{V} = eV + \lambda$. The boundary conditions for (422) are: (a) Normalization of ϱ according to (420); (b) $\bar{V} \to Z_A e^2/|\mathbf{r} - \mathbf{r}_A|$ as $\mathbf{r} - \mathbf{r}_A$ goes to zero, when a nucleus of charge $Z_A e$ is located at $\mathbf{r}_A$; (c) $V \to qe/r$ as we go to the boundary of the molecule, where q is the net charge. In this theory, it turns out that the boundary is at infinity for a neutral system and is at a finite distance from the nuclei for a positive ion. Negative ions are not stable. Other derivations of Eq. (422) are possible, such as those which derive conditions on the Fermi momentum, or invoke the semiclassical expansions. In the above derivation, the close connection between the quantum statistical theories and the one-electron picture is emphasized. Since for a neutral molecule both ϱ and V vanish at the boundaries, $\lambda = 0$. To show that (420) is satisfied for this value of λ, integrate the Poisson equation over a region bounded by the periphery of the molecule and small spheres around the nuclei. The volume integral is converted to a surface integral of ∇V. Since the electric field vanishes on the periphery only the small spheres contribute. Then, using assumption (b), $\int \varrho \, d\tau = \sum_A Z_A = N$.

For an atom of nuclear charge Z, there is spherical symmetry and $\bar{V}$ depends only on the distance r from the nucleus. We convert to dimensionless variables:

$$x = (4\pi)^{2/3} Z^{1/3} (3e^2/5\varkappa_k) r, \qquad \chi = r(Ze^2)^{-1}\bar{V}$$

to obtain the universal equation

$$\chi_{xx} = \chi^{3/2}/x^{1/2}, \qquad \chi(0) = 1. \tag{423}$$

Equation (423) is sometimes also referred to as the Thomas–Fermi equation. From it, the solution to (422) for any atom may be obtained.

The use of the kinetic energy relation (418) of the free-electron gas suggests a way of incorporating the exchange energy: Use the electron gas result locally. The wave function is a determinant of N spin orbitals formed by putting α and β spin electrons into the states ψ_k of (416). The interelectronic repulsion may be written

$$I = \tfrac{1}{2} e^2 \sum_k \sum_k \iint d\tau_1 \, d\tau_2 [\psi_k(1)\psi_{k'}(2)]^* r_{12}^{-1}$$
$$\times [4\psi_k(1)\psi_{k'}(2) - 2\psi_k(2)\psi_{k'}(1)].$$

The factors of 4 and 2 come from the different spin assignments. The first term in the second square bracket gives the Coulombic term already discussed. The second gives the exchange energy:

$$E_{ex} = -(e^2/V^2) \sum_k \sum_{k'} \iint d\tau_1 \, d\tau_2 e^{i\mathbf{k}\cdot\mathbf{R}} e^{-i\mathbf{k}'\cdot\mathbf{R}} R^{-1}$$

where $\mathbf{R} = \mathbf{r}_1 - \mathbf{r}_2$. The variables $\mathbf{r}_1$ and $\mathbf{r}_2$ are replaced by $\mathbf{R}$ and $\mathbf{r}_2$. The integration over $\mathbf{r}_2$ is easily carried out to give a factor of volume V, if we assume the result of the other integrations is independent of $\mathbf{r}_2$. The R^{-1} effectively restricts the range of the R integration, and for a sufficiently large box we may neglect the possibility of $\mathbf{r}_2$ being near the boundary. We will eventually integrate R over all space. We consider the sums over $\mathbf{k}$ and $\mathbf{k}'$ first, replacing them by integrals to get

$$E_{ex} = -(e^2 V/64\pi^6) \int d\tau_R R^{-1} \left[\int d^3 k e^{i\mathbf{k}\cdot\mathbf{R}} \right] \left[\int d^3 k' e^{-i\mathbf{k}'\cdot\mathbf{R}} \right]$$

The second square bracket is the complex conjugate of the first. For the k integration, choose polar coordinates along R so that

$$\int d^3 k e^{i\mathbf{k}\cdot\mathbf{R}} = 2\pi \int k^2 \, dk \sin\theta \, d\theta e^{ikR\cos\theta}$$

$$= (4\pi/R) \int_0^{k_F} k \sin(kR) \, dk$$

$$= (4\pi/R^3)[\sin(k_F R) - k_F R \cos(k_F R)]$$

This is substituted into E_{ex} to give

$$E_{ex} = -(e^2 V/4\pi^4) \int R^2 \, dR \sin\theta \, d\theta \, d\phi \, R^{-7}[\sin(k_F R) - k_F R \cos(k_F R)]^2. \tag{424}$$

The angular integrations in (424) are trivial. One must finally contend with

$$\mathcal{I} = \int_0^\infty dx x^{-5}[\sin x - x \cos x]^2.$$

The exchange energy is $-e^2 V \pi^{-3} k_F^4$ times this. Since

$$f(x) = \sin x - x \cos x = \sum_n (-1)^{n+1} x^{2n+1}[2n/(2n+1)!]$$

$f(x)^2$ goes as x^6 for small x and the integrand of $\mathcal{I}$ is well behaved at $x = 0$. Integration by parts three times leads to $\mathcal{I} = \frac{1}{4}$, so our expression for the exchange energy density is

$$\varrho_e = -\frac{1}{4} e^2 \pi^{-3} k_F^4 \tag{425}$$

which may be written as

$$\varrho_e = -\varkappa_e \varrho^{4/3} \qquad (426)$$

where

$$\varkappa_e = (3e^2/4\pi)(3\pi^2)^{1/3}. \qquad (427)$$

Other methods of evaluating E_{ex} are given by Gombás [1] and Slater [4].

One now may use the relation (426) for a general system, so that an exchange energy contribution $-\varkappa_e \int \varrho^{4/3} \, d\tau$ may be added to the energy expression (419). Then the variational procedure leads to a modified equation for the density, referred to as the Thomas–Fermi–Dirac equation. We will give another derivation of this later. It may be noted that the terms ϱ_K and ϱ_e are part of an expansion in decreasing powers of $\varrho^{1/3}$ of the energy of an electron gas. The next terms have also been evaluated [5] and correspond to the correlation energy. An extension of the Thomas–Fermi–Dirac theory to include a correlation energy contribution going as ϱ was given by Gombás [1] but has not so far been used for molecules.

We may introduce the notion of an "average exchange potential" $V_e = (2/e)\varkappa_e\varrho^{1/3}$ so that the exchange energy is $-\tfrac{1}{2}e \int d\tau \, V_e\varrho$, the factor of $\tfrac{1}{2}$ arising because the interaction between each pair of electrons is counted twice. (It must be noted that other coefficients of $\varrho^{1/3}$ have been suggested on various grounds and are used in the literature.) Then the exchange interaction for one electron in an orbital ϕ is the expectation value of V_e over ϕ. This idea may be used to simplify the Hartree–Fock equations by replacing the nonlocal exchange term by a local potential, as suggested by Slater [4]. The approximation has proved to be a very useful simplification in atomic structure calculations [6]. The main problems arising in the Thomas–Fermi–Dirac method can thus be ascribed to the treatment of the kinetic energy and are present in the Thomas–Fermi method as well.

Dirac's derivation of the Thomas–Fermi–Dirac equation [7, 8] utilizes a mixed density $\varrho(x, p)$ which is to be interpreted as the classical density in phase space, i.e., $\varrho(x, p) \, dx \, dp$ is the number of electrons with coordinates between x and $x + dx$ and momenta between p and $p + dp$. (Here, we symbolically put x, p, etc. for the set of three components of $\mathbf{x}$, $\mathbf{p}$, etc.) The expression for this density is

$$\varrho(x, p) = \langle x \mid \varrho \mid p \rangle\langle p \mid x \rangle \qquad (428)$$

where $\langle x \mid \varrho \mid p \rangle$ is a mixed matrix element of the density operator and $\langle p \mid x \rangle = e^{i p \cdot x/\hbar}$ the transformation matrix between position and momen-

tum representations and also the eigenfunction of position in the momentum representation. The books of Dirac and Corson [9] give good intro-
ductions to manipulation of the matrix elements and we have not the space
for a lengthy discussion here. By integrating $\varrho(x, p)$ over momentum we
obtain the electron density for the nuclear–electronic and Coulombic contributions to the energy. For the exchange energy we need off-diagonal
elements of the first-order density matrix:

$$\langle x \mid \varrho \mid x' \rangle = \int dp \langle x \mid \varrho \mid p \rangle \langle p \mid x' \rangle$$

$$= \int dp e^{ip \cdot x/\hbar} \varrho(x, p) e^{ix' \cdot p/\hbar} \qquad (429)$$

The electron–nuclear attraction energy is

$$\int dx\, dp\, eV_{\mathrm{ne}}(x)\varrho(x, p),$$

the kinetic energy

$$\int dx\, dp(p^2/2m)\varrho(x, p),$$

the Coulombic repulsion energy

$$\tfrac{1}{2}e^2 \int dx\, dx'\, dp\, dp' |\, x - x' \,|^{-1}\varrho(x, p)\varrho(x', p')$$

and the exchange energy

$$-\tfrac{1}{4}e^2 \int dx\, dx'\, dp\, dp' |\, x - x'|^{-1}\varrho(x', p)\varrho^*(x, p') \exp[i(p - p') \cdot x'/\hbar].$$

Now at absolute zero one could expect that for each position in space all
momentum cells are filled with two electrons up to some maximum value
of the momentum. The value of this Fermi momentum p_{F} depends on the
position. Thus we assume

$$\varrho(x, p) = 2/h^3, \qquad |\, p \,| \leq p_{\mathrm{F}}(x) \qquad (430a)$$

$$\varrho(x, p) = 0, \qquad |\, p \,| > p_{\mathrm{F}}(x) \qquad (430b)$$

The total number of electrons is expressed in terms of the Fermi momentum
p_{F}:

$$N = \int dx\, dp\, \varrho(x, p) = 2h^{-3} \int dx(4\pi/3)p_{\mathrm{F}}{}^3 \qquad (431)$$

Thus the density is $(8\pi/3h^3)p_F^3$. The kinetic energy is

$$2h^{-3} \int dx(4\pi/10m)p_F^5 = K \tag{432}$$

which recovers Eq. (418). The electron–nuclear attraction is

$$-2h^{-3} \int dx e V_{ne}(x)(4\pi/3)p_F^3 = E_{ne} \tag{433}$$

and the Coulombic repulsion

$$2e^2 h^{-6} \int dx\, dx' |\, x - x'\, |^{-1}(4\pi/3)^2 p_F^3(x)p_F^3(x') = I. \tag{434}$$

The exchange energy leads to a calculation like that of page 262, from which

$$E_{ex} = -(e^2/h^4) \int dx 4\pi p_F^4 \tag{435}$$

(note that $p_F = k_F\hbar$). We now attempt to minimize the energy by varying the function $p_F(x)$ with the constraint that N, Eq. (431), be constant. The density matrix would also be expected to be idempotent for a single-determinant wave function, which means the matrix used here would satisfy

$$\int dx' \langle x \,|\, \varrho \,|\, x' \rangle \langle x' \,|\, \varrho \,|\, x'' \rangle = 2 \langle x \,|\, \varrho \,|\, x'' \rangle.$$

This holds only when p_F is infinite, in which case

$$\langle x \,|\, \varrho \,|\, x' \rangle = (2/h^3)(8\pi^3\hbar^3)\,\delta(x - x').$$

This reminds us that the quantum statistical theories tend to work better for large numbers of electrons.

The minimization of $K + E_{ne} + I + E_{ex}$ with respect to variations in the function $p_F(x)$, keeping (431) as a restriction, yields the Thomas–Fermi–Dirac equation,

$$\frac{p_F^2}{2m} - eV_{ne} + 2e^2\left(\frac{4\pi}{3h^3}\right)p_F^3(x') \,|\, x - x'\, |^{-1}\, dx' - \frac{2e}{h} p_F = \lambda \tag{436}$$

The terms on the left arise from kinetic energy, nuclear–electronic attraction, Coulombic repulsion, and exchange energy. The constant λ is the Lagrange multiplier. Equation (436) is identical to Eq. (421) except for the exchange term, as is seen by using

$$\varrho = (8\pi/3h^3)p_F{}^3 \tag{437}$$

and the definition of $\varkappa_k$, Eq. (418). Like the Thomas–Fermi equation, (437) may be transformed into a differential equation by the introduction of a function V, representing the potential in which an electron moves. Even for a neutral system, the solution to the Thomas–Fermi–Dirac equation is discontinuous. Sheldon [8] shows generally that there is a boundary on which $p_F = 5e^2m/2h$, and p_F is zero outside. Thus, the electron density drops discontinuously from $(125/192\pi^5)a_0^{-3}$ to zero as we cross a bounding surface a finite distance from the nuclei.

Because the Thomas–Fermi and Thomas–Fermi–Dirac densities are derived from a variational principle, they satisfy the Hellmann–Feynman theorem [10]. The proof follows Hellmann's for the quantum mechanical case [Eq. (333)]. Let Q be a parameter entering the energy operator and ϱ_Q the density corresponding to a particular value of Q, the corresponding energy being $E(\varrho_Q, Q)$. The energy depends on Q in two ways: explicitly because of the appearance of Q in the energy operator and implicitly since it depends on ϱ_Q which depends on Q. The theorem states that

$$dE(\varrho_Q, Q)/dQ = \partial E(\varrho_Q, Q)/\partial Q \tag{438}$$

where the expression on the right means the derivative of E with respect to the explicit occurrence of Q in the energy expression. The dependence on Q through ϱ_Q is proportional to

$$\lim_{\delta Q \to 0} \frac{E(\varrho_{Q+\delta Q}, Q) - E(\varrho_Q, Q)}{\varrho_{Q+\delta Q} - \varrho_Q} = \frac{\delta E}{\delta \varrho}.$$

Now ϱ is determined by making $E(\varrho_Q, Q)$ stationary with some auxiliary condition, $A(\varrho) = \text{const}$, imposed. This means that the variation in $E(\varrho_Q, Q)$ due to any variation in ϱ which leaves $A(\varrho)$ invariant vanishes. Both ϱ_Q and $\varrho_{Q+\delta Q}$ satisfy $A(\varrho) = \text{const}$, so preceding expression represents such a variation and vanishes.

In particular, the force can be calculated either as the change in the energy with displacement of a nucleus or as the expectation value of the force operator. This operator would be $\partial V_{ne}/\partial Q$ where Q is a nuclear displacement, just as in the quantum mechanical case. Identical results must be obtained from the two recipes. This may serve as a check on whether one has come close to exact solution of the Thomas–Fermi or Thomas–Fermi–Dirac equation by an approximate method.

Another consequence of the derivation of the Thomas–Fermi (TF) and Thomas–Fermi–Dirac (TFD) equations from a variational principle on the density is that one can get approximations to the TF and TFD densities by using the variational principle directly. One assumes a form for the density, satisfying the auxiliary condition and containing one or more parameters, and makes the energy expression stationary with respect to variation of the parameters. It must be noted that little simplification is gained by using a linear variation function here since the equations are nonlinear. One may also obtain approximate energies by guessing a density and using it in the energy expression. For instance, we could use a superposition of the atomic densities for a diatomic molecule. For large internuclear distances, this is the exact solution, since the atoms have a finite extent and the densities would be truly nonoverlapping. In fact, the interatomic interaction is exactly zero until the internuclear distance is small enough for overlapping to take place.

Several relations exist [8] between the expectation values entering the energy when calculated with the correct density $\varrho = (8\pi/3h^3)p_F{}^3$. Multiply (436) by $\varrho(x)$ and integrate over all space. On using (437) we have

$$\tfrac{5}{3}K + E_{\mathrm{ne}} + 2I + \tfrac{4}{3}E_{\mathrm{ex}} = \lambda \int \varrho \, dx = \lambda N. \tag{439}$$

There is in addition a virial theorem, given by Sheldon [8]. We may derive it by introduction of a scaling factor (see Section C.1) into the density. Let $\varrho_1(x; sR)$ be a properly normalized quantum statistical density corresponding to the nuclear configuration sR. Here, R refers to a set of nuclear coordinates. A scaled density, appropriate to a different nuclear configuration, is obtained by multiplying all electronic coordinates by a scaling factor s:

$$\varrho_s(x; R) = s^3 \varrho_1(sx; sR) \tag{440}$$

The s^3 factor assures that ϱ_s is properly normalized. The expectation value of the energy using the scaled density, minimimized with respect to s, gives

$$2sK + E_{\mathrm{ne}}(sR) + (\partial/\partial s)E_{\mathrm{ne}}(sR) + I + E_{\mathrm{ex}} = 0. \tag{441}$$

If ϱ_1 is the exact density, the energy must be stationary for $s = 1$.

$$2K + E_{\mathrm{ne}} + I + E_{\mathrm{ex}} = -\left\{ \frac{\partial}{\partial s} E_{\mathrm{ne}}(sR) \right\}_{s=1}$$

$$= -\sum_i \frac{\partial E_{\mathrm{ne}}}{\partial R_i} R_i \tag{442}$$

where the sum over i is over all coordinates of the nuclei. $\partial E_{ne}/\partial R_i$ is the negative of the force on the ith nucleus. We can now include the internuclear repulsions V_{nn}. Let $\mathbf{F}_k$ be the force (nuclear plus electronic contributions) on nucleus k and $\mathbf{R}_k$ the position of nucleus k. Then (442) may be written as

$$E + K = V_{nn} + 2K + E_{ne} + I + E_{ex} = \sum_k \mathbf{F}_k \cdot \mathbf{R}_k \qquad (443)$$

where E is the total energy for the fixed-nucleus system.

2. Applications to Molecular Systems

We now have several methods to derive information about interatomic interactions from the quantum statistical calculations. Energies may be computed for several internuclear distances. Alternatively, the density may be used to evaluate the forces. Finally, the virial theorem (443) for a diatomic molecule means that a single evaluation of total and kinetic energy for some nuclear configuration gives the forces immediately, since one will have automatically $\sum_k \mathbf{F}_k = 0$. These procedures, of course, do not necessarily give the same results for an approximation to the TF or TFD solution.

Sheldon [8] found it convenient to differentiate (443) with respect to a nuclear position:

$$-\boldsymbol{\nabla}_j(\mathbf{F}_j \cdot \mathbf{R}_j) = -\boldsymbol{\nabla}_j E - \boldsymbol{\nabla}_j K = \mathbf{F}_j - \boldsymbol{\nabla}_j K \qquad (444)$$

Let nucleus A be fixed at the origin and B at a distance R. Taking $j = B$ in (444), we have for a diatomic

$$2F + R \, dF/dR = dK/dR$$

upon rearrangement. The solution is

$$F = K/R + R^{-2}\left[C - \int^R K \, dR\right] \qquad (445)$$

with C a constant of integration. For the neutral homonuclear diatomic, let K_0 be the total kinetic energy of the atoms and R_0 the sum of the radii of the atoms. When the internuclear distance equals R_0, $K = K_0$ and the forces vanish. This gives the value of C, and Eq. (445) was rewritten [8]

$$F = R^{-1}(K - K_0) + R^{-2}\int_R^{R_0} (K - K_0) \, dR. \qquad (446)$$

Differentiation of (446) gives

$$R\, d^2E/dR^2 = 2F - dK/dR$$

which is equivalent to various relations discussed in the chapter on the virial theorem.

The earliest attempt to treat a diatomic system by a quantum statistical theory was made by Hund [11] using the TF model for homonuclear diatomic molecules. Whereas we cannot go to a universal equation like (423) for atoms, for homonuclear diatomics we can reduce the number of independent parameters from two (internuclear distance and nuclear charge) to one. In Eq. (422) put $\bar{V} = 2Ze^2\Phi/R$ and measure all distances in units of R. The result is an equation for the dimensionless function Φ:

$$\bar{V}^2\Phi = \bar{x}\Phi^{3/2} \tag{447}$$

$\bar{V}^2$ means differentiation with respect to the dimensionless coordinates and

$$\bar{x} = (4Z^{1/2}/3\pi)(R/a_0)^{3/2} \tag{448}$$

with a_0 the Bohr radius $\hbar^2/me^2$. The boundary conditions are, for a neutral molecule, dependent only on $\bar{x}$. The normalization is satisfied with $\lambda = 0$, so that $\bar{V} = eV$ and $\varrho = [3eV/5\varkappa_k]^{3/2}$. The boundary condition at infinity is simply that $\Phi \to 0$ and the condition near a nucleus is

$$\Phi = (R\bar{V}/2e^2Z) \to (r_A')^{-1} \quad \text{as} \quad r_A' \to 0.$$

Here, r_A' is $|\mathbf{r} - \mathbf{r}_A|$ in dimensionless units, where $\mathbf{r}_A$ is a nuclear position. The solution to (447) for a particular value of $\bar{x}$ gives $\bar{V}$ and hence ϱ for all neutral homonuclear diatomic molecules for which R^3Z is constant.

Hund [11] introduced an approximate function satisfying the boundary conditions, approaching the atomic TF solution for charge Z near each nucleus and that for charge $2Z$ at large distances, and containing one variational parameter. The computation was still complicated, and the energy variation with changing internuclear distance (Z fixed) was not investigated. Sheldon [8] seems to have performed the first accurate quantum statistical calculation for a diatomic that considered several different values of R. He employed the TFD theory. The additional term in the equation prevents the simplification of equation (447) to a one-parameter equation, so that each molecule is a new problem. Nitrogen was chosen for investigation. (It must be noted that the change of exchange energy with distance plays a minor role in determining $U(R)$, and we can, in any case, approximate the exchange correction from the Thomas–Fermi density [10, 12].)

Sheldon converted the differential equation (in two variables because of cylindrical symmetry) to a difference equation. Unlike Hund, Sheldon had the benefit of a powerful computer. The difference equation was solved iteratively: Suppose we have values for $\bar{V}$ over a grid of points. Let the difference equation be, for the point (i, k),

$$f_{i,k} = 0 \tag{449}$$

where $f_{i,k}$ depends on the values of $\bar{V}$ at (i, k) and at neighboring points. Unless we already have the correct set of $\bar{V}_{i,k}$, (449) will not be satisfied. Then $\bar{V}_{i,k}$ is corrected by subtracting $\alpha f_{i,k}/g_{i,k}$ where $g_{i,k}$ is the derivative of $f_{i,k}$ with respect to $\bar{V}_{i,k}$ and α is between 0 and 2. We run over all the points over and over until $\bar{V}_{i,k}$ has converged. Sheldon's calculations were for internuclear distances of 1, 2.076, 3, 6, and 16 a_0. The second is close to the experimental equilibrium internuclear distance; at 16 a_0 the atoms do not overlap.

It was found that the forces from Eq. (446) were all repulsive. Sheldon [8] explained the failure to get binding in terms of the increase in kinetic energy as the nuclei approach, and concluded it was a consequence of the theory. He suggested that if $p_F(x)$ were not assumed to be spherical in momentum space, as in Eq. (430), binding could be obtained. For stable molecules, $p_F(x)$ is ellipsoidal in the bonding region. Sheldon [8] suggested, on the basis of this argument, that the quantum statistical theories describe all systems as having closed shells [where $p_F(x)$ is truly spherical] and consequently always give interatomic repulsion.

From the point of view of the Hellmann–Feynman forces, we could ascribe the repulsive behavior to the well-known failure of the quantum statistical theories to describe correctly the electron density close to the nucleus, a region emphasized by the force operator. Close to a nucleus, the density goes infinite as $r^{-3/2}$ instead of approaching a constant. Recently, Goodisman [13] performed TFD calculations on homonuclear diatomics by Sheldon's method and used the density to compute forces. These always led to repulsion even when we used the density from a modified theory, which gave improvements near the nuclei. This is not unreasonable: We generally ascribe the bonding character of an atom to the outer electrons.

The Thomas–Fermi neutral homonuclear diatomic molecule was treated by Townsend and Handler [14]. They calculated energies with a view toward determining the interatomic interaction curve. The solution to (447) was obtained by Townsend and Handler for eight values of the parameter $\bar{x}$, using a finite-difference relaxation procedure for the two-dimensional partial differential equation. The solutions for $\bar{x} = 0$ and $\bar{x} = \infty$ are just the solutions for the atoms of charges $2Z$ and Z. The normalization, used as a check on the accuracy of the solution, was tabulated, along with the other expectation values entering the energy and the exchange energy computed from this density. Now, for any value of Z one can derive, from the table, the energy for ten values of R (including 0 and ∞) by multiplying each expectation value by the proper scaling factor. Consider the pair of molecules of nuclear charges Z_1 and Z_2. If the internuclear distance for the second is $(Z_1/Z_2)^{1/3}$ times that for the first, $\bar{x}$ will be the same for both, and $\bar{V}$ for the second will be $(Z_2/Z_1)^{4/3}$ times $\bar{V}$ for the first. The energies for the second molecule will then be $(Z_2/Z_1)^{7/3}$ times that for the first (except for the exchange contribution).

Townsend and Handler [14] concluded that any binding was slight and at large distances. Subsequently, Townsend and Keller [15] introduced a correction, previously used for atomic systems, for the incorrect behavior of the Thomas–Fermi density near the nuclei and investigated its effect on the bonding properties. For the system of N electrons it may be anticipated that the error would not be much changed if the interelectronic repulsion were removed (bare nucleus problem), since the error arises from a region of space in which the interelectronic repulsion is relatively unimportant. The exact energy for the bare nucleus problem is merely a sum of one-electron energies. The bare-nucleus problem was treated exactly and by the Thomas–Fermi method, and the difference in energy used to correct the TF energy for the problem where interelectronic repulsions were considered. While Townsend and Keller found that $U(R)$ was not changed much, it was noted that it was repulsive for the rare gases and slightly attractive (at large R) for Li, N, Na, and Cu. A feature not present in the unmodified theory, periodicity or sensitivity to individual electrons, has been introduced by way of the bare nucleus problem, where energy is the sum of the $N/2$ lowest eigenvalues.

Since the electrons in purely quantum statistical theories are treated as a gas (in fact, their number can be nonintegral), such theories always give properties as smoothly varying functions of the number of electrons, whereas chemical bonding properties may vary markedly as one electron is added or removed. The failure to lead to stable molecules is related to this. As pointed out by Teller [16] the scaling behavior mentioned previously would have a strange consequence if the Thomas–Fermi model led to a stable molecule. Suppose for the homonuclear diatomic molecule with nuclear charges Z_1 the theory produced a positive binding energy D_1 and equilibrium internuclear distance R_1. Then for the molecule with nuclear charges Z_2/Z_1 at internuclear distance $(Z_1/Z_2)^{1/3}R_1$ the energy would be $(Z_2/Z_1)^{7/3}$ that for the original molecule. Since the energies of the atoms in the Thomas–Fermi theory also go as $Z^{7/3}$, the second molecule would have a binding energy at this distance equal to $(Z_2/Z_1)^{7/3}D_1$. Of course there would be no periodicity, and binding occurs for all cases; worse, binding energies of homonuclear diatomic molecules would rise monotonically (and sharply) with Z. Teller showed that in fact no molecule will be stable in this theory.

Teller [16] wrote the Thomas–Fermi equation in a slightly different form from (4??), taking the nuclear charge density as a continuous fixed distribution, $e\varrho_+$, rather than as point sources of potential. Then the Poisson equation becomes

$$\nabla^2 V = 4\pi e(\varrho - \varrho_+)$$

where ϱ refers to the electron density [equal to $(3eV/5\varkappa_k)$ in this theory]. The system is neutral: Over all space

$$\int (\varrho - \varrho_+)\, d\tau = 0.$$

By the proper choice of units, numerical coefficients may be eliminated to yield

$$\nabla^2 \varphi = \varphi^{3\,2} - \varrho_+ \tag{450}$$

with φ proportional to V. To the positive charge distribution ϱ_+ Teller let an infinitesimal positive charge distribution ζ, consisting of charge near some point in space, be added. Additional electron density was then added to maintain electrical neutrality. The equation for ε, the change in φ, may be linearized because ε is infinitesimal:

$$\nabla^2 \varepsilon = \tfrac{3}{2}\varphi^{1/2}\varepsilon - \zeta. \tag{451}$$

ε is certainly nonnegative near the location of ζ. Suppose it is negative elsewhere, in a region v. On S, the surface of v, ε vanishes, while $\zeta = 0$ within v. Then (451) shows that $\nabla^2 \varepsilon$ is always negative in v and therefore $\nabla\varepsilon$ points inward on S. Since $\varepsilon = 0$ on S, this implies $\varepsilon > 0$ within S and contradicts the hypothesis that ε is negative. Teller considered the possibility of S lying at infinity, as well as other special problems. The conclusion was the same: ε is everywhere positive.

Thus, Teller found that addition of a small amount of positive charge density anywhere in space with simultaneous addition of negative charge density to keep the system electrically neutral must increase φ and hence the potential everywhere. Starting with a neutral atom, this process may be repeated either to build up a second atom, by placing the charge infinitely far away, or to form a molecule by placing the charge close by. The energy required to form a molecule is always greater than that required to form a second atom because the potential is higher at each infinitesimal addition of ζ. Thus molecules in this theory are unstable to dissociation into atoms. Since increase of nuclear charge for a neutral system is equivalent to decrease of distances as far as the dimensionless equation is concerned, we can further argue that the energy of the Thomas–Fermi system increases as nuclei approach each other. A similar proof may be carried through for the Thomas–Fermi–Dirac case. Finally, Teller [16] extended his theorem to positive molecular ions.

Subsequently, Bálàzs [17] considered the problem of molecular stability from the point of view of the forces on the nuclei. Since the Hellmann–Feynman theorem holds, the conclusion must be the same: Molecules are not stable. It was shown by Bálàzs that this holds (definitely for homonuclear diatomics and probably for all molecules) for any model in which the electron density at any point depends *only* on the local electric potential. Schwarz [18] has pointed out that this conclusion is consistent with analyses of quantum mechanical wave functions, which show that, for the region between the nuclei, the electron density is raised compared to the sum of atomic densities while the contribution to the kinetic energy is decreased. Thus the surfaces of constant potential should not coincide with the surfaces of constant electron density, as they do for atoms.

A quantum statistical theory for which the electron density does not depend only on the potential at each point is the Fermi–Amaldi theory [1, Sect. 7]. Schwarz [18] has shown the difference between the equipotentials and constant density surfaces given by this theory for H_2^+, but Bálàzs showed [17] that only for fewer than four electrons is the Fermi–Amaldi correction important enough to give binding. Schwarz's modified statistical theory [18] makes ϱ a function of potential energy and position, so could possibly lead to binding. Another theory which could do this is the Weizsäcker correction [19] to the kinetic energy, which makes the energy depend on the derivative of the potential as well.

We briefly discuss the Weizsäcker correction here. Gombás [1, Sect. 12] argues that the usual kinetic energy expression, $\int \varkappa_k \varrho^{5/3}\, d\tau$, is a consequence of the Pauli principle, which prohibits all the electrons from going into the lowest state (which would give a low energy independent of ϱ). But even for n electrons in the lowest state, with wavefunction ψ, the kinetic energy would be

$$(n\hbar^2/2m) \int |\nabla \psi|^2 \, \partial\tau = \int \varkappa_i (|\nabla \varrho|^2/\varrho) \, d\tau$$

where ϱ is the density, $\varrho = n | \psi |^2$, and

$$\varkappa_i = \tfrac{1}{8}e^2 a_0 = \tfrac{1}{8}(\hbar^2/m). \tag{452}$$

This suggests an additional term in the kinetic energy density:

$$\varrho_W = \varkappa_i |\nabla \rho|^2/\rho. \tag{453}$$

Weizsäcker's original derivation is different. A justification for the additional term can also be given by arguments based on the WKB method [20]. All the derivations are questionable, and various modifications of the kinetic energy correction have been suggested on various grounds [21–24]. Yonei and Tomishima [23] considered corrections of the form $\lambda\varrho_W$, with λ determined experimentally by comparing atomic energies, calculated with $\lambda = 0(0.2)1$, with the correct energy. They found $\lambda = 0.2$ worked best. The addition of ϱ_W to the energy density leads to a more complicated differential equation for the density:

$$\tfrac{5}{3}\varkappa_k\varrho^{2/3} + \varkappa_i[(\nabla\varrho/\varrho)^2 - 2\nabla^2\varrho/\varrho] + V = E. \tag{454}$$

The resultant density for the atomic problem behaves more reasonably than the Thomas–Fermi density. It becomes a constant for $r \to 0$ and a decreasing exponential for $r \to \infty$.

Now (454) shows that the Weizsäcker density is exempt from Bálàzs' proof of molecular instability, since the potential depends on the gradient of the density as well as the density itself. The increased difficulty associated with the Weizsäcker equation [the Poisson equation combined with (454)] has made solutions for molecules difficult, but several results have been reported. Kołos [25], using approximate electron densities for N_2 and N atoms, reported a large decrease in the repulsive energy of N_2, compared to Sheldon's results, due to the Weizsäcker term in the energy expression. Subsequently, Gombás [26] reported a calculated binding energy for N_2 of about 20 eV, using a number of approximations. An improved calculation [27] gave 10.9 eV (experiment, 9.7 eV) and $R_e = 1.39$ Å (experiment, 1.094 Å).

Very recently Yonei [28] obtained very encouraging results from approximate computations for N_2 using the modified Weizsäcker term, $\lambda \varrho_W$, as well as the exchange energy, Eq. (426). Equation (454) became

$$\tfrac{5}{3}\varkappa_k \varrho^{2/3} - \tfrac{4}{3}\varkappa_e \varrho^{1/3} + \lambda \varkappa_i [(\boldsymbol{\nabla} \varrho/\varrho)^2 - 2\boldsymbol{\nabla}^2 \varrho/\varrho] + V = E.$$

with λ taken as $\tfrac{1}{5}$, as obtained from experiments with atoms (see page 274) [23]. Yonei employed a coordinate system introduced by Hund [11] for quantum statistical calculations on homonuclear diatomic molecules, and used by Sheldon [8] as well. In this system, the cylindrical coordinates r and z are replaced by ξ and η, where $\xi^{-1} = r_A^{-1} + r_B^{-1}$. Since the potential of the nuclei depends only on ξ, it is a reasonable approximation to assume that the electron density and total potential are functions of ξ alone. With suitable averaging over the coordinate η, the problem reduced to two differential equations in the single variable ξ. An iterative method, used previously for atoms [23], was employed to solve them.

Yonei [28] computed the energy for a series of values of internuclear distance, and found that the minimum energy was obtained for $R = 2.425a_0$. The experimental equilibrium internuclear distance is $2.0675\ a_0$. The total energy was closer to the experimental value than that from a Roothaan-type calculation, while the dissociation energy of $9.30\ \text{eV}$ was in good agreement with the experimental value. The dissociation energy was calculated as a difference between the energy at the minimum and the energy at infinite R. If $\lambda = 1$ (full Weizsäcker correction) was used, the equilibrium internuclear distance was decreased but the dissociation energy became much worse ($36.5\ \text{eV}$). The choice $\lambda = \tfrac{1}{9}$, suggested from theoretical considerations [21–24] also gave worse results [28], so that the value $\lambda = \tfrac{1}{5}$ seems close to optimum for the molecule as well as for atoms.

3. Closed-Shell Systems

The fact that all potential curves calculated by the unmodified quantum statistical theories are repulsive, and Sheldon's observation [8] that the dependence of the Fermi momentum on direction is like that for a closed-shell system, suggest that only the interaction of closed-shell systems be treated. Strictly speaking, this means interaction of rare gas atoms only. But the theories have also been used to describe parts of molecules, such as atomic cores, and ionic molecules composed of closed-shell ions. Here we discuss some applications of this kind.

Gombás [1, Sects. 18, 32] has discussed the terms in the energy when

the interaction between two ions is treated according to a quantum statistical theory, with the density taken as a sum of atomic densities. In addition
to the electrostatic interactions between the nuclei and the electronic charge
clouds, there were terms corresponding to the changes in kinetic and exchange energies, which are not linear in density. The exchange energy term is
relatively unimportant, as is any correction for correlation energy. On the
other hand, the induction and van der Waals energies may be important.
The former was computed by considering the effect of an electric field on
each spherical charge density using perturbation theory. Approximate
expressions for the latter, within the framework of the quantum statistical
theory, are also available.

Lenz and Jensen [29] investigated the results of such a calculation
for RbBr. An alkali halide was chosen for greatest ionicity, and RbBr
among these because the large number of electrons should favor a quantum
statistical theory. It was assumed that the nuclei were far enough apart so
that the electron clouds of the Rb^+ and Br^- ions were essentially unpolar-
ized. The electron density used in their work was a superposition of atomic
densities; these were obtained by a variational procedure (see page 265)
and had an exponential fall-off at large distance. For a crystal, the high
symmetry would make neglect of polarization a better approximation, and
the internuclear distances for the crystal were high by only 10%. (The
neglected van der Waals interaction was invoked to explain the discrepancy.)
For the diatomic molecule, a binding energy of 0.14 a.u. at $R_e = 5.9a_0$ was
obtained.

A reasonable potential for the interaction of two closed-shell atomic
systems, as suggested by Bohr [30, 30a], is the screened Coulomb potential.
For nuclear charges of Z_A and Z_B,

$$U(R) = (Z_A Z_B / R)e^{-R/a}. \tag{455a}$$

This function (see Chapter II, Eqs. (57) et seq.) describes the Coulombic
repulsion of the nuclei for small R and gives an interaction dying off exponentially for large R (see Section II.B). For the screening length a, Bohr
suggested

$$a = a_0(Z_A^{2/3} + Z_B^{2/3})^{-1/2}, \tag{455b}$$

where a_0 is the Bohr radius. The dependence of this characteristic length
on the $-\frac{1}{3}$ power of nuclear charge arises in the Thomas–Fermi theory
(page 261), which is known to reproduce correctly the general trend in
atomic size with atomic number. According to equations (422)–(423), the

ratio of r for an atom of charge Z to the universal variable x is $(\frac{1}{4})(9\pi^2/2Z)^{1/3}a_0 = 0.8853Z^{-1/3}a_0$. The screening length is sometimes taken as

$$a = 0.8853a_0(Z_A^{1/3} + Z_B^{1/3})^{-1} \qquad (455c)$$

or as something intermediate between (455b) and (455c). If $Z_A \gg Z_B$, the screening length is that characteristic of atom A. For $R > a$, it turns out that exponential screening is too large: $U(R)$ of (455a) falls off too rapidly with increasing R.

Using the TF theory for the dependence of electrostatic potential on radial distance from an atom Eqs. (422)–(423), the interaction of a bare nucleus of charge Z_A with a neutral atom of atomic number Z_B would be $Z_A Z_B \chi(R/a)R^{-1}$, where χ is the solution of Eq. (423) and $a = 0.8853a_0Z_B^{-1/3}$. This assumes the electrons of B do not adjust to the entry of Z_A. The "Thomas–Fermi screened potential"

$$U(R) = Z_A Z_B R^{-1}\chi(R/a)$$

is sometimes used for the interaction of closed-shell atomic systems, with a given by something like (455b) or (455c). Scattering cross sections for the screened Coulomb potential and the Thomas–Fermi screened potential have been calculated [30a]. A derivation of the Thomas–Fermi screened potential for interatomic interactions, by first-order perturbation theory, was given by Firsov [2, Sect. 18; 30a, Sect. 3.3; 32]. Subsequently, he gave a derivation [30a, Sect. 3.4; 33] using the variational principle. The intention was to derive the interaction potential between atoms for small internuclear distance, where the outer electrons play only a minor role compared to the closed-shell cores. This treatment should be applicable to closed-shell atoms for all R.

Variational principles were introduced which gave upper and lower bounds on the energy. The former is the Thomas–Fermi energy expression itself, Eq. (419). If $\varrho = \varrho_{TF} + \delta\varrho$ and $\delta\varrho$ is small compared to the Thomas–Fermi density, it is straightforward to show

$$E(\varrho) = E(\varrho_{TF}) + \tfrac{5}{9}\varkappa_k \int \varrho_{TF}^{-1/3}(\delta\varrho)^2 \, d\tau + \tfrac{1}{2}e^2 \iint \delta\varrho(\mathbf{r}_1) \, \delta\varrho(\mathbf{r}_2)r_{12}^{-1} \, d\tau_1 \, d\tau_2$$

Terms through second order in $\delta\varrho$ have been kept; the first-order terms vanish by virtue of the variational principle satisfied by ϱ_{TF}. The last term is the self-interaction of the charge distribution $\delta\varrho$ and is positive, even though $\delta\varrho$ is not everywhere of the same sign. Thus,

$$E(\varrho) > E(\varrho_{TF}) = E_{TF},$$

so $E(\varrho)$ furnishes an upper bound on E_{TF} whenever ϱ is close to ϱ_{TF} (neglect of higher terms than $(\delta\varrho)^2$ in the kinetic energy). Firsov also introduced H_1, a more complicated functional of ϱ [actually of a function $f(\varrho)$, related to the potential]. The function H_1 has the property that its value is maximized when the exact Thomas–Fermi density is inserted. The maximum value is identically equal to E_{TF}. Having the maximum and minimum principles means we can get both a lower and an upper bound on E_{TF} by insertion of approximate densities. Then the Thomas–Fermi energy is known within the upper and lower limits. The procedure is considerably simpler than attempting to solve the TF equation directly by some approximate method. Such an approximate solution, moreover, would not be accompanied by any information about the error.

For the two-center problem, a good approximate density would be a superposition of the atomic densities:

$$\varrho = \varrho_A(r_A) + \varrho_B(r_B).$$

Here, r_A and r_B are the distances from the nuclei. Correspondingly, $f(\varrho)$ for the two-center system was taken [33] as the sum of two atomic functions. With this choice, the values of E and H_1 differed by only about 8%. Thus the mean of E and H_1 differs from E_{TF} by no more than 4%. For $R = 0$, where one has a single atom, the error $[\frac{1}{2}(E + H_1) - E_{TF}]$ was actually only $1\frac{1}{2}\%$. Firsov [33], by introduction of additional approximations, derived a simple form for $U(R)$, valid for

$$R \lesssim 10a(Z_A^{1/2} + Z_B^{1/2})^{-2/3}, \qquad a = 4.7 \times 10^{-9} \text{ cm}.$$

The potential is

$$U(R) = Z_A Z_B e^2 R^{-1} \chi([Z_A + Z_B]^{2/3} R/a) \tag{456}$$

where χ is the dimensionless Thomas–Fermi function for atoms, Eq. (423). Gaspar [34] likewise used quantum statistical theories to derive a universal interaction potential for rare gas atoms. It may be shown [35] that, for large R, $U(R)$ increases as R^{-6} (no relation to the van der Waals attraction, of course).

Abrahamson et al. [36] have discussed the application of variational principles in quantum statistical theory for the diatomic problem, and have investigated in detail the resulting $U(R)$ for rare gas atoms. Abrahamson has extended [37] Firsov's work to the TFD theory. Again variational principles were used in conjunction with approximate densities to get upper and lower bounds on $U(R)$. The expression for the complete energy as a functional of the density ϱ is

$$E = K + E_{ne} + I + E_{ex} = \varkappa_k \int \varrho^{5/3}\, d\tau - e^2 \int \sum_A Z_A r_A^{-1} \varrho\, d\tau$$

$$+ \tfrac{1}{2}e^2 \iint \varrho(\mathbf{r}_1)\varrho(\mathbf{r}_2)r_{12}^{-1}\, d\tau_1\, d\tau_2 - \varkappa_e \int \varrho^{4/3}\, d\tau \tag{457}$$

Here, Z_A is the charge on nucleus A and r_A the distance from that nucleus. The exact Thomas–Fermi–Dirac density makes (457) stationary. The authors let $\varrho = \varrho_{TFD} + \delta\varrho$ with $\delta\varrho$ small compared to ϱ_{TFD} and expanded the energy through terms second order in $\delta\varrho$. Then the difference between this energy and that calculated with ϱ_{TFD} was shown [36] to be positive. The terms are

$$\tfrac{1}{9}\int (5\varkappa_k\varrho^{-1/3} - 2\varkappa_e\varrho^{-2/3})(\delta\varrho)^2\, d\tau$$

and

$$\tfrac{1}{2}e^2 \iint \frac{\delta\varrho(\mathbf{r}_1)\,\delta\varrho(\mathbf{r}_2)}{r_{12}}\, d\tau_1\, d\tau_2.$$

Since ϱ is always greater than the boundary value $(125/192\pi^5)a_0^{-3} = (\varkappa_e/2\varkappa_k)$, we have the inequality

$$\varrho^{-2/3} < (2\varkappa_k/\varkappa_a)\varrho^{-1/3}$$

so that the first term above is positive. The second term is the electrostatic self-interaction of the charge distribution $\delta\varrho$ and may be shown to be necessarily positive. Thus $E(\varrho)$ is an upper bound to the TFD energy if ϱ is not too far from ϱ_{TFD}. The potential due to the electrons is

$$-f(\mathbf{r}_1) = -e \int \varrho(\mathbf{r}_2)r_{12}^{-1}d\tau_2 \tag{458}$$

and $f \to 0$ as one gets infinitely far from the nuclei. As Firsov did for the Thomas–Fermi case, it was possible [36, 37] to construct a functional H_1 for f such that (a) when f is chosen to make H_1 stationary the value of H_1 is the TFD energy and (b) H_1 is never more than this value. Thus one can construct both upper and lower bounds to the TFD energy for a system.

Since the TFD energies for the separated atoms are supposed to be exactly known, we have an upper bound and a lower bound, from E and H_1 respectively, for $U(R)$ for each R. The TFD density, since it drops to zero at a finite distance from the nucleus, gives a vanishing interaction potential whenever R is greater than the sum of the radii of the atoms, about 8 a_0. The TF density does not do this. But for smaller values of R the TFD is more realistic, since, compared to a quantum mechanical density, the TF falls off to zero too slowly. A modified TFD potential, which should maintain the advantages of both, was introduced by Abrahamson et al. [36]. It yields an exponential fall-off for $U(R)$ over a large range of R, as expected from a quantum mechanical treatment.

Comparisons with other experimental and theoretical curves (including those of Bohr and Firsov) were carried out to assess the validity of the TFD interaction potentials for the rare gas pairs He–He, Ne–Ne, Ar–Ar, Kr–Kr, Xe–Xe, and Rn–Rn [36, 37]. Heteronuclear pairs were considered afterward [38]. In general, the modified TFD potential (actually, a variational approximation) proved very good for small internuclear distances (0.1 to 0.6 a_0), and were in reasonable agreement with empirical data at large separations (to 7.0 a_0). Abrahamson subsequently [38] gave fits of the Thomas–Fermi–Dirac $U(R)$ curves to the Born–Mayer form, $U(R) = Ae^{-Br}$.

Günther [39] has questioned Abrahamson's results because of the treatment of the boundary of the TFD electron density. This was not varied in deriving the variational principle. (In the Thomas–Fermi case, the boundary is always at infinity, and presents no problem.) According to Günther, the bounds on U_{TFD} are not valid. To demonstrate this, he derived an upper bound on U_{TFD} for Ar–Ar from Townsend and Handler's [14] Thomas–Fermi results, and showed it was considerably below (30% between $R = 0.76$ and 1.53 a_0) Abrahamson's values.

At present, applications of the quantum statistical theories to diatomic systems other than rare gas pairs is along several lines. Approximate interaction potentials, like those derived by Firsov and Abrahamson, are often used to describe closed-shell atomic cores in pseudopotential and other simple models (Volume 2, Chapter III, Section B), for such systems as the alkali-rare gas diatomics and their ions [40]. Much work using quantum statistical exchange potentials to simplify Hartree–Fock calculations has been done and such calculations should soon appear for a variety of molecules [41]. Finally, there is the possibility that, with the Weizsäcker or other modification, the quantum statistical theories may be applied to bonding situations.

REFERENCES

1. P. Gombás, "Die Statische Theorie des Atoms und Ihre Anwendungen." Springer-Verlag, Berlin and New York, 1949.
2. P. Gombás, in "Handbuch der Physik" (S. Flugge, ed.), Vol. XXXVI. Springer-Verlag, Berlin and New York, 1956.
3. N. H. March, Advan. Phys. 6, 1 (1957).

4. J. C. Slater, "Quantum Theory of Atomic Structure," Vol. II. McGraw-Hill, New York, 1960.
5. M. Gell-Mann and K. A. Brueckner, *Phys. Rev.* **106**, 364 (1957).
6. F. Herman and S. Skillman, "Atomic Structure Calculations." Prentice-Hall, Englewood Cliffs, New Jersey, 1963.
7. P. A. M. Dirac, *Proc. Cambridge Phil. Soc.* **26**, 376 (1930).
8. J. W. Sheldon, *Phys. Rev.* **99**, 1291 (1955).
9. P. A. M. Dirac, "Quantum Mechanics." Oxford Univ. Press, London, and New New York, 1958; E. M. Corson, "Perturbation Methods in the Quantum Mechanics of N-Electron Systems." Hafner, New York, 1951.
10. J. Goodisman, *Phys. Rev. A* **2**, 1 (1970).
11. F. Hund, *Z. Phys.* **77**, 12 (1932).
12. J. M. C. Scott, *Phil. Mag.* **43**, 859 (1952).
13. J. Goodisman, *Phys. Rev.* **A3**, 1819 (1971).
14. J. R. Townsend and G. S. Handler, *J. Chem. Phys.* **36**, 3325 (1962).
15. J. R. Townsend and J. M. Keller, *J. Chem. Phys.* **38**, 2499 (1963).
16. E. Teller, *Rev. Mod. Phys.* **34**, 627 (1962).
17. N. L. Bálàzs, *Phys. Rev.* **156**, 42 (1967).
18. W. H. E. Schwartz, *Theor. Chim. Acta* **23**, 21 (1971).
19. C. F. von Weizsäcker, *Z. Phys.* **96**, 431 (1935).
20. I. Fényes, *Z. Phys.* **125**, 336 (1949).
21. R. Berg and L. Wilets, *Proc. Phys. Soc. London* **68**, 229 (1955).
22. S. Golden, *Phys. Rev.* **105**, 604 (1957); **107**, 1283 (1957).
23. K. Yonei and Y. Tomishima, *J. Phys. Soc. Jap.* **20**, 1051 (1965); **21**, 142 (1966).
24. J. Goodisman, *Phys. Rev. A* **1**, 1574 (1970).
25. W. Kołos, *Phys. Rev.* **102**, 1052 (1956); *Acta Phys.* **6**, 133 (1956).
26. P. Gombás, *Z. Phys.* **152**, 397 (1958).
27. P. Gombás, *Acta Phys.* **9**, 461 (1959).
28. K. Yonei, *J. Phys. Soc. Jap.* **31**, 882 (1971).
29. W. Lenz, *Z. Phys.* **77**, 713 (1932). H. Jensen, *Z. Phys.* **77**, 722 (1932).
30. N. Bohr, *Kgl. Danske Videnskab. Selskab. Nat. Fys. Medd.* **18**, 8 (1948).
30a. I. M. Torrens, "Interatomic Potentials," Academic Press, New York and London, 1972, Sect. 4.7.
31. E. Everhart, G. Stone, and R. J. Carbone, *Phys. Rev.* **99**, 1287 (1955). J. Lindhard, V. Nielsen, and M. Scharff, *Kgl. Danske Videnskab. Mat. Fys. Medd.* **36**, 10 (1968).
32. O. B. Firsov, *Dokl. Akad. Nauk USSR* **91**, 515 (1953).
33. O. B. Firsov, *Zh. Eksp. Teor. Fiz.* **32**, 1464; **33**, 696 (1957); *Sov. Phys. JETP* **5**, 1192 (1957); **6**, 534 (1958).
34. R. Gaspár, *Acta Phys.* **11**, 71 (1960).
35. R. E. Roberts, *Phys. Rev.* **170**, 8 (1968).
36. A. A. Abrahamson, R. D. Hatcher, and G. H. Vineyard, *Phys. Rev.* **121**, 159 (1961).
37. A. A. Abrahamson, *Phys. Rev.* **130**, 693 (1963).
38. A. A. Abrahamson, *Phys. Rev. A* **133**, 990 (1964), *Phys. Rev.* **178**, 76 (1969).
39. K. Günther, *Ann. Phys.* (Leipzig) **14**, 296 (1964); *Kernenergie* **7**, 443 (1964); see also P. T. Wedepohl, *Proc. Phys. Soc. London* **92**, 79 (1967); *J. Phys.* **B2**, 307 (1968).
40. M. G. Menendez, M. J. Redman, and J. F. Aebischer, *Phys. Rev.* **180**, 69 (1969).
41. K. Schwarz and J. W. D. Connolly, *J. Chem. Phys.* **55**, 4210 (1971); J. C. Slater and J. H. Wood, *Int. J. Quantum Chem.* **4**, 3 (1971).

SUPPLEMENTARY BIBLIOGRAPHY

Below are listed, alphabetically by author, references relevant to the material of this chapter, but which came to our attention too late to be incorporated into the manuscript. We have given titles, with some clarifying information in parentheses. The letter after each reference indicates to which section it is most relevant.

W. H. Adams, Perturbation Theory for Configuration Interaction Wave Functions. *Chem. Phys. Lett.* **10**, 198 (1971). **B**

R. Ahlrichs, Error Bounds for Approximate Rayleigh-Schrödinger Perturbation Energies and Wavefunctions. *Chem. Phys. Lett.* **10**, 157 (1971). **B**

M H. Alexander and R. G. Gordon, Exact Solutions to the Coupled Hartree-Fock Perturbation Equations. *J. Chem. Phys.* **56**, 3823 (1972). **B**

M. V. Basilevsky and M. M. Berenfeld, SCF Perturbation Theory and Intermolecular Interactions. *Int. J. Quantum Chem.* **6**, 555 (1972). **B**

N. Björne, Methods for Constrained SCF–LCAO–MO Calculations; Application to N_2. *Mol. Phys.* **24**, 1 (1972). **A**

B. L. Burrows, Unrestricted Hartree–Fock Perturbation Theory. *Theor. Chim. Acta* **28**, 179 (1973). **B**

L. S. Cederbaum, On the Breakdown of Koopmans' Theorem for Nitrogen. *Chem. Phys. Lett.* **18**, 503 (1973). **A**

W. L. Clinton and L. J. Massa, The Cusp Condition: Constraint on the Electron Density Matrix Constrained Variation. *Int. J. Quantum Chem.* **6**, 519 (1972). **A**

E. R. Davidson, Selection of the Proper Canonical Roothaan–Hartree–Fock Orbitals for Particular Applications. *J. Chem. Phys.* **57**, 1999 (1972). **A**

R. G. Gordon and Y. S. Kim, Theory for the Forces Between Closed-Shell Atoms and Molecules (using quantum statistical theories). *J. Chem. Phys.* **56**, 3122 (1972). **F**

M. Kleiner, A Note on the Relation Between CI and Rayleigh–Schrödinger Perturbation Theory. *Theoret. Chim. Acta* **25**, 121 (1972). **B**

J. Koutecký and V. Bonačić-Koutecký, Direct Minimization of the Hartree–Fock Expectation Value of the Hamiltonian, *Chem. Phys. Lett.* **15**, 558 (1972). **B**

M. S. Ostlund, Complex and Unrestricted Hartree–Fock Functions, *J. Chem. Phys.* **57**, 2994 (1972). **A**

J. C. Slater, Hellmann–Feynman and Virial Theorems in the $X\alpha$ Method. *J. Chem. Phys.* **57**, 2389 (1972). **F**

B. Swanstrøm, K. Thomsen, and P. B. Yde, Analytical *ab initio* Computation of Force Constants and Dipole Derivatives: LiH, Li_2, and BH (by perturbationlike treatment of Roothaan equations). *Mol. Phys.* **20**, 1135 (1971); **23**, 691 (1972). **B**

J. H. Walker, T. E. H. Walker and H. P. Kelly, Ground and Low-Lying Excited Electronic States of FeH (by many-body perturbation theory). *J. Chem. Phys.* **57**, 2094 (1972). **B**

T. E. H. Walker and H. P. Kelly, Energy of CH Calculated by Many-Body Perturbation Theory. *Phys. Rev.* **A5**, 1986 (1972). **B**

K. Yonei, errata for *J. Phys. Soc. Japan* **31**, 882 (1971). *J. Phys. Soc. Japan* **32**, 293, 586 (1972). **F**

General Bibliography

The discussions of the preceding chapters assume a knowledge of quantum mechanics such as would be obtained from a one-semester course. Material which is standard or elementary has not been included, or merely summarized. Instead, reference has been made to one or more quantum chemistry texts. These basic works are listed in Section 1 below. In the present volume, all sources in Section 1 of this Bibliography are referred to by the name of the first author, in upper-case letters and *without* a reference number. In Section 2, we give some books closely related to the subject matter discussed here, to which we have probably not referred as often as we should have.

Our coverage of the literature of diatomic molecule calculations emphasizes recent work. In particular, an attempt was made to mention all papers published in 1970, 1971 and 1972, since the most recent of the bibliographies listed below (Richards, Walker, and Hinkley) includes papers through 1969. The articles and books listed under Bibliographies of Calculations (Section 3) and Reviews of the Literature (Section 4) provide complete coverage of the published literature.

1. *Basic Texts in Quantum Chemistry*

H. Eyring, J. Walter, and G. E. Kimball, "Quantum Chemistry." Wiley, New York, 1944.
W. Kauzmann, "Quantum Chemistry." Academic Press, New York, 1957.

F. L. Pilar, "Basic Quantum Chemistry." McGraw-Hill, New York, 1968.

R. McWeeny and B. T. Sutcliffe, "Methods of Molecular Quantum Mechanics." Academic Press, New York, 1969.

I. N. Levine, "Quantum Chemistry," Vol. I. Allyn and Bacon, Boston, 1970.

2. *Books discussing the Calculation of* $U(R)$

G. Herzberg, "Molecular Spectra and Molecular Structure, I. Spectra of Diatomic Molecules." Van Nostrand, New York, 1950.

J. O. Hirschfelder, C. F. Curtiss and R. Byron Bird, "Molecular Theory of Gases and Liquids." Wiley, New York, 1954.

J. O. Hirschfelder, ed., "Intermolecular Forces" (Advances in Chemical Physics, Vol. 12). Interscience, New York, 1967.

H. Margenau and N. R. Kestner, "Theory of Intermolecular Forces." Pergamon, Oxford, 1969 (2nd Ed. 1971).

J. C. Slater, "Quantum Theory of Molecules and Solids, Vol. I, Electronic Structure of Molecules." McGraw-Hill, New York, 1963.

I. M. Torrens, "Interatomic Potentials." Academic Press, New York and London, 1972.

3. *Bibliographies of Calculations*

H. Yoshizumi, *Advan. Chem. Phys.* **2**, 323 (1959). A bibliographical survey of methods for treating interelectronic correlation. Papers dealing with SCF and other calculations for simple molecules are listed.

J. C. Slater, "Quantum Theory of Molecules and Solids," Vol. I. McGraw-Hill, New York, 1963. A bibliography is given (p. 401 *et seq.*) of papers and books giving a "fairly complete account of the literature" on molecular electronic structure. Articles are listed by author, with cross references, and titles are given.

R. G. Parr, "Quantum Theory of Molecular Electronic Structure." Benjamin, New York, 1963. In section 7 is a summary of a variety of *ab initio* calculations on small molecules.

R. K. Nesbet, *Advan. Quantum Chem.* **3**, 1 (1967). Methods for approximate Hartree–Fock calculations on small molecules are discussed, together with the results of such calculations.

A. D. McLean and M. Yoshimine, Suppl. to *IBM J. Res. Devel.* **12**, 206 (1968). Tables of Linear Molecule Wave Functions.

M. Krauss, Natl. Bur. Stds. Technical Note 438 (U. S. Dep't. of Commerce, Dec. 1967). Gives references to *ab initio* molecular electronic calculations from 1960 to 1967, including material unpublished at the time of the report. For each molecule, a list of calculations is given, with references, together with information on the method used, the properties calculated, and whether part of the potential curve was derived. In separate tables, the best calculated values for dissociation energies, spectroscopic constants, and other quantities are given.

G. Klopman and B. O'Leary, *Fortschritte der Chem. Forsch.* **15**, 445 (1970). A book-length discussion and critique of semi-empirical molecular orbital calculations, with a complete bibliography and tables of results.

W. G. Richards, T. E. H. Walker, and R. K. Hinkley, "A Bibliography of ab initio Molecular Wave Functions." Oxford Univ. Press, London, 1971. Includes all *ab initio* calculations published through the end of 1969. For each molecule, calculations are listed with references, the nature of each calculation being given. The energy obtained by each (a rough measure of accuracy) is indicated, together with indications of which properties were calculated in addition to the energy. Properties considered include potential curve, spectroscopic constants, and dissociation energies.

4. Reviews of the Literature

Reviews of the literature and surveys of recent work appear regularly as chapters in *Annual Reviews of Physical Chemistry*. These are generally complete (each covering the period since the previous review of the same subject), with the amount of discussion variable. Below are listed the chapters that have appeared, since 1958, dealing with molecular calculations.

M. Kotani, Y. Mizuno, K. Kayama, and H. Yoshizumi, Quantum Theory of Electronic Structure of Molecules. **9**, 245 (1958).
J. A. Pople, Quantum Theory. Theory of Molecular Structure, and Valence. **10**, 331 (1959).
P.-O. Löwdin, Quantum Theory of Electronic Structure of Molecules. **11**, 107 (1960).
O. Sinanoğlu and D. F.-T. Tuan, Quantum Theory of Atoms and Molecules. **15**, 251 (1964).
B. M. Gimarc and R. G. Parr, Quantum Theory of Valence. **16**, 451 (1965).
F. Prosser and H. Shull, Quantum Chemistry. **17**, 37 (1966).
A. Golebiewski and H. S. Taylor, Quantum Theory of Atoms and Molecules. **18**, 353 (1967).
C. Schlier, Intermolecular Forces. **20**, 191 (1969).
L. C. Allen, Quantum Theory of Structure and Dynamics. **20**, 315 (1969).
A. D. Buckingham and B. D. Utting, Intermolecular Forces. **21**, 287 (1970).
F. E. Harris, Quantum Chemistry, **23**, 415 (1972).

Author Index

Numbers in parentheses are reference numbers and indicate that an author's wokr is referred to although his name is not cited in the text. Numbers in italics show the page on which the complete reference is listed.

A

Abrahamson, A. A., 72, *86*, 278 (36, 37), 280 (37, 38), *281*
Adams, W. H., *282*
Aebischer, J. F., 280 (40), *281*
Ahlrichs, R., *282*
Alder, B. J., 82, *87*
Alexander, M., 233 (34), *241*
Alexander, M. H., 98, *149*, *282*
Allen, L. C., *285*
Amdur, I., 39 (47), *51*, *52*
Anderson, A. B., 234, *241*
Andriesse, C. D., *52*
Aquilanti, V., *52*

B

Bachrach, V. L., *88*
Bader, R. F. W., 99, *149*, 184, *194*, 219, 225 (17), 229 (26), 230, 233 (27, 33), *240*, *241*
Bagus, P. S., 111 (46), *149*
Bálàzs, N. L., 273, *281*
Balfour, W. J., 36 (31), *50*
Bandrauk, A. D., 229 (26), 233, *241*
Bargmann, V., 77 (23), *87*
Barker, J. A., 47 (68), *51*, *52*, *53*, 82, *87*
Barnett, G. P., 134 (76b), *150*
Barnett, M. P., 145 (110), *151*

Barron, T. H. K., 48 (73, 74), *52*
Bartlett, J. H., 241 (1), *256*
Bar-ziv, E., *52*
Basilevsky, M. V., *282*
Bates, D. R. W., 41 (54), 42 (54), *51*
Battali-Cosmarici, C., *52*
Battino, R., 47 (69), *51*
Baumann, H., 211, *214*
Bazley, N. W., 94 (11, 12), 95 (11, 12), 97 (24), *148*, *149*
Beck, D. E., 36 (30), *50*, 81, 82 (36), *87*
Beckel, C. L., 34, *50*, *54*, 73, *86*
Bell, D. H., 247 (13), *257*
Bell, R. P., 198 (4a), 211, *214*
Bender, C. F., 133, 134 (74), 148, *150*, *151*
Bennewitz, H. G., *52*
Benston, M. L., 100 (37), *149*, 184, 187, *194*, 210, *214*, 229, 234 (35), 235 (35), *240*, *241*
Berenfeld, M. M., *282*
Berg, R., 274 (21), 275 (21), *281*
Berkowitz, J., *52*
Berlin, T., 219, *240*
Bernstein, R. B., 31 (7), 36, 37 (38), 38 (43), 40 (43, 49), 41 (53), *50*, *51*, 77 (24), *87*
Berthier, G., 124, 134 (69), *150*
Bezzub, L. I., 212 (25), *214*
Bird, R. B., 38 (44), 43 (44), 44 (44), 45 (44), 46 (44), *51*, 55 (1), 56 (1), *61*, 72 (1), 77 (1), 79 (1), 81 (1), *86*, *284*
Birnbaum, G., 36 (29), *50*

287

Subject Index

A

Adiabatic theorem, 2, 15
Adiabatic treatment, 14–19, 26, 42
Allowed variation, 90, 196, 203, 217, 221
Anharmonicity, 73, 184, 206, 233, 234
Antibonding orbital, 69
Antisymmetrization, 60, 61, 105, 122, 168, 169
Ar–Ar, 47, 48, 49, 82
Atomic orbital, 68, 71
Average energy denominator, 156, 209

B

Badger's Rule, 78
Basis functions, 99, 102, 103, 156, 221
Binding, 219, 270
Binding energy, 64, 68, 211, 272
Bonding orbital, 69, 137
Born–Mayer potential, 280
Born–Oppenheimer separation, 7–14, 32, 33, 178, 179, 181
Born separation, 17, 26
Bracketing method, *see* partitioning method
Brillouin's theorem, 114, 115, 124, 128, 129, 134, 174, 176
Brillouin-Wigner expansion, 127, 158–160, 161
Buckingham–Corner potential, 79
Buckingham potential, 43, 74, 81

C

Centrifugal potential, 23, 29, 30, 32, 41
CI, *see* configuration interaction

Closed shells, 65, 66, 67, 270, 277, 280
Closure approximation, *see* average energy denominator
Cluster expansion, 141–143
CO, 206
Configuration, 122, 126, 129, 133
Configuration Interaction, 121–136, 137, 138, 140, 143, 146, 172, 226
Constrained variation, 99–101, 213, 230
Core repulsion, 62, 63
Correlated wave function, 138–140, 142, 148, 171, 172, 253, 255, 256
Correlation, 123, 134, 135, 137, 138, 144, 162
Correlation energy, 123, 136, 139, 141, 142, 167, 177, 227, 228, 263, 276
Coulomb hole, 136, 139
Coulomb integral, 63, 64, 65, 144, 168
Coulomb operator, 110, 164, 175
Coulombic energy, *see* electrostatic energy
Cross section, 37–39, 42, 43
Cusp condition, 138, 222, 248–249, 253

D

Density matrix, 101, 116–121, 130, 131, 145, 177, 181, 258, 264, 265
Determinantal, wave function, 106, 111, 115, 119, 124, 132, 162, 177, 211, 226, 251, 258, 265
Differential overlap, 60, 61, 65, 129
Dispersion forces, *see* van der Waals forces
Dissociation energy, 31, 85, 228, 238
Distinguishable Electron Method, 174

296

Physical Chemistry

A Series of Monographs

Ernest M. Loebl, Editor

Department of Chemistry, Polytechnic Institute of

Brooklyn, Brooklyn, New York